Glycoprotein Analysis in Biomedicine

Methods in Molecular Biology

John M. Walker, SERIES EDITOR

14. **Glycoprotein Analysis in Biomedicine,** edited by ***Elizabeth F. Hounsell, 1993***
13. **Protocols in Molecular Neurobiology,** edited by ***Alan Longstaff and Patricia Revest, 1992***
12. **Pulsed-Field Gel Electrophoresis:** *Protocols, Methods, and Theories,* edited by ***Margit Burmeister and Levy Ulanovsky, 1992***
11. **Practical Protein Chromatography,** edited by ***Andrew Kenney and Susan Fowell, 1992***
10. **Immunochemical Protocols** , edited by ***Margaret M. Manson, 1992***
9. **Protocols in Human Molecular Genetics,** edited by ***Christopher G. Mathew, 1991***
8. **Practical Molecular Virology,** edited by ***Mary K. L. Collins, 1991***
7. **Gene Transfer and Expression Protocols,** edited by ***E. J. Murray, 1991***
6. **Plant Cell and Tissue Culture,** edited by ***Jeffrey W. Pollard and John M. Walker, 1990***
5. **Animal Cell Culture,** edited by ***Jeffrey W. Pollard and John M. Walker, 1990***
4. **New Nucleic Acid Techniques,** edited by ***John M. Walker, 1988***
3. **New Protein Techniques,** edited by ***John M. Walker, 1988***
2. **Nucleic Acids,** edited by ***John M. Walker, 1984***
1. **Proteins,** edited by ***John M. Walker, 1984***

Methods in Molecular Biology • 14

Glycoprotein Analysis in Biomedicine

Edited by

Elizabeth F. Hounsell

Medical Research Council,
Clinical Research Centre, Harrow, UK

Humana Press **Totowa, New Jersey**

Dedication

To my father

999 Riverview Drive
Totowa, New Jersey 07512

Printed in the United States of America

Library of Congress Cataloging in Publication Data
Main entry under title:

Methods in molecular biology.

Glycoprotein analysis in biomedicine / edited by Elizabeth F. Hounsell.
p. cm. – (Methods in molecular biology ; 14)
Includes bibliographical references and index.
ISBN 0-89603-226-4
1. Glycoproteins–Analysis. I. Hounsell, Elizabeth F.
II. Series: Methods in molecular biology (Totowa, N.J.) ; 14.
QP552.G59G598 1993
574.19'2454–dc20 92-31163
CIP

Preface

As the editor of *Glycoprotein Analysis in Biomedicine* I have tried to bring together a comprehensive range of detailed protocols that will render carbohydrate biochemistry accessible to all, and not just to the dedicated few. I wish to thank the authors, all of whom have striven significantly to realize this aim.

There is now a growing interest in glycoprotein analysis from researchers in many disciplines who realize that protein glycosylation can no longer be ignored. Oligosaccharide chains of glycoproteins form a diverse array of molecules with an inherently large content of chemical information that is functional in biological recognition. In addition to their specific interactions with antibodies and carbohydrate binding proteins, the carbohydrate moieties of glycoproteins have profoundly important effects on the conformations, stability, antigenicity, and function of the protein components.

Some specialized techniques, i.e., NMR and MS, are introduced at the beginning of the book since these are capable of defining exactly the diversity of oligosaccharide sequences, and are used to detect the growing number of new structures as well as to indentify, in new scenarios, those previously documented. Together with the chromatographic methods of purification and analysis discussed in Chapters 1–12, there now exists a powerful armamentarium for glycoprotein characterization. The more chemical techniques are documented in detail in the earliest chapters, whereas the enzymatic methods of releasing oligosaccharides from protein are introduced later. Analysis of glycoproteins from mammalian, parasitic, bacterial, and viral origins are each discussed.

The second half of the book deals with the more medical aspects of glycoprotein analysis, introducing sensitive staining techniques and radioactive labeling methods for several different groups of glycoproteins—cytoplasmic, membrane bound, mucinous, proteoglycan-type, and the like. Lastly, two chapters deal with the interactions of oligosaccharides, i.e., with the identification of endogenous lectins and the extension of electron microscopic techniques, accessible in every laboratory, to STM with the potential to visualize glycoproteins and their ligands.

All the glycoproteins discussed have oligosaccharide sequences that interact with lectins, antibodies, and enzymes to aid in their characterization. The same oligosaccharide sequence can be found on different protein backbones and on different carbohydrate–protein core regions (*see* Glosssary). The overall patterns of glycosylation are highly ordered with diagnostic alterations that can be detected during development and disease progression. The different glycoforms of a protein will also vary when that protein is expressed in different cells and at different stages of cell growth which, together with the function of carbohydrate in glycoprotein activity and antigenicity, makes it imperative that adequate analytical techniques are available to define glycoprotein structure. The rewards for such attention include better drugs and diagnostics in fields ranging through the immune pathologies, infection, and oncology.

Elizabeth F. Hounsell

Contents

Contributors

JACQUES BARA • *Mucin Immunochemistry Laboratory, Institut de Recherche sur le Cancer, Villejuif Cedex, France*

CHI KONG CHING • *Department of Medicine, University of Liverpool, UK*

ANTHONY P. CORFIELD • *Department of Medicine Laboratories, Bristol Royal Infirmary, Bristol, UK*

MICHAEL A. J. FERGUSON • *Department of Biochemistry, University of Dundee, Scotland*

GORDON FORSTNER • *Research Institute, Department of Biochemistry, The Hospital for Sick Children, and University of Toronto, Ontario, Canada*

JANET FORSTNER • *Research Institute, Department of Biochemistry, The Hospital for Sick Children, and University of Toronto, Ontario, Canada*

HANS-JOACHIM GABIUS • *Institut für Pharmazeutische Chemie der Philipps-Universität, Marburg, Germany*

SIGRUN GABIUS • *Abteilung Hamatologie-Onkologie, Medizinische Universitätsklinik, Göttingen, Germany*

HILDEGARD GEYER • *Institute of Biochemistry, University of Giessen, Germany*

RUDOLF GEYER • *Institute of Biochemistry, University of Giessen, Germany*

MARY GREGORIOU • *Laboratory of Molecular Biophysics, University of Oxford, UK*

M. LUCIA GÜTHER • *Biologia Celular, Escola Paulista de Medicina, Sao Paulo, Brazil*

ROBERT S. HALTIWANGER • *Department of Biological Chemistry, Johns Hopkins University School of Medicine, Baltimore, MD (Present address: Department of Biochemistry, State University of New York at Stony Brook, NY)*

GUNNAR C. HANSSON • *Department of Medical Biochemistry, University of Göteborg, Sweden*

GERALD W. HART • *Department of Biological Chemistry, Johns Hopkins University School of Medicine, Baltimore, MD*

Sumihiro Hase • *Department of Chemistry, Osaka University College of Science, Toyonaka, Osaka, Japan*
Anton Haselbeck • *Boehringer Mannheim GmbH, Biochemica Research Center, Tutzing/Obb., Germany*
Susumu Honda • *Faculty of Pharmaceutical Sciences, Kinki University, Kowakae, Higashi-Osaka, Japans*
Wolfgang Hösel • *Boehringer Mannheim GmbH, Biochemica Research Center, Tutzing/Obb., Germany*
Kyoko Hotta • *Department of Biochemistry, School of Medicine, Kitasato University, Sagamihara, Kanagawa, Japan*
Elizabeth F. Hounsell • *Clinical Research Centre, Harrow, UK*
Hitoo Iwase • *Department of Biochemistry, School of Medicine, Kitasato University, Sagamihara, Kanagawa, Japan*
Kazuaki Kakehi • *Faculty of Pharmaceutical Sciences, Kinki University, Kowakae, Higashi-Osaka, Japan*
Hasse Karlsson • *Department of Medical Biochemistry, University of Göteborg, Sweden*
Ismat Khatri • *Research Institute, Department of Biochemistry, The Hospital for Sick Children, and University of Toronto, Ontario, Canada*
Johann Lechner • *Lehrstuhl für Biochemie, Regensburg, Germany*
Terence J. McMaster • *H. H. Wills Physics Laboratory, University of Bristol, UK*
Tsuguo Mizuochi • *Institute for Comprehensive Medical Science, Fujita Health University School of Medicine, Toyoake, Aichi, Japan*
Victor J. Morris • *AFRC Institute of Food Research, Norwich Laboratory, Norwich Research Park, Colney, Norwich, UK*
Bo Nilsson • *BioCarb Technology, Lund, Sweden*
Rafael Oriol • *Biochimie, Faculté de Pharmacie, Chatenay-Malabry, France*
Toshiaki Osawa • *Division of Chemical Toxicology and Immunohistochemistry, Faculty of Pharmaceutical Sciences, University of Tokyo, Japan*
Christos Paraskeva • *Department of Pathology, University of Bristol, UK*
Jonathan M. Rhodes • *Department of Medicine, University of Liverpool, UK*
Christopher C. Rider • *Department of Biochemistry, Royal Holloway and Bedford New College, University of London, Egham Hill, UK*
Anne P. Sherblom • *Department of Biochemistry, Microbiology, and Molecular Biology, University of Maine, Orono, ME (Present name and address: Anne P. Clark, Lung Biology and Pathology Study Section, Division of Research Grants, National Institutes of Health, Bethesda, MD)*

ROSALITA M. SMAGULA • *Department of Biochemistry, Microbiology, and Molecular Biology, University of Maine, Orono, ME*
TSUTOMU TSUJI • *Division of Chemical Toxicology and Immunohistochemistry, Faculty of Pharmaceutical Sciences, University of Tokyo, Japan*
FELIX WIELAND • *Institut für Biochemie, Heidelberg, Germany*
KAZUO YAMAMOTO • *Division of Chemical Toxicology and Immunohistochemistry, Faculty of Pharmaceutical Sciences, University of Tokyo, Japan*

CHAPTER 1

A General Strategy for Glycoprotein Oligosaccharide Analysis

Elizabeth F. Hounsell

1. Introduction

There is a large diversity of oligosaccharide sequences linked to proteins. The molecules classically called glycoproteins comprise serum and cell membrane glycoproteins of approximate mol-wt range 20–200 kDa, having oligosaccharide chains linked to Ser/Thr or Asn, i.e., *O*- and *N*-linked, respectively, making up between 10 and 60% by weight. The structures of these chains are introduced in Chapters 2–12 of this book and the variety of glycosylation assessed in the Glossary. Additionally covered are (1) mucins, which are traditionally defined as larger mol-wt glycoproteins of 10^6 kDa and upwards having >60% oligosaccharide, mainly *O*-linked via GalNAc-containing oligosaccharide cores, and (2) proteoglycans or glycosaminoglycans, which also have a high carbohydrate/protein ratio, but classically have disaccharide repeating units with an alternate uronic acid residue and a large degree of sulfation. The distinction between these categories is becoming increasingly blurred, and they can now be seen as a spectrum of the aforementioned glycosylation patterns occurring on high- and low-mol-wt, secreted, and cell-surface glycoproteins. As examples of this diversity, classical mucin and proteoglycan sequences can occur on cell-membrane-attached proteins of relatively low mol-wt *(1);* glycoproteins and proteoglycans are found in forms attached to the membrane by lipid-linked anchors (ref. *1* and Chapter 8), and cytoplasmic glycoproteins having GlcNAc linked to Ser and Thr (ref. *2* and Chapter

From: *Methods in Molecular Biology, Vol. 14: Glycoprotein Analysis in Biomedicine*
Edited by: E. F. Hounsell

Table 1
Examples of Analysis Techniques that Detect Carbohydrates in a Highly Specific Way with Very Few Cross-reactions to Other Biological Molecules

Chemical	• Oxidation with sodium metaperiodate, which cleaves specifically between two adjacent hydroxyl groups (as in the periodic acid Schiff reagent, PAS). • Phenol/sulfuric acid charring of mono- or oligosaccharides having a hydroxyl group at C2. • Reduction of mono- or oligosaccharides having a free reducing end after release from protein or hydrolysis of glycosidic bonds. • Binding of reduced oligosaccharides (alditols) to PBA. • Addition of a chemical label by reductive amination. • Nitrous acid cleavage of oligosaccharides at non-*N*-acetylated hexosamine residues. • Detection of polysulfated oligosaccharides by Alcian blue staining.
Biological	• Release of monosaccharides by exoglycosidases. • Release of oligosaccharides by endoglycosidases. • Metabolic labeling with ^{35}S or ^{3}H monosaccharides. • Addition of monosaccharides by glycosyl transferases. • Binding to lectins or anticarbohydrate antibodies.
Physicochemical	• Characteristic mol wt by mass spectrometry (MS). • Characteristic chromatographic profile. • Characteristic signals in a nuclear magnetic resonance spectrum (NMR).

14) are now also known. Additional oligosaccharide-to-protein linkages are found in proteoglycans *(1)* and bacterial glycoproteins (ref. *3* and Chapter 9). Because of the variety of glycoproteins, monosaccharides, and their derivatives found in nature, the assessment of glycosylation status of a protein, requires four levels of questioning.

1.1. Is It a Glycoprotein?

There are several methods for assessing the presence of mono- to oligosaccharide molecules linked to protein, or of free oligosaccharides and other oligosaccharide conjugates. These are summarized in Table 1. Oxidation with periodate is a classical method, e.g., the periodate-Schiff reagent (PAS) and Smith degradation, more recently adopted (Section 3.1.2. and Chapter 3) as part of microsequencing strategies for structural analysis and as commercial kits for glycoprotein detection in conjunction with lectins or antibodies (Chapter 13). The phenol sulfuric acid assay (Chapter 12) can be carried out at microscale in a multiwell titer plate and read by an ELISA plate reader to detect down to 50 ng of monosaccharides having a C2 hydroxyl group (e.g., galactose, mannose, glucose). Reduction methods (concomitant with oligosaccharide release for *O*-linked chains) are discussed in several chapters. Reduction can be used to detect oligosaccharides specifically by introduction of a radioactive label and purification on a phenylboronic acid (PBA) column (Section 3.1.2. and refs. *2,5*). High sensitivity analysis can also be achieved by addition of a chemical label by a related technique called reductive amination. This relies on the fact that a reduced chain can be oxidized by periodate to give a reactive aldehyde for linkage to an amine-containing compound or that free reducing sugars exist for part of the time in the open chain aldehyde form, which can react equally. Derivatives chosen include lipid for TLC overlay assays and TLC-MS analysis *(4,6)* UV-absorbing groups, which also give sensitive MS detection *(6,7,* and Chapter 6 herein), and sulfated aromatic amines for electrophoretic detection *(8,9)*. These can be detected down to picomole level.

1.2. What Type of Oligosaccharide Sequences Are Present?

Essential in any analysis strategy is an initial screen for the types of oligosaccharide chain present, e.g., *O*- or *N*-linked chains, and also for the presence of any labile chemical linkages that might be destroyed by the subsequent analytical techniques used. High-sensitivity monosaccharide analysis by HPLC (*see* Chapter 7) or HPAE (Section 3.1.3. and ref. *10*) can be achieved. However, the GC analysis method described in the present chapter using trimethylsilyl ethers of methyl glycosides is the most widely applicable, being able in one run to identify pentoses (e.g., ribose), deoxyhexoses (e.g., fucose, xylose, ribose, rhamnose, arabinose), hexoses (e.g., mannose, glucose, galactose, talose), hexosamines (e.g., glucosamine, galactosamine), uronic acids, and sialic acids. Related analyses are discussed in Chapters 8 and 9. GC of chiral derivatives *(11)* can be additionally used to determine the D and L configurations of monosaccharides. The technique of GC-MS analysis of partially methylated alditol acetates derivatized as described in Chapters 3, 4, and 8 is also a very useful technique that can identify the hydroxyl groups through which each monosaccharide is linked, thus establishing their presence in a chain and giving vital structural information. This type of analysis can now be conveniently performed on bench top GC–MS equipment at the picomole to nanomole levels.

Obtaining a high-field MS analysis of released oligosaccharide chains in their native form, e.g., by fast atom bombardment (FAB) or liquid secondary ion (LSI) MS, is a very useful procedure in discovering any labile groups removed by derivatization. Permethylated oligosaccharides, available as part of the route to partially methylated alditol acetates, can also be analyzed by this technique to give additional sequence information. Alternative derivatives are peracetylated oligosaccharides, which are readily formed and extracted to give very clean samples for MS analysis *(12,13)*. High-sensitivity detection of high-mol-wt molecules down to a few picomole of material can be achieved by the largest mass spectrometers, particularly of oligosaccharides derivatized at the reducing end as discussed earlier.

1.3. What Is the Best Strategy for Release of Oligosaccharide Chains?

When initial clues as to oligosaccharide types have been gained, confirmatory evidence can be obtained by specific chemical or enzymatic release. Both types of method have been researched extensively over the past decade to achieve a high degree of perfection minimizing any nonspecific side reactions while maximizing oligosaccharide yield. To obtain typical *N*- and *O*-linked oligosaccharides, chemical release can be best achieved by hydrazinolysis or alkali treatment. Hydrazinolytic cleavage of *N*-linked chains is discussed in Chapter 5. A commercial instrument for this has recently become available from Oxford Glycosystems (Oxford, UK) that includes a step using milder conditions, which is claimed to release intact *O*-linked chains as well. The latter step has been classically achieved by mild alkali treatment, e.g., 0.05*M* sodium hydroxide at 50°C for 16 h in the presence of 0.5–1*M* $NaBH_4$ (*see* Chapters 4, 9, and 10) to yield intact oligosaccharide alditols. The mild hydrazinolysis and alkali/reduction conditions result in some peptide breakdown, whereas hydrazinolysis for release of *N*-linked chain cleaves the majority of peptide bonds. Enzymatic release leaves the peptide intact and obviates possible chemical breakdown of oligosaccharides. However, occasionally it may be necessary to protease digest first to achieve complete oligosaccharide release *(14)*, and the enzymes may not cleave all possible structures (e.g., *see* Chapters 10 and 12 for a discussion of *N*- and *O*-linked chains, respectively). The extent of deglycosylation can be readily judged by the detection methods discussed in Table 1. Specific identification of different forms of *N*-linked chains by enzymatic release are discussed in Chapters 10 and 11.

1.4. What Does My Glycoprotein Look Like?

The oligosaccharide chains of glycoproteins are fashioned by a series of enzymes acting in specific sequence in different subcellular compartments. The end product is dependent on a number of factors, including the initial protein message and its processing, availability of enzymes, substrate levels, and so on, factors that can vary among different cell types, different species, and at different times in the cell cycle. It is therefore important to address the question of glycoprotein structure to specific glycosylation sites and have profiling methods capable of detecting minor changes in structure that may be important in function and antigenicity. The following route is discussed in the present chapter and succeeding contributions in the first half of this book.

1. Initial characterization of type and amount of each monosaccharide (picomole to nanomole).
2. Protease digestion and analysis of the complete digest by high-field MS (peptides in 20 pmol digest detected).
3. HPLC peptide mapping and microassay for glycopeptides (*see* Table 1) followed by peptide *N*-terminal analysis of identified glycopeptides.
4. Release of oligosaccharides and chromatographic profiling giving composition, sequence, and linkage information.
5. NMR analysis of >50 μg chromatographically pure oligosaccharide.
6. Conformational analysis of oligosaccharides by computer graphics molecular modeling.

The second half of the book looks at methods for *in situ* analysis of glycoproteins and glycoprotein interactions. This is particularly important in the biomedical field where specific patterns of oligosaccharide sequences on different cell types are markers of differentiation and disease. This is exemplified by the latest papers describing the potential of oligosaccharide analysis in diagnosis in a special issue of *Carbohydrate Research* on lectins *(15)*. By having detailed knowledge of oligosaccharide sequences, we can devise analogs as inhibitors of endogenous carbohydrate-binding proteins (lectins, selectins, lecams, and so on) such as ELAM-1, which mediates neutrophil–endothelial cell interactions *(16)*. Analysis of glycoproteins, glycosyl-phosphatidyl inositol glycans, and lipopolysaccharide structure by the methods discussed here will lead to an enhanced understanding of processes in infection and immunity where oligosaccharides are often in the front line. Given the present difficulties of visualizing oligosaccharides in the crystal structure of glycoproteins (discussed in ref. *17*), the physicochemical and microscopic methods described in this and subsequent chapters are an essential part of any biochemistry and molecular biology laboratory.

2. Materials

1. 0.5*M* Methanolic HCl (Supelco, Bellefont, PA).
2. Screw-top PTFE septum vials.
3. Phosphorus pentoxide.
4. Silver carbonate.
5. Acetic anhydride.
6. Trimethylsilylating (TMS) reagent (Sylon HTP kit, Supelco: pyridine hexamethyldisilazane, trimethylchlorosilane; **Use Care: corrosive**).
7. Toluene stored over 3-Å molecular sieve.
8. GC apparatus fitted with flame ionization or MS detector and column, e.g., for TMS ethers 25*M* × 0.22 mm id BP10 (SGE, Austin, TX) and for partially methylated alditol acetates 25*M* × 0.22 mm id 5% PhMe silicone (Hewlett Packard, Bracknell, Berks., UK).
9. 1*M* $NaBH_4$ in 0.05*M* NaOH made up fresh.
10. Cation-exchange column.
11. Phenyl boronic acid (PBA) Bond Elut columns (Jones Chromatography, Hengoed Glamorgan, UK) activated with MeOH.
12. 0.2*M* NH_4OH.
13. HCl (0.01, 0.1, and 1*M*).
14. HPLC apparatus fitted with UV detector (approx 1 nanomole mono- and oligosaccharides containing *N*-acetyl groups can be detected at 195–210 nm) and pulsed electrochemical detector (oligo- and monosaccharides ionized at high pH can be detected at picomole level). Columns: reversed phase (RP) C18, amino bonded silica, porous graphitized carbon (Shandon, Runcorn, Cheshire, UK), glycopak N (Waters, Harrow, UK), CarboPac PAl (Dionex, Camberley, UK).
15. Eluents for RP-HPLC: eluent A, 0.1% aqueous TFA; eluent B acetonitrile containing 0.1% TFA.
16. Eluents for high-pH anion-exchange (HPAE) Dionex Chromatography: 12.5*M* NaOH (BDH) diluted fresh each day to 200, 100, and 1 m*M*; 500 m*M* sodium acetate (Aldrich, Gillingham, UK). After chromatography and detection, salt needs to be removed by a Dionex

micro membrane suppressor or by cation-exchange chromatography before further analysis, e.g., by methylation.

17. NMR tubes (5 mL).
18. D_2O (99.90% for repeated evaporation and 99.96% for the final solution for NMR).
19. Acetone.
20. Access to 270–600 MHz NMR.
21. High-resolution computer graphics terminal plus data processing, e.g., Silicon Graphics Personal IRIS (Reading, Berks, UK), and supported graphics software.

3. Methods

3.1. Initial Characterization of the Type and Amount of Sugar

3.1.1. GC Composition Analysis (see *Note 1)*

1. Concentrate glycoproteins or oligosaccharides containing 1–100 µg carbohydrate and 10 µg internal standard (e.g., arabinose or perseitol) in screw-top septum vials. Dry under vacuum in a dessicator over phosphorus pentoxide.
2. Place the sample under a gentle stream of nitrogen, and add 200 µL methanolic HCl (*see* Note 2).
3. Cap immediately, and heat at 80°C for 18 h (*see* Note 3).
4. Cool the vial, open, and add approx 50 mg silver carbonate.
5. Mix the contents, and test for neutrality (*see* Note 4).
6. Add 50 µL acetic anhydride, and stand at room temperature for 4 h in the dark (*see* Note 5).
7. Spin down the solid residue (*see* Note 6), and remove the supernatant to a clean vial.
8. Add 100 µL methanol, and repeat step 7, adding the supernatants together.
9. Repeat step 8 and evaporate the combined supernatants under a stream of nitrogen.
10. Dry over phosphorus pentoxide before adding 20 µL trimethylsilylating reagent.
11. Heat at 60°C for 5 min, evaporate remaining solvent under a stream of nitrogen, and add 20 µL dry toluene.
12. Inject onto a standard or capillary GC column. (A typical chromatogram is shown in Fig. 1.)
13. Calculate the total peak area of each monosaccharide by adding individual peaks and dividing by the peak area ratio of the internal standard. Compare to standard curves for molar calculation determination.

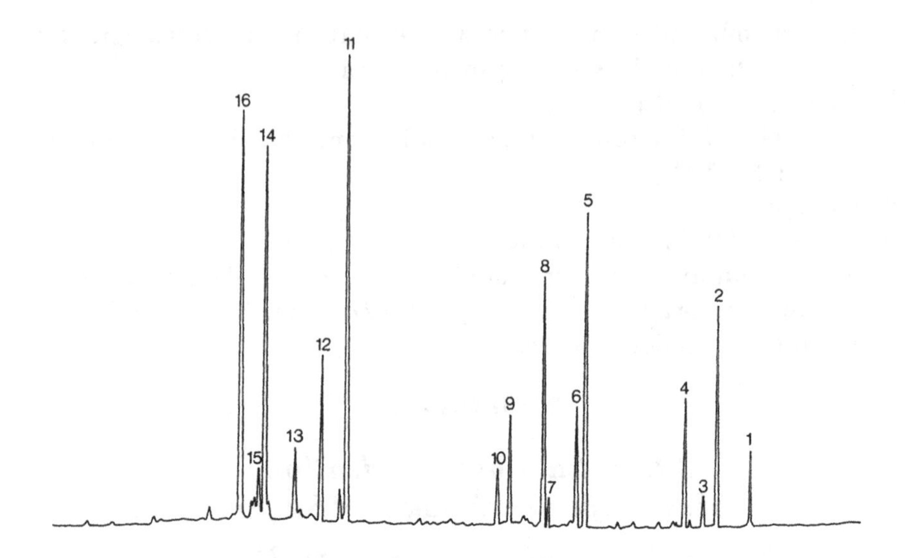

Fig. 1. GC of trimethylsilyl ethers of methyl glycoside derivatives of monosaccharides: peaks 1–4 fucose; 5,7, mannose; 6,8,9, galactose; 10, glucose; 11, inositol; 12,14, *N*-acetylgalactosamine; 13,15,16, *N*-acetylglucosamine. (*N*-acetylneuraminic acid occurs as a single peak with longer retention than the hexosacetamido residues; glucuronic and galacturonic acids chromatograph close to glucose and galactose, respectively; rhamnose, xylose and arabinose chromatograph between fucose and mannose).

3.1.2. Characterization of O*-Linked Glycosylation*

1. Release *O*-linked chains by treatment with 0.05*M* sodium hydroxide in the presence of 1*M* $NaBH_4$ or $NaB[^3H]_4$ for 16 h at 50°C, and desalt on a cation-exchange column (*see* Chapters 4, 9, and 10).
2. Dissolve the samples in 200 µL 0.2*M* NH_4OH and add to the top of a PBA mini column prewashed with MeOH water and 0.2*M* NH_4OH.
3. Wash the column with 2 × 100 µL 0.2*M* NH_4OH and 2 × 100 µL water.
4. Elute the sample in 1*M* acetic acid (*see* Note 7).
5. Evaporate the sample and reevaporate with water 2 × 100 µL.
6. Carry out periodate oxidation as described in Chapters 3, 15, and 21 or ref. *4* using conditions suitable for alditol oxidation, e.g., 6 m*M* periodate for 5 min at 0°C or 1 m*M* for 48 h at 4°C.
7. Couple the reactive aldehyde to an organic amine of choice as discussed in Section 1.1. and Chapter 6.
8. Analyze by TLC or HPLC, and compare to standards (*see* Note 8).

3.1.3. Characterization of Sialic Acid Residues (see *Note 9*)

1. Hydrolyze oligosaccharides or glycoproteins with 0.01*M* HCl for 1 h at 70°C to remove *N*-glycolyl or *N*-acetylneuraminic acid with mostly intact *O*- and *N*-acyl groups.
2. Hydrolyze with 0.025*M* (2 h) to 0.1*M* (1 h) HCl at 80°C to remove the majority of sialic acids, but with some *O*- and *N*-acyl degradation.
3. Hydrolyze with 0.5*M* HCl at 80°C for 1 h to remove all sialic acids and fucose.
4. Analyze the released sialic acids by HPAE chromatography *(18)* on a Dionex CarboPac PAl column with pulsed electrochemical detection and gradient 100 m*M* NaOH 0–4 min to 70% 100 m*M* NaOH/30% 1*M* NaOAc 4–30 min.

3.2. Glycopeptide Preparation

1. Choose a protease or proteases that will cleave the glycoprotein into peptides and glycopeptides of ideally 5–15 amino acids (*see* Note 10).
2. Set up the protease digestion in an appropriate buffer (*see* Note 11).
3. Stop the reaction by storage at 4°C, evaporate in a Gyrovap, and reevaporate with water 3 × 100 µL.
4. Take up the samples in HPLC eluent A and load a 1-µL aliquot containing a 200-pmol sample onto the FAB- or LSI-MS probe tip containing the matrix thioglycerol:TFA (10:1) or thioglycerol:glycerol (1:1). A typical spectrum is shown in Fig. 2 (*see* Note 12).

3.3. Glycopeptide Purification

1. Chromatograph 200-picomol glycopeptide mixture by RP-HPLC using a gradient of 98% eluent A/2% eluent B to 18% eluent A and 82% eluent B in 80 min at a flow rate of 1 mL/min and UV detection at 210 nm.
2. Pool the fractions of each peak.
3. Analyze the peaks for sugar by one of the microtechniques described in Table 1.
4. Take 20 pmol glycopeptides for peptide *N*-terminal analysis (*see* vol. 1 of this series).

3.4. Oligosaccharide Profiling

1. Release oligosaccharides by enzymic or chemical methods.
2. Oligosaccharides removed by enzymes as described in Chapters 10 and 11 can be recovered in the supernatant after ethanol precipitation *(19)*.

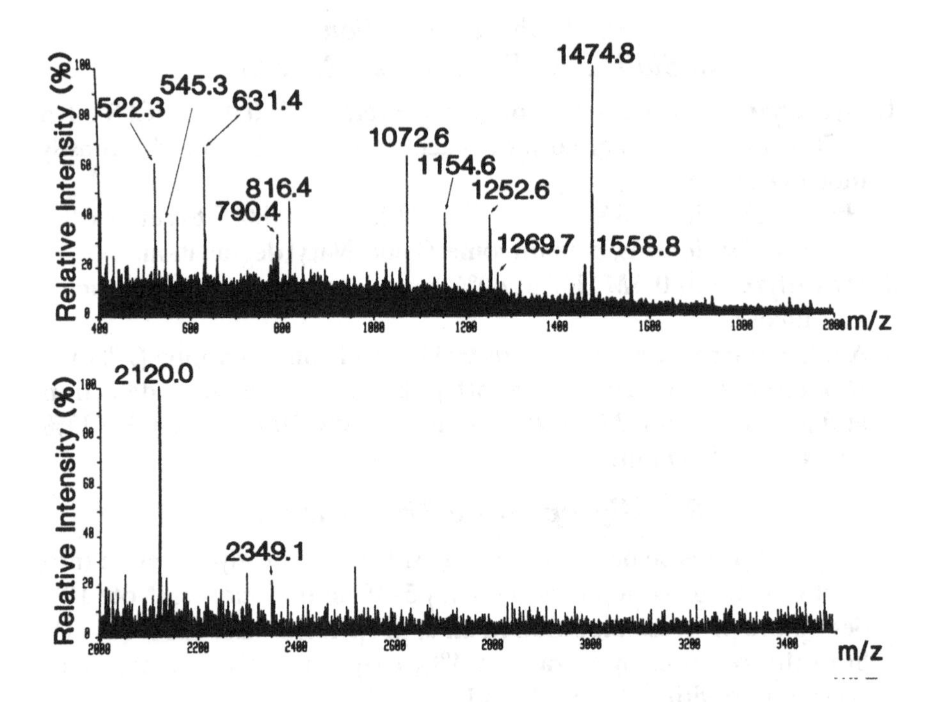

Fig. 2. The LSIMS spectrum of a trypsin digest of the glycoprotein fetuin *(19)* showing mass/charge (*m/z* where *z* = 1) values for nonglycosylated peptides. Reproduced from ref. *19* by permission of John Wiley and Sons.

3. Analyze by HPLC or HPAE as described in Chapters 6–12 and in ref. *20*.
4. Methylate 5 nmol purified desalted oligosaccharide as described in Chapters 3, 4, and 8.
5. If a high-field MS instrument is available, analyze half of the sample by FAB-MS or LSIMS.
6. Derivatize the remainder of the sample to partially methylated alditol acetates, and analyze by GC or preferably GC–MS (Chapter 8).
7. Carry out additional structural analysis techniques as described in Chapters 2–12.

3.5. ^{1}H-NMR Analysis of Oligosaccharides or Glycopeptides

1. Evaporate the purified and desalted sample three times from D_2O (*see* Note 13).
2. Take up the sample in 400 µL D_2O, and add 1 µL of 5% acetone in D_2O for each 50-µg sample present.

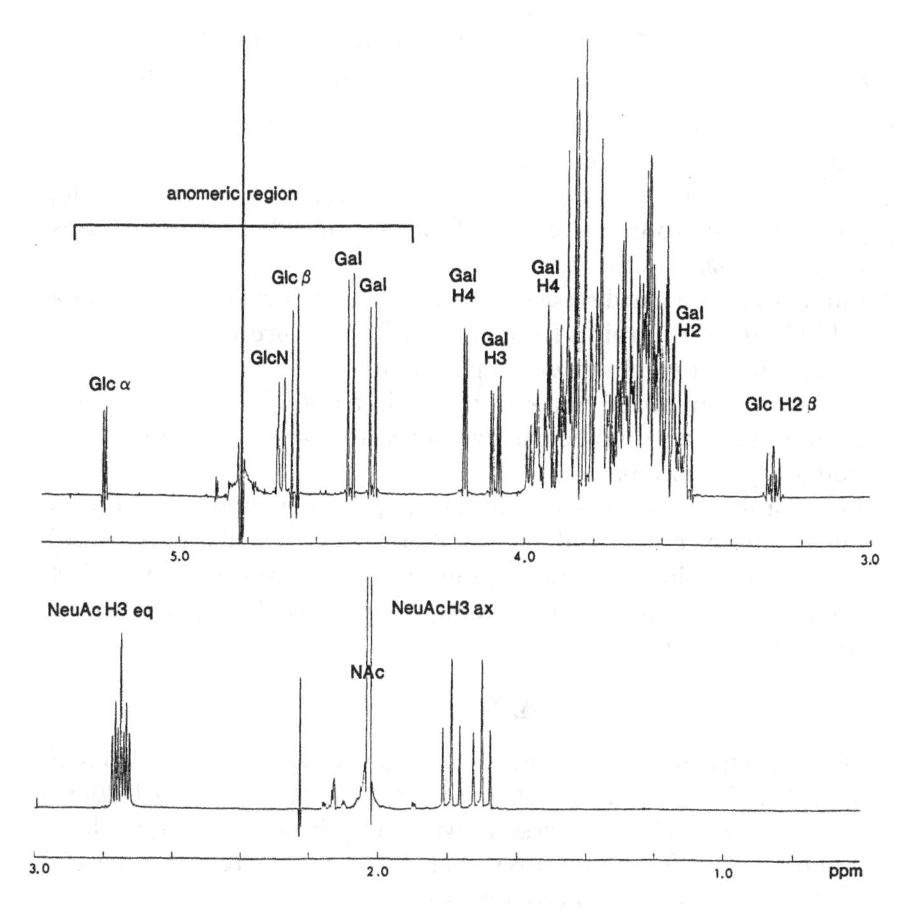

Fig. 3. The 1D proton (^1H) NMR spectrum in D_2O at 22°C of the oligosaccharide Galβ1-3GlcNAcβ1-3Galβ1-4Glcα/β

| 2,3 | 2,6

NeuAcα NeuAcα

(ppm referenced from acetone at 2.225 ppm)

3. Transfer the sample to an NMR tube, and store capped at 4°C (*see* Note 14).
4. Carry out standard 1D and 2D ^1H-NMR experiments at 22°C, and assign chemical shifts to specific proton signals (*see* Note 15; Fig. 3).
5. Carry out rotating frame NOE experiments using 500–100 ms mixing times, e.g., 200 ms for tri- to heptasaccharides (*see* Note 16) in order to obtain information of through-space interactions.

3.6. Computer Graphics Molecular Modeling of Oligosaccharides, Glycopeptides, and Glycoproteins

1. Establish a monosaccharide library based on Cambridge crystallographic coordinates (*see* Note 17).
2. Make 3D ϕ/ψ/energy plots of disaccharides using force fields parameterized to take account of oligosaccharide specific features, e.g., the exoanomeric effect *(21,22)*.
3. Model oligosaccharides based on ϕ,ψ angles at local energy minima.
4. Add distance constraints from NMR NOE measurements.
5. Carry out energy minimization procedures.
6. Set up the minimized oligosaccharide in a water box, and carry out minimization and molecular dynamics simulated annealing under boundary conditions.
7. Link minimized oligosaccharides to peptides or modeled proteins, and reminimize. *O*-linked oligosaccharides in particular have considerable interaction with the peptide backbone, and therefore, NOE distance constraints from NMR experiments of *O*-linked glycopeptides should be included.

4. Notes

1. Monosaccharide compositonal analysis can also be carried out by HPLC- and HPAE-based methods (*see* Chapter 7 and refs. *10, 20*) if a GC is not available or extra sensitivity is required. However, if a large number of different monosaccharides are expected, the present method gives unambiguous analysis.
2. The use of methanolic HCl for cleavage of glycosidic bonds and oligosaccharide-peptide cleavage yields methyl glycosides and carboxyl group methyl esters, which gives acid stability to the released monosaccharides, and thus, monosaccharides of different chemical lability can be measured in one run. If required as free reducing monosaccharides (e.g., for HPLC), the methyl glycoside can be removed by hydrolysis (Chapter 9). The reagent can be obtained from commercial sources or made in laboratory by bubbling HCl gas through methanol or by adding acetyl chloride to methanol (Chapter 9).
3. An equilibrium of the α and β methyl glycosides of monosaccharide furanose (f) and pyranose (p) rings is achieved after 18 h so that a characteristic ratio of the four possible (fα, fβ, pα, pβ) molecules is formed to aid in unambiguous monosaccharide assignment.
4. Solid-silver carbonate has a pink hue in an acidic environment, and therefore, neutrality can be assumed when green coloration is achieved.

5. The acidic conditions remove *N*-acetyl groups, which are replaced by acetic anhydride. This means that the original status of *N*-acetylation of hexosamines and sialic acids is not determined in the analysis procedure.
6. Direct re-*N*-acetylation by the addition of pyridine-acetic anhydride 1:1 in the absence of silver carbonate can be achieved, but this gives more variable results.
7. If ^3H $NaBH_4$ is used, this method serves to give a very sensitive detection of the presence of oligosaccharides after reduction of free reducing ends.
8. Released *O*-linked chains having the GalNAc-ol at the reduced end are cleaved by periodate at the C4–C5 bond, thus giving a characteristic product for sequences linked at C3 and/or C6 (*see* Chapter 3 and ref. 4).
9. The term sialic acid refers to a family of monosaccharides, of which the major types are *N*-acetyl or *N*-glycolyl, having various *O*-acyl groups. Other hydrolysis conditions removes these *O*- and *N*-acyl groups, but the method described in Section 3.1.1. can serve to identify the presence of any one of the family.
10. The presence of disulfide bonds and glycosylation sites restricts peptide cleavage by proteases; therefore, the protease(s) chosen should have consensus cleavage sites away from these features. Protease digestion after reduction and carboxymethylation can make other sites available.
11. Sialic acid residues may be partially destroyed by basic conditions above pH 8.0.
12. At the 20 pmol level, only peptides are detected in a glycopeptide/peptide mixture owing to suppression of peptide ionization by the presence of oligosaccharide. Glycopeptides once identified (Section 3.3.) are therefore analyzed for amino acid sequence by peptide *N*-terminal analysis.
13. This serves to exchange all OH and NH groups to OD and ND (where D is deuterium), and therefore, only the CH protons are detected. The experiments can be repeated in H_2O containing 1% D_2O as a spin lock. If it is required to cool the sample to <4°C, acetone/H_2O can be used as the solvent.
14. NMR tubes can crack if stored containing D_2O at <4°C for long periods. At high concentrations, bacterial contamination can be avoided by making up the samples in 0.02% sodium azide evaporated from D_2O.
15. Chemical shifts from 1D experiments can be assigned with the aid of computer-assisted interpretation as described in ref. *23*. The computerized data base and the majority of the literature data for ^1H-NMR of oligosaccharides are for analysis at 295 K (22°C). Additional experiments can be performed at different temperatures to reveal

signals near the HOD resonance. Chemical shifts are usually given with respect to the acetone signal at 2.225 ppm at 295 K.

16. Oligosaccharides in the size range tetra- to heptasaccharide cannot be detected by standard NOE experiments because of cancellation of positive and negative signals at around 22°C.
17. The first generation of computer graphics software for molecular modeling does not adequately cater to monosaccharide and oligosaccharides. The required data can now be obtained from the Institut National de la Recherche Agronomique, Nantes, France.

References

1. Poole, A. R. (1986) Proteoglycans in health and disease: structure and functions. *Biochem. J.* **236,** 1–14.
2. King, I. A. and Hounsell, E. F. (1989) Cytokeratin 13 contains *O*-glycosidically linked *N*-acetylglucosamine residues. *J. Biol. Chem.* **264,** 14,022–14,028.
3. Messner, P. and Sleytr, U. B. (1991) Bacterial surface layer glycoproteins, *Glycobiology,* **1,** 545–551, and references in Chapter 9 of this book.
4. Stoll, M. S., Hounsell, E. F., Lawson, A. M., Chai, W., and Feizi, T. (1990) Microscale sequencing of *O*-linked oligosaccharides using mild periodate oxidation of alditols, coupling to phospholipid and TLC-MS analysis of the resulting neoglycolipids. *Eur. J. Biochem.* **189,** 499–507.
5. Stoll, M. S. and Hounsell, E. F. (1988) Selective purification of reduced oligosaccharides using a phenylboronic acid bond elut column: potential application in HPLC, mass spectrometry, reductive amination procedures and antigenic/serum analysis. *Biomed. Chromatogr.* **2,** 249–253.
6. Lawson, A. M., Chai, W., Cashmore, G. C., Stoll, M. S., Hounsell, E. F., and Feizi, T. (1990) High-sensitivity structural analyses of oligosaccharide probes (neoglycolipids) by liquid-secondary-ion mass spectrometry. *Carbohyd. Res.* **200,** 47–57.
7. Poulter, L., Karrer, R., and Burlingame, A. L. (1991) n-Alkyl p-amino benzoates as derivatising agents in the isolation, separation and characterisation of submicrogram quantities of oligosaccharides by LSIMS. *Anal. Biochem.* **195,** 1–13.
8. Lee, K. B., Lindhart, R. J., and Al-Hakim, A. (1991) Gel and capillary electrophoresis based oligosaccharide sequencing. *Glycoconj. J.* **8,** 250–251.
9. Sullivan, M. T., Klock, J., and Stack, R. J. (1991) Characterisation of N-linked glycoprotein oligosaccharides by fluorescence assisted carbohydrate electrophoresis (FACE). *Glycoconj. J.* **8,** 249.
10. Hardy, M. R. (1989) Monosaccharide analysis of glycoconjugates by high-performance anion-exchange chromatography with pulsed amperometric detection. *Adv. Enzymol.* **179,** 77–83.
11. Gerwig, G. J., Kamerling, J. P., and Vliegenthart, J. F. G. (1979) Determination of the absolute configuration of monosaccharides in complex carbohydrates by capillary G.L.C. *Carbohydr. Res.* **77,** 1–7.

12. Hounsell, E. F., Madigan, M. J., and Lawson, A. M. (1984) Fast-atom-bombardment mass spectrometry and electron-impact direct-probe mass spectrometry in the structural analysis of oligosaccharides with special reference to the poly(N-acety-lactosamine) series. *Biochem. J.* **219,** 947–952.
13. Dell, A. (1987) FAB-MS of carbohydrates. *Adv. Carbohydr. Chem. Biochem.* **45,** 19–72.
14. Carr, S. A., Barr, J. R., Roberts, G. D., Anumula, K. R., and Taylor, P. B. (1991) Identification of attachment sites and structural classes of asparagine-linked carbohydrates in glycoproteins. *Methods Enzymol.* **193,** 501–517.
15. A collection of papers on lectins in honour of Professor Toshiaki Osawa and Professor Nathan Sharon (1991) *Carbohydr. Res.* **213,** 1–359.
16. Lowe, J. B., Stoolman, L. M., Nair, R. P., Larsen, R. D., Berhand, T. L., and Marks, R. M. (1990) ELAM-1-dependent cell adhesion to vascular endothelium determined by a transfected human fucosyltransferase DNA. *Cell* **63,** 475–484.
17. Hounsell, E. F. (1993) Physicochemical analysis of oligosacchride determinants of glycoproteins. *Adv. Carbohydr. Chem. Biochem.*, **50.**
18. Manzi, A. E., Diaz, S., and Varki, A. (1990) High-pressure liquid chromatography of sialic acids on a pellicular resin anion-exchange column with pulsed amperometric detection: a comparison with six other systems. *Anal. Biochem.* **188,** 20–32.
19. Smith, K. D., Harbin, A-M., Carruthers, R. A., Lawson, A. M., and Hounsell, E. F. (1990) Enzyme degradation, high performance liquid chromatography and liquid secondary ion mass spectrometry in the analysis of glycoproteins. *Biomed. Chromatogr.* **4,** 261–266.
20. Hounsell, E. F. (1993) Glycoprotein carbohydrates, in *HPLC of Small Molecules* (Lim, C. K., ed.), Oxford University Press, Oxford, UK.
21. Lemieux, R. U. and Bock, K. (1983) The conformational analysis of oligosaccharides by ^{1}H-NMR and HSEA calculations. *Arch. Biochem.* **221,** 125–134.
22. French, A. D. and Brady, J. W., eds. (1990) *Computer Modeling of Carbohydrate Molecules.* ACS Symposium Series 430, American Chemical Society, Washington, DC.
23. Hounsell, E. F. and Wright, D. J. (1990) Computer-assisted interpretation of ^{1}H-nmr spectra in the analysis of the structure of oligosaccharides. *Carbohydr. Res.* **205,** 19–29.

CHAPTER 2

Analysis of Asparagine-Linked Oligosaccharides by Sequential Lectin-Affinity Chromatography

Kazuo Yamamoto, Tsutomu Tsuji, and Toshiaki Osawa

1. Introduction

Sugar moieties on the cell surface play one of the most important roles in cellular recognition. In order to elucidate the molecular mechanism of these cellular phenomena, assessment of the structure of sugar chains is indispensable. However, it is difficult to elucidate the structures of cell-surface oligosaccharides because of two technical problems. First is the difficulty in fractionating various oligosaccharides heterogeneous in the number, type, and substitution patterns of outer sugar branches. The second problem is that very limited amounts of material can be available, which makes it difficult to perform detailed structural studies. Lectins are proteins with sugar-binding activity. Each lectin binds specifically to a certain sugar sequence in oligosaccharides and glycopeptides. To overcome the problems just described, lectins are very useful tools. Recently, many attempts have been made to fractionate oligosaccharides and glycopeptides on immobilized lectin columns. The use of a series of immobilized lectin columns, whose sugar-binding specificities have been precisely elucidated, enables us to fractionate a very small amount of radioactive oligosaccharides or glycopeptides (ca. 10 ng, depending on the specific activity) into structurally distinct groups. In this chapter, we summarize the serial lectin-

From: *Methods in Molecular Biology, Vol. 14: Glycoprotein Analysis in Biomedicine*
Edited by: E. F. Hounsell

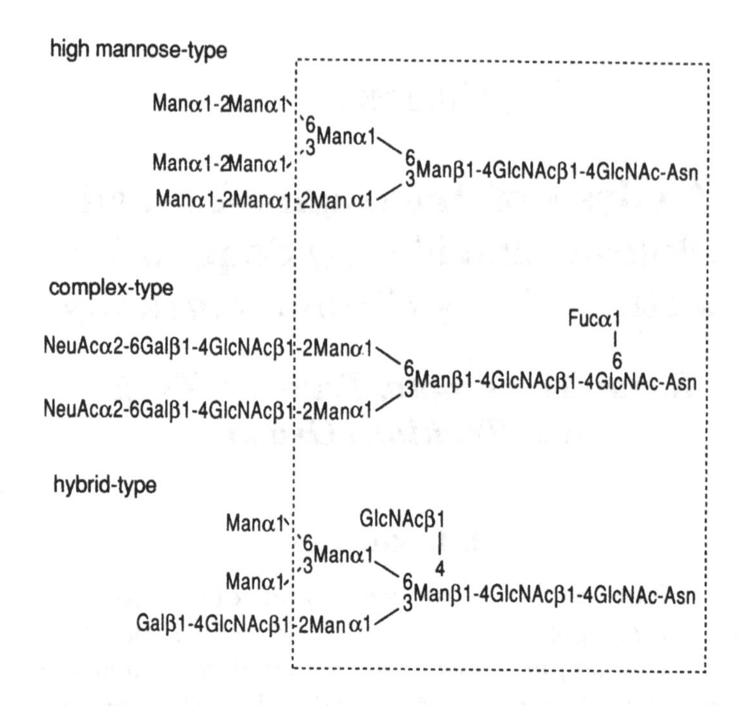

Fig. 1. Structures of major types of asparagine-linked oligosaccharides. The boxed area encloses the core structure common to all asparagine-linked structures.

Sepharose affinity chromatographic technique for rapid, sensitive, and specific fractionation and analysis of asparagine-linked oligosaccharides of glycoproteins.

Structures of asparagine-linked oligosaccharides fall into three main categories termed high mannose-type, complex-type and hybrid-type *(1)*. They share the common core structure Manα1-3(Manα1-6)Manβ1-4GlcNAcβ1-4GlcNAc-Asn, but differ in their outer branches (Fig. 1). High mannose-type oligosaccharides have two to six additional α-mannose residues to the core structure. Typical complex-type oligosaccharides contains two to four outer branches with a sialyllactosamine sequence. Hybrid-type structures have the features of both high mannose-type and complex-type oligosaccharides, and most of them contain bisecting *N*-acetylglucosamine, which is linked β1–4 to the β-linked mannose residue of the core structure. An additional type of carbohydrate chain, so-called poly-*N*-acetyllactosamine-type, has also been described *(2–4)*. Its outer branches have a characteristic structure composed of *N*-acetyllactosamine repeating units. It may be classified to be of complex-type; however, it is antigenically and functionally distinct from standard complex-type sugar chains *(5)*. Some poly-*N*-acetyllactosamine-type oligosaccharides have branched sequences containing Galβ1-4GlcNAcβ1-3(Galβ1-4GlcNAcβ1-6)Gal units *(2,3)*, which is the determinant of the I-antigen.

Glycopeptides or oligosaccharides can be prepared from glycoproteins by enzymatic digestions or chemical methods, as discussed in subsequent chapters in this book. The most widely used means for preparing glycopeptides is to digest the material with pronase completely. Oligosaccharide can be prepared from glycoproteins or glycopeptides by treating samples with anhydrous hydrazine (*6* and Chapter 5) or endoglycosidases (Chapters 10–12). Since the released oligosaccharides retain their reducing termini, they can be radiolabeled by reduction with NaB^3H_4 *(7)*. The primary amino group of the peptide backbone of glycopeptides can be labeled by acetylation with [^{3}H]- or [^{14}C]-acetic anhydride *(8)*. Before employing columns of immobilized lectins for analyses, oligosaccharides or glycopeptides should be separated on a column of QAE- or DEAE-cellulose based on anionic charge derived from sialic acid, phosphate, or sulfate residues. Acidic oligosaccharides thus separated should be converted to neutral ones for simplifying the following separation. To simplify discussion, the oligosaccharides discussed here do not contain sialic acid, phosphate, or sulfate residues, although these acidic residues, especially sialic acid residues, are found in many oligosaccharides. In most cases, the influence of these residues on the interaction of oligosaccharides with immobilized lectins is weak, but where documentation of the influence of these residues is available, it will be mentioned in the appropriate sections. In this chapter, we describe the general procedure of serial lectin-affinity chromatography of glycopeptides and oligosaccharides using several well-defined immobilized lectins.

2. Materials

1. Mono Q HR5/5, DEAE-Sephacel, Sephadex G-25 (Pharmacia, Uppsala, Sweden).
2. High-performance liquid chromatograph, two pumps, with detector capable of monitoring UV absorbance at 220 nm.
3. Neuraminidase: 1 U/mL of neuraminidase from *Streptococcus sp.* (Seikagaku Kogyo, Tokyo, Japan) in 50 m*M* acetate buffer, pH 6.5.

4. Dowex 50W-X8 (50–100 mesh, H^+ form).
5. Bio-Gel P-4 minus 400 mesh (Bio-Rad, Richmond, CA).
6. HPLC mobile phase for Mono Q: A, 2 m*M* Tris-HCl, pH 7.4; B, 2 m*M* Tris-HCl, pH 7.4, 0.5*M* NaCl.
7. Mobile phase for Bio-Gel P-4: distilled water.
8. Standard for Bio-Gel P-4: partial hydrolysate of chitin prepared according to Rupley *(9)*; 10 µg mixed with 50 µL distilled water. Store frozen.
9. Concanavalin A, *Ricinus communis* lectin, wheat germ lectin, *Datura stramonium* lectin, *Maackia amurensis* lectin, *Allomyrina dichotoma* lectin, *Amaranthus caudatus* lectin (EYLaboratories, San Mateo, CA), *Phaseolus vulgaris* erythroagglutinin, *Phaseolus vulgaris* leukoagglutinin (Seikagaku Kogyo). Immobilized lectins were prepared at a concentration of 1–5 mg lectin/mL of gel (*see* Notes 1–3) or obtained commercially (e.g., Pharmacia, EY Laboratories, Bio-Rad, Seikagaku Kogyo): *Galanthus nivalis* lectin, *Lens culinaris, Pisum sativum, Vica fava* lectin, pokeweed mitogen, *Sambucus nigra L* lectin, *Tetracarpidium conophorum* lectin (*see* Notes 1–3).
10. ^3H-$NaBH_4$: 100 mCi of ^3H-$NaBH_4$ (sp. act. 5–15 Ci/mmol; NEN, Boston, MA) mixed with 2 mL 10 m*M* NaOH; store –80°C.
11. Tris-buffered saline (TBS): 10 m*M* Tris-HCl, pH 7.4, 0.15*M* NaCl.
12. Lectin column buffer: 10 m*M* Tris-HCl, pH 7.4, 0.15*M* NaCl, 1 m*M* $CaCl_2$, 1 m*M* $MnCl_2$.
13. Methyl-α-mannoside: 100 m*M* in TBS; store refrigerated.
14. Methyl-α-glucoside: 10 m*M* in TBS; store refrigerated.
15. Lactose: 50 m*M* in TBS; store refrigerated.
16. *N*-acetylglucosamine: 200 m*M* in TBS; store refrigerated.

3. Methods

3.1. Separation of Acidic Sugar Chains on Mono Q HR5/5 or DEAE-Sephacel, and Removal of Sialic Acids

3.1.1. Ion-Exchange Chromatography

1. Equilibrate the Mono Q HR5/5 or DEAE-Sephacel column with 2 m*M* Tris-HCl, pH 7.4, at a flow rate of 1 mL/min at room temperature.
2. Dissolve the oligosaccharides or the glycopeptides in 0.1 mL of 2 m*M* Tris-HCl, pH 7.4, and apply to the column.
3. Elute with 2 m*M* Tris-HCl, pH 7.4, for 10 min, and then with a linear gradient (0–20%) of 2 m*M* Tris-HCl, pH 7.4, 0.5*M* NaCl in 60 min at a flow rate of 1 mL/min.

4. Neutral oligosaccharides are recovered in the pass through fraction. Acidic monosialo-, disialo-, trisialo-, and tetrasialooligosaccharides are eluted out successively by the linear gradient of NaCl.

3.1.2. Removal of Sialic Acid Residues

1. To 10–100 μg oligosaccharides free of buffers or salts, add 100 μL of neuraminidase buffer and 100 μL of neuraminidase, and incubated at 37°C for 18 h.
2. Heat-inactivate the neuraminidase by immersion in a boiling water bath for 3 min.
3. Apply to the column of Dowex 50W-X8 (0.6 cm id × 2.5 cm), wash the column with 1 mL of distilled water, and concentrate the eluates under vacuum.

Alternatively, add 500 μL of 0.1*M* HCl and heat at 80°C for 30 min, and dry up the sample using evaporator.

3.2. Separation of Poly-N-Acetyllactosamine-Type Sugar Chains from Other Types of Sugar Chains

Poly-*N*-acetyllactosamine-type sugar chains vary as to the number of *N*-acetyllactosamine repeating units and the branching mode and structural characterization of this type sugar chains has been quite difficult. The structural characterization of poly-*N*-acetyllactosamine-type sugar chains has been quite difficult *(10)*. This type of sugar chain has a higher mol wt compared to other high mannose-type, complex-type, or hybrid-type chains. Thus, poly-*N*-acetyllactosamine-type sugar chains are easily separated from others on a column of Bio-Gel P-4 having a mol mass of more than 4000 excluded from the Bio-Gel P-4 column chromatography *(11,12)*.

1. Equilibrate two coupled columns (0.8 cm id × 50 cm) of Bio-Gel P-4 in water at 55°C by use of a water jacket.
2. Elute the oligosaccharides at a flow rate of 0.3 mL/min, and collect fractions of 0.5 mL. Monitor absorbance at 220 nm.
3. Collect poly-*N*-acetyllactosamine-type oligosaccharides that are eluted at the void volume of the column. Other types of oligosaccharides included in the column are subjected to the next separations (Sections 3.3.–3.6.) illustrated in Fig. 2. The specificity of the lectins used is summarized in Fig. 3 and Table 1.

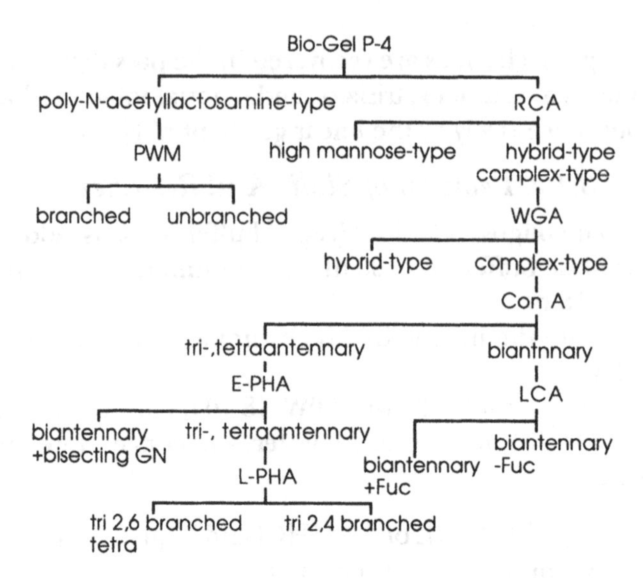

Fig. 2. Scheme of fractionation of asparagine-linked sugar chains by combining affinity chromatography on immobilized lectins.

3.3. Separation of High Mannose-Type Sugar Chains from Complex-Type and Hybrid-Type Sugar Chains

3.3.1. Affinity Chromatography on Immobilized RCA

After the separation of high-mol-wt poly-*N*-acetyllactosamine-type oligosaccharides, a mixture of other types of sugar chains can be separated on a column of *Ricinus communis* lectin (RCA), which recognizes the Galβ1-4GlcNAc sequence *(13,14)*.

1. Equilibrate the RCA-Sepharose column (0.6 cm id × 5.0 cm) in TBS.
2. Dissolve oligosaccharides or glycopeptides in 0.5 mL of TBS, and apply to the column.
3. Elute (1.0 mL fractions) successively with 3 column vol of TBS, and then with 3 column vol of 50 m*M* lactose at flow rate 2.5 mL/h at room temperature.
4. Bind both complex-type and hybrid-type sugar chains to the RCA-Sepharose (*see* Note 4).
5. Collect high mannose-type oligosaccharides, which pass through the column.
6. Purify the oligosaccharides (or glycopeptides) from salts and haptenic sugar by gel filtration on Sephadex G-25 column (1.2 cm id × 50 cm) equilibrated with distilled water.

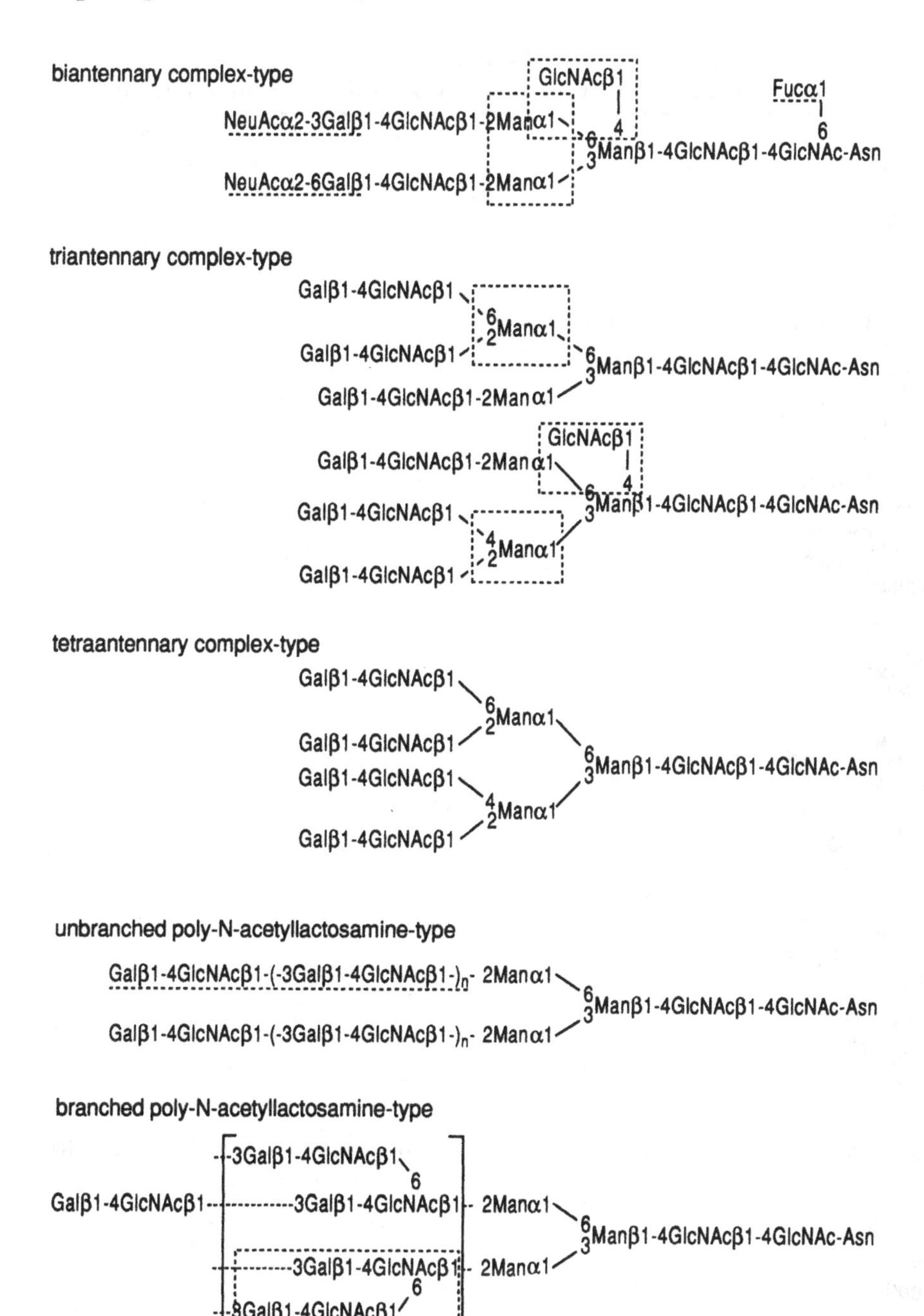

Fig. 3. Structures of several complex-type oligosaccharides. The boxed area indicates the characteristic structures recognized by several immobilized lectins as summarized in Table 1.

Table 1
Characteristic Structures Recognized by Several Immobilized Lectins

	RCA	GNA	WGA	Con A	LCA	PSA	VFA	E-PHA	L-PHA	DSA	TCA
M-M 6 M-M 3 >M 6 M-M-M 3 >M-GN-GN	-	R	-	+	-	-	-	-	-	-	-
M >M GN M 6 4 M-GN-GN G-GN-M 3	+	-	R	+	-	-	-	-	-	-	-
G-GN 2 -M >M-GN-GN G-GN 2 -M	+	-	-	+	-	-	-	-	-	-	+
G-GN-M F 6 >M-GN-GN G-GN-M	+	-	-	+	+	+	+	-	-	-	+
G-GN-M GN F >M-GN-GN G-GN-M	+	-	-	+	-	-	-	R	-	-	+
G-GN-M G-GN 4 >M-GN-GN G-GN 2 >M	+	-	-	-	-	-	-	-	-	R	++
G-GN 6 G-GN 2 >M >M-GN-GN G-GN-M	+	-	-	-	-	-	-	-	R	+	+
G-GN-M F G-GN >M-GN-GN G-GN >M	+	-	-	-	-	-	-	-	-	R	N.D.
G-GN >M F G-GN >M-GN-GN G-GN-M	+	-	-	-	-	-	-	-	R	+	+
G-GN-M GN G-GN >M-GN-GN G-GN >M	+	-	-	-	-	-	-	R	-	R	N.D.
G-GN >M GN G-GN >M-GN-GN G-GN-M	+	-	-	-	-	-	-	-	R	N.D.	N.D.
G-GN 6 >M G-GN 2 G-GN 4 >M-GN-GN G-GN 2 >M	+	-	-	-	-	-	-	-	R	+	+
G-GN >M F G-GN G-GN >M-GN-GN G-GN >M	+	-	-	-	-	-	-	-	R	+	N.D.

+, bound; R, retarded; –, not bound; N.D., not determined.

3.3.2. Affinity Chromatography on Immobilized Snowdrop Lectin

High mannose-type glycopeptides, which carry Manα1-3Man units, are specifically retarded on the immobilized snowdrop *(Galanthus nivalis)* lectin (GNA) *(15)*.

1. Equilibrate the GNA-Sepharose column (0.6 cm id × 5.0 cm) in TBS.
2. Dissolve oligosaccharides or glycopeptides in 0.5 mL of TBS, and apply to the column.
3. Elute (0.5-mL fraction) with 5 column vol of TBS, to collect sugar chains lacking Manα1-3Man units or hybrid-type, which are not retarded.
4. Elute with 3 column vol of 100 m*M* methyl-α-mannoside at flow rate 2.5 mL/h at room temperature to obtain the specifically retarded high mannose-type glycopeptides that carry Manα1-3Man units

3.4. Separation of Hybrid-Type Sugar Chains from Complex-Type Sugar Chains

3.4.1. Affinity Chromatography on Immobilized WGA

Wheat germ lectin (WGA)-Sepharose has a high affinity for the hybrid-type sugar chains. It is demonstrated that the sugar sequence GlcNAcβ1-4Manβ1-4GlcNAcβ1-4GlcNAc-Asn structure is essential for tight binding of glycopeptides to WGA-Sepharose column *(16)*.

1. Equilibrate the WGA-Sepharose column (0.6 cm id × 5.0 cm) in TBS.
2. Dissolve glycopeptides in 0.5 mL of TBS, and apply to the column.
3. Elute (0.5-mL fractions) successively with 5 column vol of TBS.
4. Collect hybrid-type glycopeptides with a bisecting *N*-acetylglucosamine residue, which are retarded on WGA column.
5. Collect sugar chains having the typical complex-type (and also high mannose-type) sugar chains eluted at the void volume of the column with TBS.

3.5. Separation of Complex-Type Biantennary Sugar Chains (Fig. 3)

3.5.1. Affinity Chromatography on Immobilized Concanavalin A (Con A)

Oligosaccharides and glycopeptides with tri- and tetraantennary complex-type sugar chains pass through Con A-Sepharose, whereas biantennary complex-type, hybrid-type, and high mannose-type oligosaccharides bind to the Con A and can be differentially eluted from the column *(17,18)*.

1. Equilibrate the Con A-Sepharose column (0.6 cm id × 50 cm) in lectin column buffer.
2. Pass the oligosaccharide mixture of complex-type chain from WGA column through Con A-Sepharose column.
3. Elute (1-mL fractions) successively with 3 column vol of lectin column buffer.
4. Collect oligosaccharides with tri- and tetraantennary complex-type sugar chains, which pass through the column. Complex-type biantennary oligosaccharide having bisecting GlcNAc also pass through the column.
5. Elute (1-mL fractions) successively with 3 column vol of 10 m*M* methyl-α-glucoside and finally with 3 column vol of 100 m*M* methyl-α-mannoside.
6. Collect complex-type biantennary sugar chains, which are eluted after the addition of methyl-α-glucoside.
7. Collect high mannose-type and hybrid-type oligosaccharides (glycopeptides) eluted after the addition of 100 m*M* methyl-α-mannoside.

3.5.2. Affinity Chromatography on Immobilized LCA, PSA, or VFA

The biantennary complex-type sugar chains bound to Con A-Sepharose column and eluted with 10 m*M* methyl-α-glucoside contain two-types of oligosaccharides, which will be separated on a column of lentil lectin *(Lens culinaris,* LCA) pea lectin *(Pisum sativum,* PSA), or fava lectin *(Vica fava,* VFA) *(19–21)*.

1. Equilibrate the LCA-, PSA-, or VFA-Sepharose column (0.6 cm id × 5.0 cm) in lectin column buffer.
2. Pass the biantennary complex-type sugar chains from the Con A column through the LCA-, PSA-, or VFA-Sepharose column.
3. Elute (1.0-mL fraction) successively with 3 column vol of lectin column buffer, then with 3 column vol of 100 m*M* methyl-α-mannoside at flow rate 2.5 mL/h at room temperature.
4. Collect biantennary complex-type sugar chains without fucose that pass through the column.
5. Elute bound biantennary complex-type sugar chains having fucose residue attached to the innermost *N*-acetylglucosamine to the column (*see* Note 5).

3.5.3. Affinity Chromatography on Immobilized E-PHA

Complex-type biantennary sugar chains having outer galactose residues and "bisecting" *N*-acetylglucosamine are retarded by *Phaseolus vulgaris* erythroagglutinin (E-PHA)-Sepharose *(14,22)*.

1. Equilibrate the E-PHA-Sepharose column (0.6 cm id × 5.0 cm) in lectin column buffer.
2. Apply the pass-through fraction from the Con A column on the E-PHA-Sepharose column.
3. Elute (0.5-mL fraction) successively with 5 column vol of lectin column buffer at flow rate 2.5 mL/h at room temperature.
4. Collect biantennary complex-type sugar chains having bisecting *N*-acetylglucosamine residue retarded on the E-PHA column *(see* Note 6). When the elution of the column is performed at 4°C, biantennary complex-type oligosaccharide without bisecting *N*-acetylglucosamine is also retarded by the E-PHA-Sepharose column (data not shown).

3.6. Separation of Complex-Type Triantennary and Tetrantennary Sugar Chains

3.6.1. Affinity Chromatography on Immobilized E-PHA

E-PHA-Sepharose interacts with high affinity with triantennary (having 2,4 branched mannose) oligosaccharides containing both outer galactose residues and bisecting a *N*-acetylglucosamine residue *(22)*.

1. Equilibrate the E-PHA-Sepharose column (0.6 cm id × 5.0 cm) in lectin column buffer.
2. Apply the pass-through fraction from the Con A column on the E-PHA-Sepharose column.
3. Elute (0.5-mL fraction) successively with 5 column vol of lectin column buffer at flow rate 2.5 mL/h at room temperature.
4. Collect retarded triantennary (having 2,4 branched mannose) oligosaccharides containing both outer galactose and bisecting *N*-acetylglucosamine. Other tri- and tetraantennary oligosaccharides pass through the column.

3.6.2. Affinity Chromatography on Immobilized L-PHA

Phaseolus vulgaris leukoagglutinin (L-PHA), which is an isolectin of E-PHA, interacts with triantennary and tetraantennary complex-type glycopeptides having α-linked mannose residues substituted at positions C-2 and C-6 with Galβ1-4GlcNAc *(23)*.

1. Equilibrate the L-PHA-Sepharose column (0.6 cm id × 5.0 cm) in lectin column buffer.
2. Apply the pass-through fraction from the E-PHA column on L-PHA-Sepharose column.

3. Elute (0.5-mL fraction) successively with 5 column vol of lectin column buffer at flow rate 2.5 mL/h at room temperature.
4. Collect retarded triantennary and tetraatermry complex-type glycopeptides having both 2,6-branched α-mannose and outer galactose on the L-PHA column (*see* Note 7).

Other tri-and tetraantennary oligosaccharides pass through the column.

3.6.3. Affinity Chromatography on Immobilized DSA

Datura stramonium lectin (DSA) shows high affinity with tri- and tetraantennary complex-type oligosaccharides. Triantennary complex-type oligosaccharides containing 2,4 substituted α-mannose are retarded by DSA-Sepharose column. Triantennary and tetraantennary complex-type oligosaccharides having α-mannose residue substituted at the C-2,6 positions bind to the column and are eluted by the GlcNAc oligomer *(24,25)*.

1. Equilibrate the DSA-Sepharose column (0.6 cm id × 5.0 cm) in TBS.
2. Apply the pass-through fraction from the E-PHA column on the DSA-Sepharose column.
3. Elute (0.5-mL fraction) successively with 3 column vol of TBS to obtain retarded triantennary complex-type glycopeptides having 2,4-branched α-mannose on the DSA column.
4. Elute with 3 column vol of 5 mg/mL *N*-acetylglucosamine oligomer at flow rate 2.5 mL/h at room temperature to obtain bound triantennary and tetraantennary oligosaccharides having α-mannose residue substituted at the C-2,6 positions.

3.6.4. Affinity Chromatography on Immobilized TCA

The specificity of *Tetracarpidium conophorum* (Nigerian walnut) lectin (TCA) is the reverse of that of L-PHA and DSA. It is indicated that biantennary, tetraantennary glycan, and a triantennary glycan carrying Galβ1-4GlcNAc units substituted to C-2 and C-6 of α-mannosyl residue bind much less tightly than does triantennary glycan substituted on C-2 and - 4 of α-mannosyl residue *(26)*.

1. Equilibrate the NWA-Sepharose column (0.6 cm id × 5.0 cm) in lectin column buffer.
2. Apply the pass-through fraction from the E-PHA column on the TCA-Sepharose column.
3. Elute (0.5-mL fraction) successively with 3 column vol of lectin column buffer and then with 3 column vol of 10 m*M* lactose at flow rate 2.5 mL/h at room temperature to obtain the bound triantennary and

tetraantennary oligosaccharides having 2,6-branched α-mannose on the TCA column.

4. Collect triantennary complex-type oligosaccharide having α-mannose substituted at the C-2,4 positions eluted after the addition of 100 m*M* lactose at 37°C.

3.7. Separation of Poly-N-Acetyllactosamine-Type Sugar Chains

High-mol-wt poly-*N*-acetyllactosamine-type oligosaccharides are classified into two groups. One is branched poly-*N*-acetyllactoseminoglycan containing Galβ1-4GlcNAcβ1-3(Galβ1-4GlcNAcβ1-6)Gal unit, and the other is the linear poly-*N*-acetyllactosamine structure, which lacks galactose residues substituted at the C-6 position.

3.7.1. Affinity Chromatography on Immobilized PWM

Branched poly-*N*-acetyllactosamine-type oligosaccharides can be separated by the use of a pokeweed mitogen (PWM, Pa-1, Pa-2, and Pa-4)-Sepharose column *(27)*. Since the sugar sequence Galβ1-4GlcNAcβ1-6Gal firmly binds to the PWM-Sephrose column, the branched poly-*N*-acetyllactosamine chains can be retained by the column, whereas unbranched oligosaccharides are recovered without retardation *(28)*.

1. Equilibrate the PWM-Sepharose column (0.6 cm id × 5.0 cm) in TBS.
2. Apply the poly-*N*-acetyllactosamine-type sugar chains separated on Bio-Gel P-4 (*see* Section 3.2.) on a PWM-Sepharose column.
3. Elute (1.0-mL fraction) successively with 3 column vol of TBS and then with 3 column vol of 0.1*M* NaOH at flow rate 2.5 mL/h at room temperature.
4. Collect unbranched poly-*N*-acetyllactosamine-type sugar chains, which pass through the column.
5. Collect bound branched poly-*N*-acetyllactosamine-type sugar chains in the 0.1*M* NaOH eluant.

3.7.2. Affinity Chromatography on Immobilized DSA

Immobilized DSA lectin interacts with high affinity with sugar chains having the linear, unbranched poly-*N*-acetyllactosamine sequence. For the binding to DSA-Sepharose, more than two intact *N*-acetyllactosamine repeating units is essential *(25)*.

1. Equilibrate the DSA-Sepharose column (0.6 cm id × 5.0 cm) in TBS.
2. Apply the poly-*N*-acetyllactosamine-type sugar chains separated on Bio-Gel P-4 (*see* Section 3.2.) on PWM-Sepharose column.

3. Elute (0.5-mL fraction) successively with 3 column vol of TBS and then with 3 column vol of 5 mg/mL GlcNAc oligomer at flow rate 2.5 mL/h at room temperature.
4. Collect branched poly-*N*-acetyllactosamine-type sugar chains, which pass through the column, separated from unbranched poly-*N*-acetyllactasamine-type sugar chains, which bind.

3.8. Separation of Sialylated Sugar Chains

The basic Galβ1-4GlcNAc sequence present in complex-type sugar chains may contain sialic acids in α2,6 or α2,3 linkage to outer galactose residues.

3.8.1. Affinity Chromatography on Immobilized MAL

Maackia amurensis lectin (MAL) *(29,30)* interacts with high affinity with complex-type tri- and tetraantennary glycopeptides containing outer sialic acid residue linked α2,3 to penultimate galactose. Glycopeptides containing sialic acid linked only α2,6 to galactose do not interact detectably with the immobilized MAL.

1. Equilibrate the MAL-Sepharose column (0.6 cm id × 5.0 cm) in lectin column buffer.
2. Apply the acidic oligosaccharides or glycopeptides separated on Mono Q HR5/5, or DEAE-Sephacel (*see* Section 3.1.1., step 1) on MAL-Sepharose column.
3. Elute (0.5-mL fraction) successively with 3 column vol of lectin column buffer and then with 3 column vol of 50 m*M* lactose at flow rate 2.5 mL/h at room temperature.
4. Collect glycopeptides (oligosaccharides) containing α2,6-linked sialic acid(s), which pass through the column.
5. Elute bound glycopeptides (oligosaccharides) having α2,3-linked sialic acid(s) in the 50-m*M* lactose eluant.

3.8.2. Affinity Chromatography on Immobilized Allo A

Allomyrina dichotoma lectin (allo A) *(31,32)* recognizes the other isomer of sialyllactosamine compared to MAL. Mono-, di-, and triantennary complex-type oligosaccharides containing terminal sialic acid(s) in α2,6 linkage bind to allo A-Sepharose, whereas complex-type sugar chains having isomeric α2,3-linked sialic acid(s) do not bind to the immobilized allo A.

1. Equilibrate the allo A-Sepharose column (0.6 cm id × 5.0 cm) in TBS.
2. Apply the acidic oligosaccharides or glycopeptides separated on Mono

Q HR5/5, or DEAE-Sephacel (*see* Section 3.1.1., step 1) on an allo A-Sepharose column.
3. Elute (0.5-mL fraction) successively with 3 column vol of TBS and then with 3 column vol of 50 m*M* lactose at flow rate 2.5 mL/h at room temperature.
4. Collect glycopeptides (oligosaccharides) containing α2,3-linked sialic acid(s), which pass through the column.
5. Elute bound glycopeptides (oligosaccharides) having α2,6-linked sialic acid(s) (*see* Note 8).

3.8.3. *Affinity Chromatography on Immobilized SNA*

Elderberry (*Sambucus nigra L.*) bark lectin (SNA) (*33,34*) shows the same sugar-binding specificity as allo A. All types of oligosaccharides that contain at least one NeuAcα2-6Gal unit in the molecule bound firmly to the SNA-Sepharose.

1. Equilibrate the SNA-Sepharose column (0.6 cm id × 5.0 cm) in TBS.
2. Apply the acidic oligosaccharides or glycopeptides separated on Mono Q HR5/5, or DEAE-Sephacel (*see* Section 3.1.1., step 1) on the SNA-Sepharose column.
3. Elute (0.5-mL fraction) successively with 3 column vol of TBS and then with 3 column vol of 50 m*M* lactose at flow rate 2.5 mL/h at room temperature.
4. Collect glycopeptides (oligosaccharides) containing α2,3-linked sialic acid(s), which pass through the column.
5. Elute bound glycopeptides (oligosaccharides) having α2,6-linked sialic acid(s) in the 50 m*M* lactose eluant.

3.9. Summary

Various immobilized lectins can be successfully used for fractionation and for structural studies of asparagine-linked sugar chains of glycoproteins (*see* Note 9). This method needs <10 ng of a radiolabeled oligosaccharide prepared from a glycoprotein by hydrazinolysis or by digestion with endo-β-*N*-acetylglucosaminidases. The fractionation and the structural assessment through the use of immobilized lectins make the subsequent structural studies much easier.

4. Notes

1. During the coupling reactions, sugar-binding sites of lectins must be protected by the addition of the specific haptenic sugars.
2. Immobilized lectin is stored at 4°C. In most cases, immobilized lectin is stable for several years.

3. Some lectins, especially legume lectins, need Ca and Mn ions for carbohydrate binding, so that the buffers used for the affinity chromatography on the lectin column must contain 1 m*M* $CaCl_2$ and $MnCl_2$.
4. Complex-type or hybrid-type oligosaccharides are retarded on a column of RCA-Sepharose rather than tightly bound when their sugar sequences are masked by sialic acids.
5. Intact *N*-acetylglucosamine and asparagine residues at the reducing end are required for tight binding of complex-type oligosaccharides to both LCA, PSA, or VFA-Sepharose column.
6. High-affinity interaction with E-PHA-Sepharose is prevented if both outer galactose residues on a bisected sugar chain are substituted at position C-6 by sialic acid.
7. L-PHA-Sepharose does not retard the elution of sugar chains lacking outer galactose residues.
8. Complex-type oligosaccharides without sialic acid of mono-, di-, tri-, and tetraantennary complex-type are retarded by the allo A lectin column.
9. More detailed reviews on the separation of oligosaccharides and glycopeptides by means of affinity chromatography on immobilized lectin columns have been published *(35,36)*.

References

1. Kornfeld, R. and Kornfeld, S. (1985) Assembly of asparagine-linked oligosaccharides. *Ann. Rev. Biochem.* **54,** 631–664.
2. Tsuji, T., Irimura, T., and Osawa, T. (1981) The carbohydrate moiety of Band 3 glycoprotein of human erythrocyte membrane. *J. Biol. Chem.* **256,** 10,497–10,502.
3. Fukuda, M., Dell, A., Oates, J. E., and Fukuda, M. N. (1984) Structure of branched lactosaminoglycan, the carbohydrate moiety of Band 3 isolated from adult human erythrocytes. *J. Biol. Chem.* **259,** 8260–8273.
4. Merkle, R. K. and Cummings, R. D. (1987) Relationship of the terminal sequences to the length of poly-*N*-acetyllactosamine chains in asparagine-linked oligosaccharides from the mouse lymphoma cell ine BW5147. *J. Biol. Chem.* **262,** 8179–8189.
5. Fukuda, M. (1985) Cell surface glycoconjugates as onco-differentiation markers in hematopoietic cells. *Biochim. Biophys. Acta.* **780,** 119–150.
6. Fukuda, M., Kondo, T., and Osawa, T. (1976) Studies on the hydrazinolysis of glycoproteins. Core structures of oligosaccharides obtained from porcine thyroglobulin and pineapple stem bromelain. *J. Biochem.* **80,** 1223–1232.
7. Takasaki S. and Kobata, A. (1978) Microdetermination of sugar composition by radioisotope labeling. *Methods Enzymol.* **50,** 50–54.

8. Tai, T., Yamashita, K., Ogata, M. A., Koide, N., Muramatsu, T., Iwashita, S., Inoue, Y., and Kobata, A. (1975) Structural studies of two ovalbumin glycopeptides in relation to the endo-β-*N*-acetylglucosaminidase specificity. *J. Biol. Chem.* **250**, 8569–8575.
9. Rupley, J. A. (1964) The hydrolysis of chitin by concentrated hydrochloric acid, and the preparation of low-molecular-weight substrates for lysozyme. *Biochim. Biophys. Acta.* **83**, 245–255.
10. Krusius, T., Finne, J., and Rauvala, H. (1978) The poly(glycosyl) chains of glycoproteins. Characterizaion of a novel type of glycoprotein saccharides from human erythrocyte membrane. *Eur. J. Biochem.* **92**, 289–300.
11. Yamamoto, K., Tsuji, T., Tarutani, O., and Osawa, T. (1984) Structural changes of carbohydrate chains of human thyroglobulin accompanying malignant transformations of thyroid grands. *Eur. J. Biochem.* **143**, 133–144.
12. Tsuji, T., Irimura, T., and Osawa, T. (1980) The carbohydrate moiety of Band-3 glycoprotein of human erythrocyte membranes. *Biochem. J.* **187**, 677–686.
13. Baenziger, J. U. and Fiete, D. (1979) Structural determinants of *Ricinus communis* agglutinin and toxin specificity for oligosaccharides. *J. Biol. Chem.* **254**, 9795–9799.
14. Irimura, T., Tsuji, T., Yamamoto, K., Tagami, S., and Osawa, T. (1981) Structure of a complex-type sugar chain of human glycophorin A. *Biochemistry* **20**, 560–566.
15. Shibuya, N., Goldstein, I. J., Van Damme, E. J. M., and Peumans, W. J. (1988) Binding properties of a mannose-specific lectin from the snowdrop *(Galanthus nivalis)* bulb *J. Biol. Chem.* **263**, 728–734.
16. Yamamoto, K., Tsuji, T., Matsumoto, I., and Osawa, T. (1981) Structural requirements for the binding of oligosaccharides and glycopeptides to immobilized wheat germ agglutinin *Biochemistry* **20**, 5894–5899.
17. Ogata, S., Muramatsu, T., and Kobata, A. (1975) Fractionation of glycopeptides by affinity column chromatography on concanavalin A-Sepharose. *J. Biochem.* **78**, 687–696.
18. Krusius, T., Finne, J., and Rauvala, H. (1976) The structural basis of the different affinities of two types of acidic N-glycosidic glycopeptides from concanavalin A-Sepharose. *FEBS Lett.* **71**, 117–120.
19. Kornfeld, K., Reitman, M. L., and Kornfeld, R. (1981) The carbohydrate-binding specificity of pea and lentil lectins. *J. Biol. Chem.* **256**, 6633–6640
20. Katagiri, Y., Yamamoto, K., Tsuji, T., and Osawa, T. (1984) Structural requirements for the binding of glycopeptides to immobilized *vicia faba* (fava) lectin. *Carbohydr. Res.* **129**, 257–265.
21. Yamamoto, K., Tsuji, T., and Osawa, T. (1982) Requirement of the core structure of a complex-type glycopeptide for the binding to immobilized lentil- and pea-lectins. *Carbohydr. Res.* **110**, 283–289.
22. Yamashita, K., Hitoi, A., and Kobata, A. (1983) Structural determinants of *Phaseolus vulgaris* erythroagglutinating lectin for oligosaccharides. *J. Biol. Chem.* **258**, 14,753–14,755.

23. Cummings, R. D. and Kornfeld, S. (1982) Characterization of the structural determinants required for the high affinity interaction of asparagine-linked oligosaccharides with immobilized *Phaseolus vulgaris* leukoagglutinating and erythroagglutinating lectins. *J. Biol. Chem.* **257,** 11,230–11,234.
24. Cummings, R. D. and Kornfeld, S. (1984) The distribution of repeating [Galβ1,4GlcNAcβ1,3] sequences in asparagine-linked oligosaccharides of the mouse lymphoma cell line BW5147 and PHA^{R}2.1. *J. Biol. Chem.* **259,** 6253–6260.
25. Yamashita, K., Totani, K., Ohkura, T., Takasaki, S., Goldstein, I. J., and Kobata, A. (1987) Carbohydrate binding properties of complex-type oligosaccharides on immobilized *Datura stramonium* lectin. *J. Biol. Chem.* **262,** 1602–1607.
26. Sato, S., Animashaun, T., and Hughes, R. C. (1991) Carbohydrate-binding specificity of *Tetracarpidium conophorum* lectin. *J. Biol. Chem.* **266,** 11,485–11,494.
27. Irimura, T. and Nicolson, G. L. (1983) Interaction of pokeweed mitogen with poly(*N*-acetyllactosamine)-type carbohydrate chains. *Carbohydr. Res.* **120,** 187–195.
28. Kawashima, H., Sueyoshi, S., Li, H., Yamamoto, K., and Osawa, T. (1990) Carbohydrate binding specificities of several poly-*N*-acetyllactosamine-binding lectins. *Glycoconjugate J.* **7,** 323–334.
29. Wang, W.-C. and Cummings, R. D. (1988) The immobilized leukoagglutinin from the seeds of *Maackia amurensis* binds with high affinity to complex-type Asn-linked oligosaccharides containing terminal sialic acid-linked α-2,3 to penultimate galactose residues. *J. Biol. Chem.* **263,** 4576–4585.
30. Kawaguchi, T., Matsumoto, I., and Osawa, T. (1974) Studies on hemagglutinins from *Maackia amurensis* seeds. *J. Biol. Chem.* **249,** 2786–2792.
31. Sueyoshi, S., Yamamoto, K., and Osawa, T. (1988) Carbohydrate binding specificity of a beetle *(Allomyrina dichotoma)* lectin. *J. Biochem.* **103,** 894–899.
32. Yamashita, K., Umetsu, K., Suzuki, T., Iwaki, Y., Endo, T., and Kobata, A. (1988) Carbohydrate binding specificity of immobilized *Allomyrina dichotoma* lectin II. *J. Biol. Chem.* **263,** 17,482–17,489.
33. Shibuya, N., Goldstein, I. J., Broekaen, W. F., Lubaki, M. N., Peeters, B., and Peumans, W. J. (1987) Fractionation of sialylated oligosaccharides, glycopeptides, and glycoproteins on immobilized elderberry *(Sambucus nigra L.)* bark lectin. *Arch. Biochem. Biophys.* **254,** 1–8.
34. Shibuya, N., Goldstein, I. J., Broekaert, W. F., Lubaki, M. N., Peeters, B., and Peumans, W. J. (1987) The elderberry *(Sambucus nigra L.)* bark lectin recognizes the Neu5Ac(α2-6)Gal/GalNAc sequence. *J. Biol. Chem.* **262,** 1596–1601.
35. Osawa, T. and Tsuji, T. (1987) Fractionation and structural assessment of oligosaccharides and glycopeptides by use of immobilized lectins. *Ann. Rev. Biochem.* **56,** 21–42.
36. Osawa, T. (1989) Recent progress in the application of plant lectins to glycoprotein chemistry. *Pure Appl. Chem.* **61,** 1283–1292.

CHAPTER 3

Sequence and Linkage Analysis of *N*- and *O*-Linked Glycans by Fast Atom Bombardment Mass Spectrometry

Bo Nilsson

1. Introduction

Most proteins are in fact glycoproteins, which means substitution of certain amino acids with carbohydrates in a covalent linkage. Asparagine, serine, and threonine are the most commonly glycosylated amino acids. Only in rare cases, other amino acids have been shown to carry carbohydrates. Asparagine is substituted in an *N*-glycosidic linkage with *N*-acetylglucosamine, whereas serine and threonine are glycosylated, usually with *N*-acetylgalactosamine in an *O*-glycosidic linkage. All *N*-linked glycans have a common branched trimannosyl-chitobiose core structure. Depending on substitution of the mannose residues, the *N*-linked glycans can roughly be divided into three groups: complex, high-mannose, and hybrid type of structures. Attachment of two, three, or four *N*-acetyllactosamine (Galβ1-4GlcNAc) sequences makes the bi-, tri-, and tetraantennary complex type of structures, respectively. The complex types contain, to a variable degree, sialic acid, fucose, and α-linked galactose as terminating sugars. The high-mannose type contains multiple mannose residues and the hybrid-type mannose on one branch of the core structure, and a lactosamine sequence on the other. In *O*-linked glycans, the common *N*-acetylgalactosamine is either 3-substituted or 3,6-disubstituted. The substituents are usually galactose, *N*-acetylglucosamine, and sialic acid as a terminal residue.

From: *Methods in Molecular Biology, Vol. 14: Glycoprotein Analysis in Biomedicine*
Edited by: E. F. Hounsell

In order to carry out structural analysis, the glycans have to be released from the protein and fractionated into pure compounds. Liberation of *N*- and *O*-linked glycans can be accomplished by chemical or enzymatic methods. For separation of released glycans, several methods are available, such as gel filtration, affinity chromatography, ion-exchange chromatography, and high-performance liquid chromatography. Pure compounds are subjected to structural analysis using, for example, enzymatic, chemical, and spectroscopic methods, e.g., mass spectrometry and NMR. When fast atom bombardment ionization was introduced, it became possible to analyze high-mol-wt glycoprotein oligosaccharides by mass spectrometry. Fast atom bombardment mass spectrometry (FAB-MS) of permethylated glycoconjugates gives information about the mol wt and the monosaccharide sequence in terms of sialic acid, hexose, deoxyhexose, and hexosamine. Binding positions of glycosidic linkages can, in general, not be determined by FAB-MS. The present chapter describes a method including the following reaction steps:

1. Periodate oxidation;
2. Reduction with $NaBD_4$; and
3. Permethylation.

The product is analyzed by FAB-MS, and from the primary and secondary sequence ions, binding positions between monosaccharide residues can be determined.

2. Materials

1. 0.1*M* Acetate buffer, pH 5.5, containing 8 m*M* sodium periodate.
2. Ethylene glycol.
3. 0.1*M* Sodium hydroxide.
4. Sodium borodeuteride.
5. Concentrated acetic acid.
6. Methanol.
7. PD 10 column (Pharmacia, Uppsala, Sweden) for gel filtration in water.
8. Pyridine.
9. Acetic anhydride.
10. Dimethyl sulfoxide (DMSO).
11. 1*M* Potassium hydride in DMSO.
12. Methyl iodide. **Carcinogen.**
13. Chloroform. **Potential Carcinogen, Liver Poison.**

14. Acetone.
15. Column (1 × 15 cm) packed with LH-20 (Pharmacia) for gel filtration in acetone/chloroform 2/1 (v/v).
16. Thin-layer chromatography (TLC) silica plates.
17. Anisaldehyde reagent: 10 mL of anisaldehyde, 10 mL of concentrated sulfuric acid, and 180 mL of ethanol.
18. Thioglycerol (1-thio-2,3-propanediol).
19. Sodium iodide.

3. Methods

3.1. Periodate Oxidation

1. Dissolve the oligosaccharide alditols (50–100 µg) in 3 mL of acetate buffer containing sodium periodate.
2. Carry out the periodate oxidation in the dark at 4°C for 48 h (*see* Note 1).
3. Decompose excess periodate by the addition of 25 µL of ethylene glycol, and leave the sample at 4°C overnight.
4. Add 0.1*M* sodium hydroxide (about 1.5 mL) until pH 7 is reached (*see* Note 2).
5. Reduce the oxidized compound with 25 mg of sodium borodeuteride at 4°C overnight.
6. Add acetic acid (about 0.3 mL) to pH 4, and concentrate the sample to dryness on a rotary evaporator.
7. Remove boric acid by evaporations with 3 × 3 mL methanol (*see* Note 3).
8. Remove inorganic salts on a PD 10 column eluted with water. (Alternative—*see* Note 4.)
9. Lyophilize the sample.

3.2. Permethylation

1. Dissolve the periodate oxidized and reduced compound in 0.5 mL of DMSO in a glass vessel that is sealed.
2. Flush the vessel with nitrogen, add 0.5 mL of a potassium hydride solution in DMSO, and reseal. Leave at room temperature for 4 h.
3. Freeze the sample, and add 0.3 mL of methyl iodide slowly by a syringe during venting of the vessel with a needle.
4. Sonicate for 15 min.
5. Remove excess methyl iodide by evaporation.
6. Add water (3 mL) and extract the methylated product three times with chloroform (3 × 1 mL).
7. Wash the combined chloroform phases three times with water (3 × 5 mL).
8. Purify the permethylated product on an LH-20 column (*see* Note 5).

9. Test fractions from the column for sugar by spotting an aliquot on a TLC plate, spray with the anisaldehyde reagent, and heat in an oven at 100°C for 5 min. Fractions containing carbohydrates give a green color.

3.3. *Fast Atom Bombardment Mass Spectrometry* (see *Note 6*)

1. Dissolve the permethylated samples in chloroform, and load on the stainless-steel target, which is mounted on a probe (*see* Note 7).
2. Evaporate the chloroform, and add one drop of thioglycerol matrix.
3. Insert the probe in the inlet of the mass spectrometer.
4. Perform the ionization using xenon atoms with a kinetic energy of 8 keV.
5. Accelerate the ions using a voltage of 10 kV, and record them in the positive mode.

3.4. *Interpretation of Mass Spectra* (see *Note 8*)

1. A guide for interpretating mass spectra of periodate degraded *N*-and *O*-linked glycoprotein oligosaccharides is shown in Tables 1 and 2. FAB-mass spectra of permethylated *N*-acetylhexosamine containing oligosaccharides are characterized by intense ions formed by cleavage of the 2-acetamido-2-deoxyhexosyl linkages. Binding positions between monosaccharide residues are determined from the primary and secondary sequence ions. The monosaccharide sequence from the nonreducing terminal is determined from the primary ions formed after cleavage of the glycosidic bonds. The secondary ions are formed from the primary by eliminations. A sequence of Gal-GlcNAc is determined from the primary sequence ion of *m/z* 424 (Table 1). For a 1-3 linkage between these residues, an intense secondary ion of *m/z* 228 is formed by elimination of the oxidized Gal residue, whereas for a 1-4 linkage, a secondary ion of *m/z* 392 is formed by elimination of methanol *(1)*. The complex type of structures contains branches with a sequence of Neu5Ac-Gal-GlcNAc. A nonreducing terminal sialic acid is recognized by the ions of *m/z* 289 and 257, whereas the latter is formed from the former by elimination of methanol. A sequence of Neu5Acα2-3Galβ1-4GlcNAc is determined by the primary and secondary ions of *m/z* 289, 257, 493, 738, and 706. The ion of *m/z* 493 differs from *m/z* 289 by 204 mass units and represents addition to a nonreducing terminal sialic acid by a periodate-resistant Gal residue, which therefore must be 3-substituted. The substitution of the GlcNAc residue in the 4-position is determined from the secondary fragment of *m/z* 706 formed by elimination of methanol from *m/z* 738 in analogy with the Galβ1-4GlcNAc sequence as discussed earlier. Another

Table 1
Ions Determining Linkages in Nonreducing Terminal Sequences

Nonreducing terminal sequence	Primary and secondary sequence ions
Galβ1-3GlcNAcβ1-	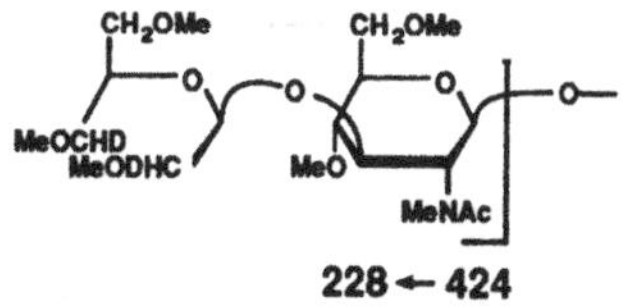 CH_2OMe CH_2OMe O O O MeOCHD MeODHC MeO MeNAc O 228 ← 424
Galβ1-4GlcNAcβ1-	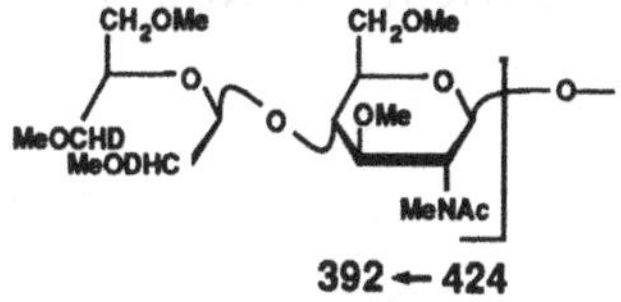 CH_2OMe CH_2OMe O O O MeOCHD MeODHC OMe MeNAc O 392 ← 424
Galα1-3Galβ1-4GlcNAcβ1-	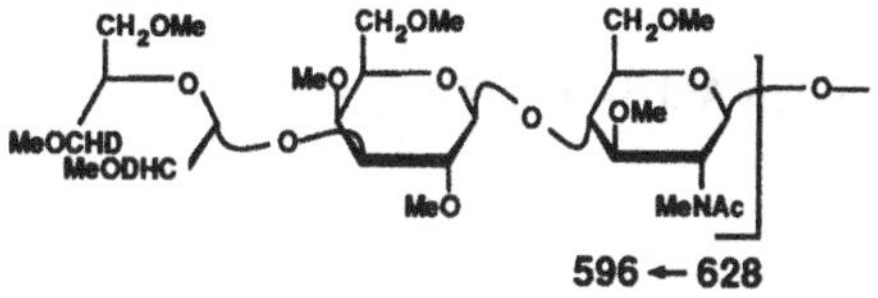 CH_2OMe CH_2OMe CH_2OMe O MeO O O O MeOCHD MeODHC OMe MeO MeNAc O 596 ← 628
Neu5Acα2-3Galβ1-4GlcNAcβ1-	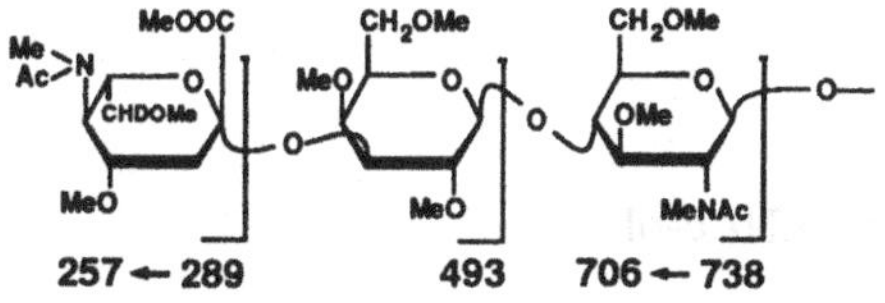 MeOOC CH_2OMe CH_2OMe Me Ac N O MeO O CHDOMe O O OMe MeO MeO MeNAc O 257 ← 289; 493; 706 ← 738
Neu5Acα2-6Galβ1-4GlcNAcβ1-	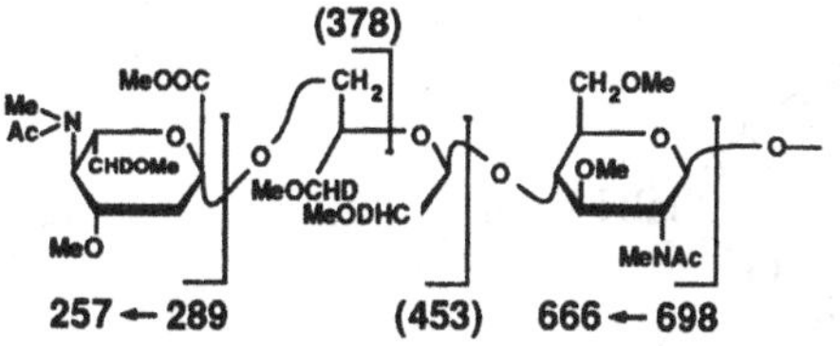 (378) MeOOC CH_2 CH_2OMe Me Ac N O O O O CHDOMe MeOCHD MeODHC OMe MeO MeNAc O 257 ← 289; (453); 666 ← 698
Neu5Acα2 | 6 Neu5Acα2-3Galβ1-3GlcNAcβ1-	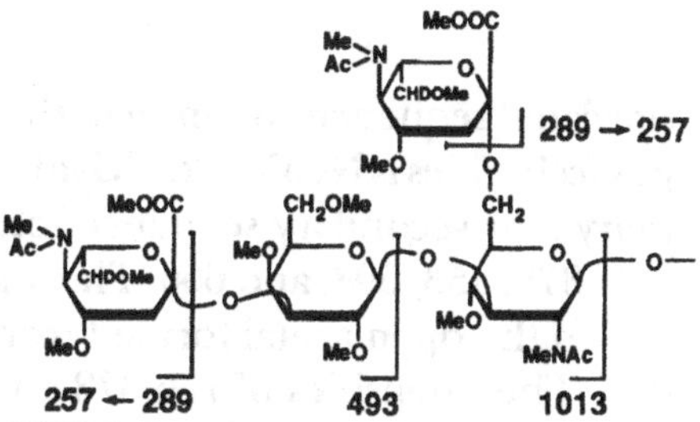 MeOOC Me Ac N O CHDOMe MeO O CH_2 289 → 257 MeOOC CH_2OMe Me Ac N O MeO O O O CHDOMe O MeO MeO MeO MeNAc O 257 ← 289; 493; 1013

Table 2
Ions Determining Linkages in Reducing Terminal Sequences

Reducing terminal sequence	Primary sequence ions
-4GlcNAcβ1-4GlcNAc-ol	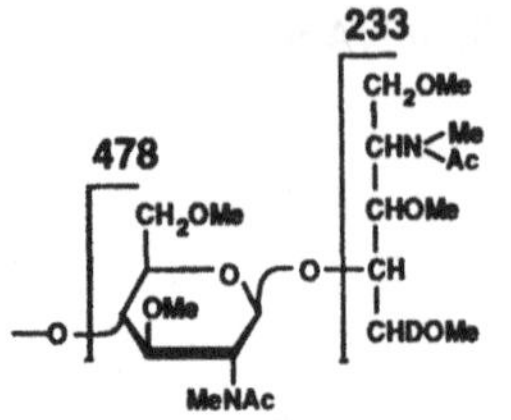 478, 233
Fucα1 – 6 -4GlcNAcβ1-4GlcNAc-ol	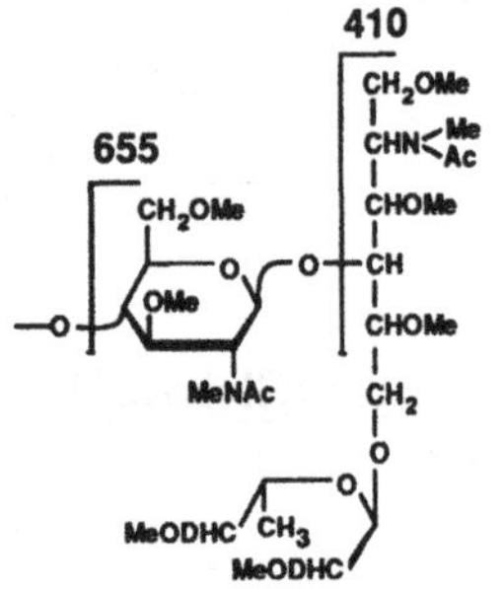 655, 410
-3GalNAc-ol	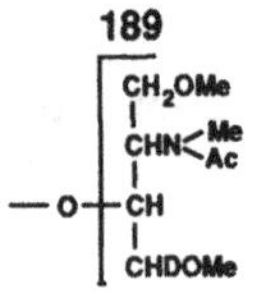 189
R_2-O-6, R_1-O-3 GalNAc-ol	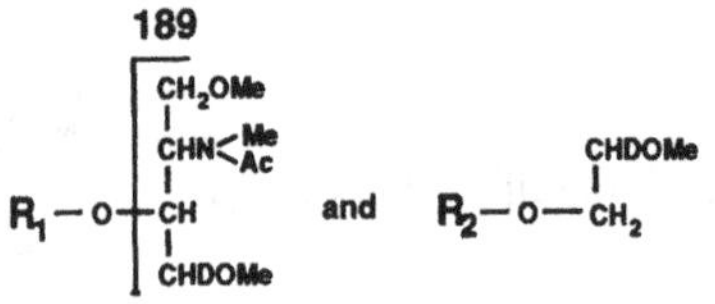 189 and R_2–O–CH_2–CHDOMe

sialylated sequence common in the complex type of glycoprotein oligosaccharides is Neu5Acα2-6Galβ1-4GlcNAc. The corresponding primary and secondary sequence ions for this sequence are *m/z* 289, 257, 378, 453, 698, and 666. The ion of *m/z* 378 is formed by cleavage within the open acetal formed from the periodate-oxidized Gal residue. The intensities of *m/z* 378 and 453 are, however, low as can be seen in the spectrum shown in Fig 1.

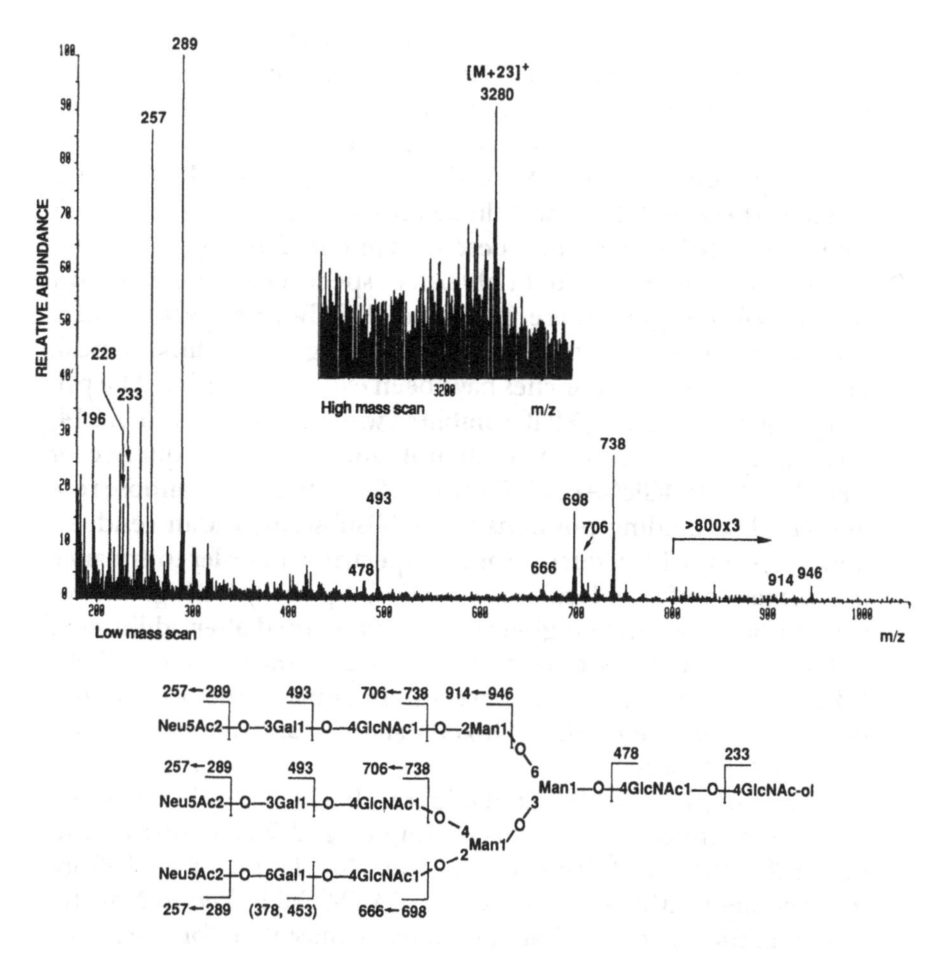

Fig. 1. FAB-mass spectrum of a triantennary structure, isolated from bovine fetuin, after periodate oxidation, $NaBD_4$ reduction, and permethylation. The high mass scan was recorded separately in the presence of sodium iodide. For interpretation, *see* Tables 1 and 2.

A disialylated branch in a sequence of Neu5Acα2-3Galβ1-3[Neu5Acα2-6]GlcNAc is recognized by the sequence ions of *m/z* 289, 493, and 1013. In order to determine the binding position to the reducing terminal, the periodate oxidation has to be carried out on the reduced oligosaccharide. *N*-linked glycans all have a reducing chitobiose sequence, GlcNAcβ1-4GlcNAc, which is determined by the ions of *m/z* 233 and 478. Substitution of the reducing GlcNAc in the 6-position by fucose gives the ions of *m/z* 410 and 655. *O*-linked glycans contain a reducing terminal GalNAc, which can be 3-substituted or 3,6-disubstituted. A 3-substituted reduced GalNAc is degraded by periodate to *N*-acetylthreosaminitol and recognized by an ion of *m/z* 189. A 3,6-disubstituted residue is cleaved by periodate between carbons 4 and 5, and two products are formed. The branch connected to the 3-position will have *N*-acetylthreosaminitol as alditol (as discussed earlier), and the branch linked to the 6-position will be terminated with ethylene glycol (as described in refs. *2* and *3*).

2. To illustrate the type of data obtained, spectra of some common *N*- and *O*-linked glycans are presented. A FAB-mass spectrum of a complex triantennary structure is shown in Fig. 1. The linkage positions in the sialylated branches have been explained earlier. The primary sequence ion of *m/z* 946 combined with a secondary of *m/z* 914, formed by elimination of methanol, shows that a sequence of Neu5Ac2-3Gal1-4GlcNAc is linked to the 2-position of a mannose residue *(1)*. The binding positions to the disubstituted Man residues, however, cannot be deduced from the spectrum. In order to enhance the formation of a molecular ion species, especially for high-mol-wt compounds, a separate high-mass scan is recorded after addition of sodium iodide to the sample and an $[M + 23]^+$ ion is formed *(4)*. An $[M + 23]^+$ ion of *m/z* 3280 supports a triantennary structure with two branches terminated with 2-3-linked sialic acid and one branch with 2-6-linked sialic acid.

   An example of an *O*-linked oligosaccharide alditol containing a 3,6-disubstituted GalNAc-ol is shown in Fig. 2. The sequence ions of *m/z* 289, 493, and 189 together with an $[M + 1]^+$ ion of *m/z* 699 are in agreement with a sequence of Neu5Ac2-3Gal linked to *N*-acetylthreosaminitol, showing that this sequence must therefore be linked to the 3-position of the GalNAc in the native compound. Another $[M + 1]^+$ ion of *m/z* 815 combined with the primary and secondary ions of *m/z* 289, 493, 738, and 706 is consistent with a sequence of Neu5Ac2-3Gal1-4GlcNAc linked to ethylene glycol. This sequence must therefore occupy the 6-position of the GalNAc in the native compound.

   The last example of application of this method is demonstrated on a high-mannose type of structure. This compound was isolated from human urine with one GlcNAc residue as the reducing terminal. The FAB-mass spectrum obtained is shown in Fig. 3. Since no internal *N*-acetyl hexosamine residues are present the sequence ions are of low abundance or absent. The ion of *m/z* 179 represent a periodate degraded nonreducing terminal mannose. A secondary sequence ion of *m/z* 355, which is formed from a primary of *m/z* 387 (not seen) by elimination of methanol, determines a sequence of Man1-2Man. No

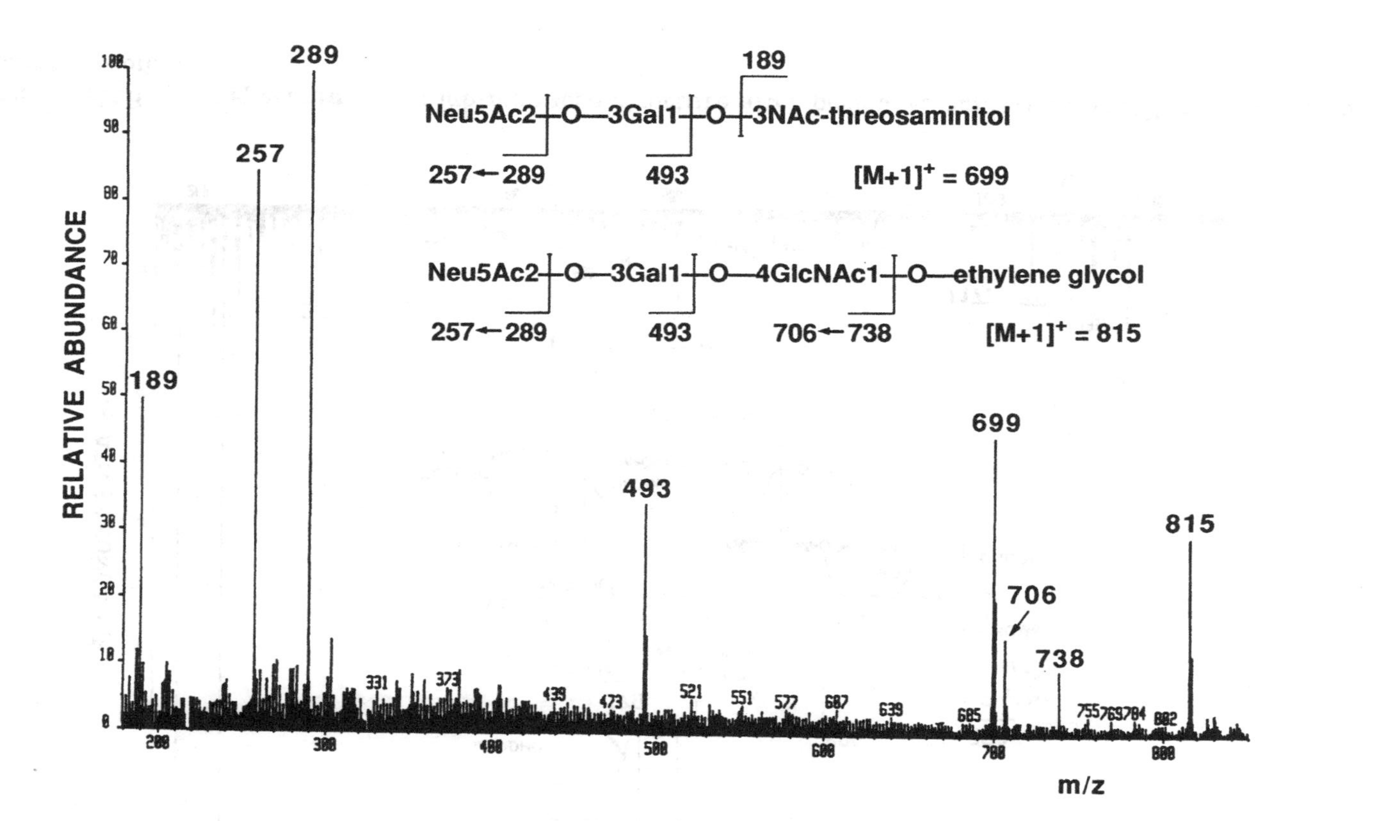

Fig. 2. FAB-mass spectrum of an *O*-linked oligosaccharide alditol, isolated from bovine fetuin, Neu5Acα2-3Galβ1-3[Neu5Acα2-3Galβ1-4GlcNAcβ1-6] GalNAc-ol after periodate oxidation, $NaBD_4$-reduction, and permethylation. For interpretation, *see* Tables 1 and 2.

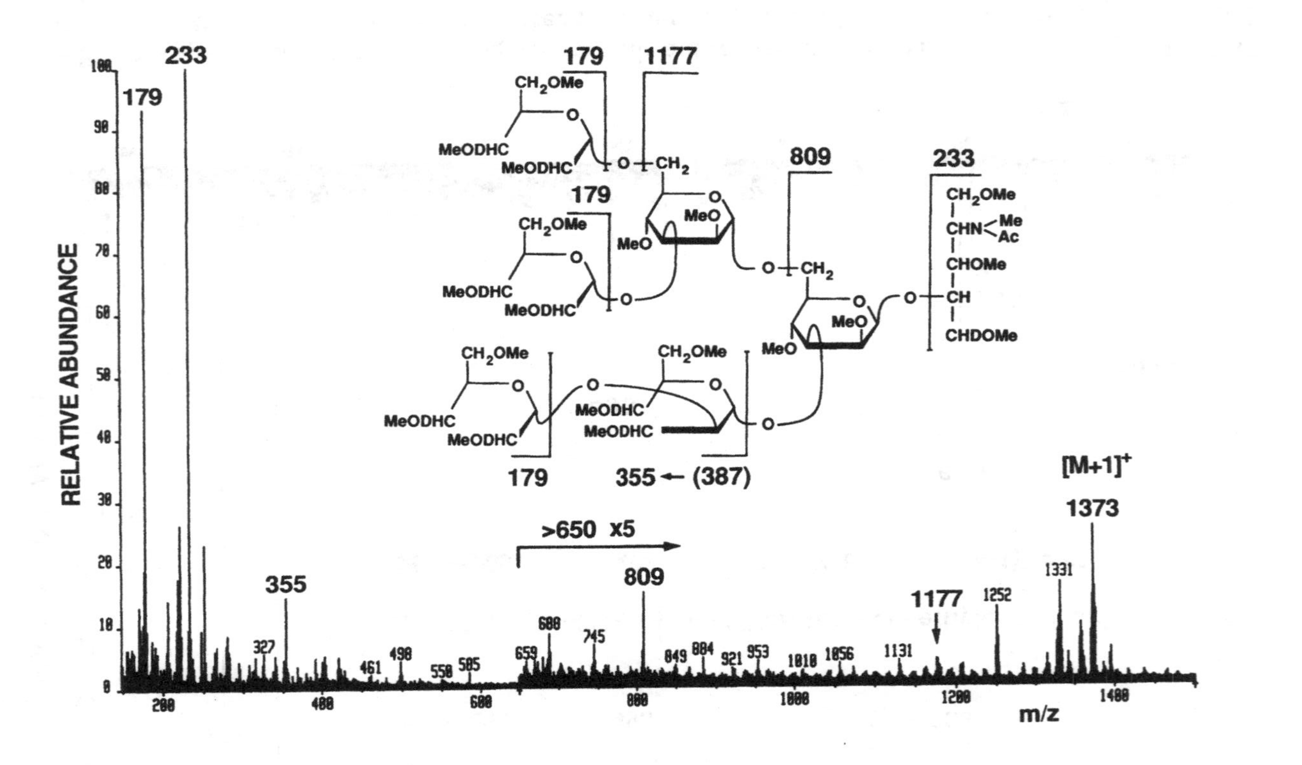

Fig. 3. FAB-mass spectrum of a high-mannose structure after periodate oxidation, $NaBD_4$-reduction, and permethylation.

other significant sequence ions from the nonreducing terminal are seen. High-mannose oligosaccharides often have 3,6-disubstituted Man residues. Rather intense alditol-containing ions are formed by a specific cleavage of the bond to the 6-position of the disubstituted Man residue with retention of the positive charge on the alditol part. The ions at *m/z* 809 and 1177 are such linkage-specific ions. The two branches linked to the 3- and 6-positions of the Man residue must therefore be Man1-2Man and Man1-3[Man1-6]Man, respectively. An $[M + 1]^+$ ion of *m/z* 1373 gives further evidence for the linkage positions.

4. Notes

1. It is important that the periodate oxidation is carried out in the dark to avoid unspecific oxidation. The periodate reagent has to be prepared fresh, since it is degraded when exposed to light.
2. Periodate oxidation is one of the most reliable chemical reactions in carbohydrate chemistry. Sugar residues with hydroxyl groups in vicinal positions are cleaved quantitatively between the carbons, and aldehydes are formed. The aldehydes are subsequently reduced with $NaBD_4$. Since the periodate oxidation is carried out at pH 5.5, it is important to adjust the pH to approx 7 prior to addition of $NaBD_4$. Otherwise, the $NaBD_4$ will be consumed without reduction of the aldehydes.
3. Addition of methanol in an acidic environment leads to the formation of volatile methyl borate.
4. An alternative to desalting on a PD 10 column, especially when low-mol-wt products are formed, is the following procedure: The oxidized and reduced sample is concentrated to dryness and acetylated in 3 mL of acetic anhydride/pyridine 1/1 (v/v) at 100°C for 30 min. Ethanol is added, and the sample is concentrated to dryness. Water (3 mL) is added, and the acetylated product is extracted three times with chloroform (3 × 1 mL). The combined chloroform phases are washed with water (3 × 1 mL) and concentrated to dryness. Residual pyridine is removed by evaporation with 1 mL of toluene. The sample can be analyzed as peracetylated derivative or permethylated as previously described. The *O*-acetyl groups will be removed in the alkaline conditions used during the methylation procedure.
5. The quality of the FAB-mass spectra depends to a great extent on the purity of the material obtained after the derivatization procedure. It is therefore necessary to remove contaminants, introduced from the solvents by gel filtration on an LH-20 column. By mixing in some NaB^3H_4, in the reduction step (Section 3.1., step 5), a small amount of sample can be purified by monitoring the radioactivity.

6. The requirements for mass spectrometry are fast atom bombardment ionization and an instrument with a mass range up to approx 4000 mass units. These derivatives (permethylated or peracetylated) can also be used in electron ionization mass spectrometry with a similar fragmentation pattern, except for the mol-wt information, which is not obtained in electron ionization.
7. The amount of material loaded on the FAB probe is about 1–10 µg. It is sometimes advantageous to record a separate high-mass scan in the presence of sodium iodide, to obtain mol-wt information.
8. The FAB-mass spectra presented contain several ions not discussed. Most of these ions can be explained, but they add no further structural information.

References

1. Angel, A-S., Lindh, F., and Nilsson, B. (1987) Determination of binding positions in oligosaccharides and glycosphingolipids by fast atom bombardment mass spectrometry. *Carbohydr. Res.* **168,** 15–31.
2. Krotkiewski, H., Lisowska, E., Angel, A-S., and Nilsson, B. (1988) Structural analysis by fast atom bombardment mass spectrometry of the mixture of alditols derived from *O*-linked oligosaccharides of murine glycophorins. *Carbohydr. Res.* **184,** 27–38.
3. Angel, A-S. and Nilsson, B . (1990) Linkage positions in glycoconjugates by periodate oxidation and fast atom bombardment mass spectrometry, in *Methods in Enzymology,* vol. 193 (McCloskey, J., ed.), Academic, New York, pp. 587–607.
4. Angel, A-S. and Nilsson, B. (1990) Analysis of glycoprotein oligosaccharides by fast atom bombardment mass spectrometry. *Biomed. Env. Mass Spect.* **19,** 721–730.

CHAPTER 4

Gas Chromatography and Gas Chromatography–Mass Spectrometry of Glycoprotein Oligosaccharides

Gunnar C. Hansson and Hasse Karlsson

1. Introduction

The composition of the oligosaccharide side chains of glycoproteins is often very complex. This demands analytical techniques with a high resolution. We have adapted the development of high-temperature capillary GC and GC–MS using thin-film thermostable bonded stationary phases for the analysis of large permethylated oligosaccharides up to about decasaccharides *(1,2)*. The technique is suitable for the analysis of most of the *O*-linked type of oligosaccharides, but the *N*-linked type of oligosaccharides are usually too large.

The approach presented here is suited for compositional and comparative analyses especially when limited amounts of material are available. Other techniques, for example, isolation of individual components and analysis with NMR spectroscopy, give more detailed information about the individual oligosaccharides, but will need more starting material and take a longer time. The GC and GC–MS approaches presented here also give a good estimate of the amounts of individual components. However, one has to be aware of the upper chromatographic limit. All unknown samples should therefore also be analyzed with FAB-MS. This chapter includes a description of the release of *O*-linked oligosaccharides using alkaline-borohydride and permethylation, methods applicable also to other techniques described in this book.

From: *Methods in Molecular Biology, Vol. 14: Glycoprotein Analysis in Biomedicine*
Edited by: E. F. Hounsell

2. Materials

2.1. Release and Isolation of O-Linked Oligosaccharides from Glycoproteins

1. KOH (0.1*M*) containing 2*M* $NaBH_4$: Add $NaBH_4$ (store dry) to a stock solution of KOH. Make a new solution every time.
2. Acetic acid (2.0*M*).
3. Ion-exchange resin, AG50Wx8 (Bio-Rad, Richmond, CA) 100–200 mesh, hydrogen form: Wash in distilled water and methanol on a sintered funnel. Store in methanol, and wash with water before use.
4. Pyridinium-acetate (1.0*M*), pH 5.4: 135 mL pyridine, 57 mL acetic acid, and water to 1.0 L; adjust pH with pyridine or acetic acid.
5. DEAE-Sephadex A-25 (Pharmacia, Uppsala, Sweden): Swell the gel for at least 3 h at room temperature.

2.2. Permethylation

1. Sodium hydroxide: Finely grain NaOH pellets in a mortar. Store in closed tubes. Can be used for at least 1 yr.
2. Dimethyl sulfoxide: Store with molecular sieves added.
3. Iodomethane: Store in the dark. **Carcinogen.**
4. Hydrochloric acid (0.1*M*).
5. Chloroform. **Potential Carcinogen, Liver Poison.**

2.3. High-Temperature Capillary Gas Chromatography

2.3.1. Column Preparation

Fused silica columns for high-temperature GC with film thicknesses of <0.1 μm are not yet commercially available. However, they can be ordered on special request from Chrompack, Middelburg, the Netherlands. The following scheme for a static coating of a fused silica column can be followed by the slightly experienced:

1. Fused silica columns (2–10 m × 0.25 mm id); HT-polyimide coated on the outside and deactivated can be purchased from Chrompack, Middelburg, The Netherlands.
2. The stationary phase (PS 264 from Fluka, Buchs, Switzerland) is dissolved in dichloromethane:*n*-pentane (1:1) at a concentration of 0.16–0.80 mg/mL to give a film thickness of 0.01–0.05 μm for a 0.25 mm id column. Equation for film thickness:

$$d_f[\mu m] = c[mg/mL]\ d[mm]/4$$

where *c* is the concentration of the stationary phase solution and *d* the inner diameter of the column. Let the stationary-phase solution stand for at least 6 h. Immediately before filling the column, add a solution of dicumyl peroxide (crosslinking agent) in dichloromethane (concentration 15 mg/mL) to give a final concentration of 2% (w/w). Filter the solution through a 0.5 µm filter (Millex FH, Millipore, Bedford, MA) down into a pear-shaped retort. Polish one end of the column with a microflame, and thread a piece of Teflon™ tubing onto the column. Fill the capillary column with coating solution using pressure (nitrogen gas) from the other end. Seal the Teflon™ tubing end with a clip, and begin to evaporate the solvent with the column placed in a water bath at room temperature using a vacuum pump.

3. After the solvent has been evaporated, the column is flushed with dry nitrogen gas for 30 min. Connect both ends to vacuum and evacuate. Close both ends (melting using a microtorch gas flame) and crosslink the stationary phase in a GC oven at 145°C for 30 min.
4. Cut the ends of the column, and extract slowly (3–4 h) with 5 mL of dichloromethane using pressure (nitrogen gas).
5. Conditioning of the column is performed in a GC oven by running a slow-temperature program from 70–400°C at a rate of 1°C/min using hydrogen as a carrier gas. Leave at 400°C for 90 min.

2.3.2. Reconditioning of Fused Silica Columns

After several months of use, a column may have lost its efficiency as well as become active. To recondition the column, slowly extract it from the detector end with 5 mL of dichloromethane. Flush the column overnight with dry nitrogen gas (0.1 bar pressure) followed by conditioning as described earlier in Section 2.3.1., step 5.

2.3.3. Gas Chromatography

Gas chromatography can be performed in any chromatograph equipped for capillary columns that allows temperatures up to at least 400°C. A gas chromatograph (as for example the Hewlett-Packard 5890A, Avondale, PA) allowing fast temperature programs (30°C/min) is needed for chromatographing larger oligosaccharides on shorter columns. The chromatograph should be equipped with an on-column injector and oxygen traps in the carrier gas line. Hydrogen is preferred as carrier gas using velocities from 80–250 cm/s at 70°C. The detector normally used is a flame ionization detector kept at the highest temperature reached during the run, usually 400°C.

2.3.4. Gas Chromatography–Mass Spectrometry

GC–MS can be performed on several mass spectrometers, although sector instruments are usually preferred, since they allow good sensitivity at higher masses. The major problem is the interface between the GC and the MS, because these have not been designed to allow high temperatures (up to 400°C) without having cold or hot spots. The length of the interface must be as short as possible. Currently, we use an in-house-built interface to a VG ZAB-HF. The capillary column should be introduced directly into the electron impact ion source, and the tip positioned 1–2 mm from the electron beam. Helium is used as carrier gas with an in-line high-capacity gas purifier and an indicating purifier (both Supelco, Bellefonte, PA). The scan rate of the mass spectrometer ought to be fast (1 s/decade), although we have used 2 s/decade scanning from *m/z* 1600 to 160, giving a total cycle time of 3.5 s with satisfactory results. Other conditions for the mass spectrometer have been interface temperature 400°C, ion-source temperature 380°C, electron energy 70 eV, trap current 500 μA, accelerating voltage 8 kV, and resolution 1400.

3. Methods

3.1. Release and Isolation of O-Linked Oligosaccharides from Glycoproteins

1. Dissolve the glycoprotein to a concentration of 2 mg/mL in water (*see* Note 1). Use a relatively long tube with a screw cap. Adjust the pH using pH test paper to 8–9 using 0.1*M* KOH.
2. Add an equal amount of freshly made 2*M* $NaBH_4$ in 0.10*M* KOH, and incubate for 16 h (or up to 45 h) at 45°C in a fume hood *(3)*. Unscrew the cap just to allow for the formed hydrogen gas to escape.
3. Freeze the sample in a freezer before gently adding 2*M* acetic acid until no more bubbles are formed and the pH is approx 6 (pH test paper). Take care that the sample does not bubble out of the tube. This can be stopped by the addition of a few drops of methanol.
4. Remove salts by ion-exchange chromatography or gel filtration chromatography on Bio-Gel P-2 (Bio-Rad) or Sephadex G-10 (Pharmacia). The ion-exchange procedure is described here. Pack a column with AG 50Wx8 (1.5 mL resin/mL of KOH-$NaBH_4$-solution), and rinse the column with water. Add the neutralized sample, and elute the oligosaccharides with water (5 mL/mL resin).
5. Evaporate the eluate in a rotary evaporator at 40°C. Dissolve the residue in methanol, add one to two drops of acetic acid, and evaporate. Repeat the last step another three times (*see* Note 2).

6. The oligosaccharides are separated into neutral and acidic ones using ion-exchange chromatography. Dissolve the sample in water containing 1% butanol, and apply to a column packed with DEAE-Sephadex A-25 (0.5 g gel/10 mg starting glycopeptide). Allow it to equilibrate for 2–4 h, and elute the neutral oligosaccharides with water containing 1% butanol (50 mL/0.5 mg gel) followed by the acidic ones with 1.0*M* pyridinium-acetate (50 mL/0.5 mg gel).
7. Lyophilize the oligosaccharides.

3.2. Permethylation (see *Note 3*)

The permethylation protocol using solid NaOH *(4)* has quickly gained acceptance and is now used instead of the dimethylsulfinyl carbanion technique *(5)* for most applications (*see* Note 4). The procedure described here is based on minor modifications *(6)* of the original solid NaOH procedure. The procedure should be performed in a fume hood using gloves.

1. Transfer the oligosaccharides (dissolved in water) to clean tubes with Teflon™-faced screw caps. Evaporate the water by a gentle stream of nitrogen gas at 50°C. Dry the tubes in a vacuum desiccator connected to a vacuum pump for 30 min.
2. Add a small magnetic bar.
3. Add 0.5 mL (for up to 0.2 mg of sample) or 1.0 mL (for 0.3–2 mg of sample) of dimethyl sulfoxide. The oligosaccharides can be helped to dissolve by a short (seconds) treatment in an ultrasonic cleaning bath.
4. Add 0.1 mL or 0.2 mL of iodomethane.
5. Add about 25 or 50 mg of NaOH powder. Use a spoon previously calibrated to give about the required amounts.
6. Stir the sample for 10 min at room temperature.
7. Stop the reaction by the addition of 2 or 4 mL of 0.1*M* HCl. Add 1 or 2 mL of chloroform.
8. Gently shake the tube to mix the two phases. To speed up the phase separation, spin the tube in a small table-top centrifuge for 5 min and remove the upper water phase by a Pasteur pipet.
9. Add 2 or 4 mL of water and centrifuge again. Remove the water phase. Repeat this another two times.
10. Add 2 or 4 mL of water, centrifuge, and by using a Pasteur pipet, carefully remove the lower (chloroform phase) to a clean tube. Add 1 or 2 mL of chloroform, centrifuge, and remove the lower phase once more to the same tube.
11. Dry the permethylated oligosaccharides in a stream of nitrogen gas at 40–50°C. Dry in a desiccator connected to a vacuum pump for 30 min.

3.3. High-Temperature Gas Chromatography and Gas Chromatography–Mass Spectrometry

Dissolve the permethylated oligosaccharides in ethyl acetate (1–100 ng oligosaccharide component/µL). Usually, approx 1 µL is injected on-column at 70°C. For oligosaccharides of up to eight sugar residues, a 10-m column with a film thickness of 0.03–0.05 µm can be used. A head pressure of 0.7 bar is used, giving a linear velocity of approx 90 cm/s at 70°C. After injecting the sample at 70°C and a 1-min delay, a linear temperature program of 10°C/min is used up to 400°C, where it is held for 5 min. For the larger permethylated oligosaccharides (more than seven to eight sugar residues), a shorter column (2 m) with a thin film thickness (0.03 µm) can be used and a rapid temperature program of 30°C/min up to 400°C. A higher linear gas velocity is then used (approx 250 cm/s at 70°C with a head pressure of 0.4 bar). Also a longer, 10 m, column can be used for the larger oligosaccharides, but then a film thickness down to 0.01 µm is needed.

So far, GC–MS can only be performed on the longer (i.e., 10 m) columns, and the larger permethylated oligosaccharides can be tried on the columns with the 0.01-µm phase mentioned earlier (*see* Note 5).

4. Notes

1. The amounts used depend on the starting material and on the analyses wanted. For obtaining gas chromatograms, glycoproteins with 50–80% carbohydrates, such as in mucins, can be analyzed down to under 1 mg of starting material.
2. This is done to remove borate.
3. The permethylation protocol gives good yields for most compounds. However, oligosaccharides containing glucitols (reduced glucose) give lower yields. This can be improved by increasing the amount of iodomethane to the same as that of dimethyl sulfoxide *(1)*.
4. Sialic acid containing oligosaccharides having a reduced *N*-acetylgalactosaminitol, i.e., obtained from *O*-linked glycans, gives poor yield when permethylated with the described protocol.
5. Gas chromatography using high temperatures and thin film columns as described here is relatively easy and will, by the high resolution obtained and comparison with known compounds, by itself give very useful information. The structural information obtained from GC–MS will be more informative, but the technique is at the same time more complicated. Our current mass range is approx 9 sugar residues for GC–MS and 11 for GC.
6. An example of the analysis of the permethylated neutral oligosaccharides obtained from rat intestinal mucin glycopeptides by GC is shown in Fig. 1.

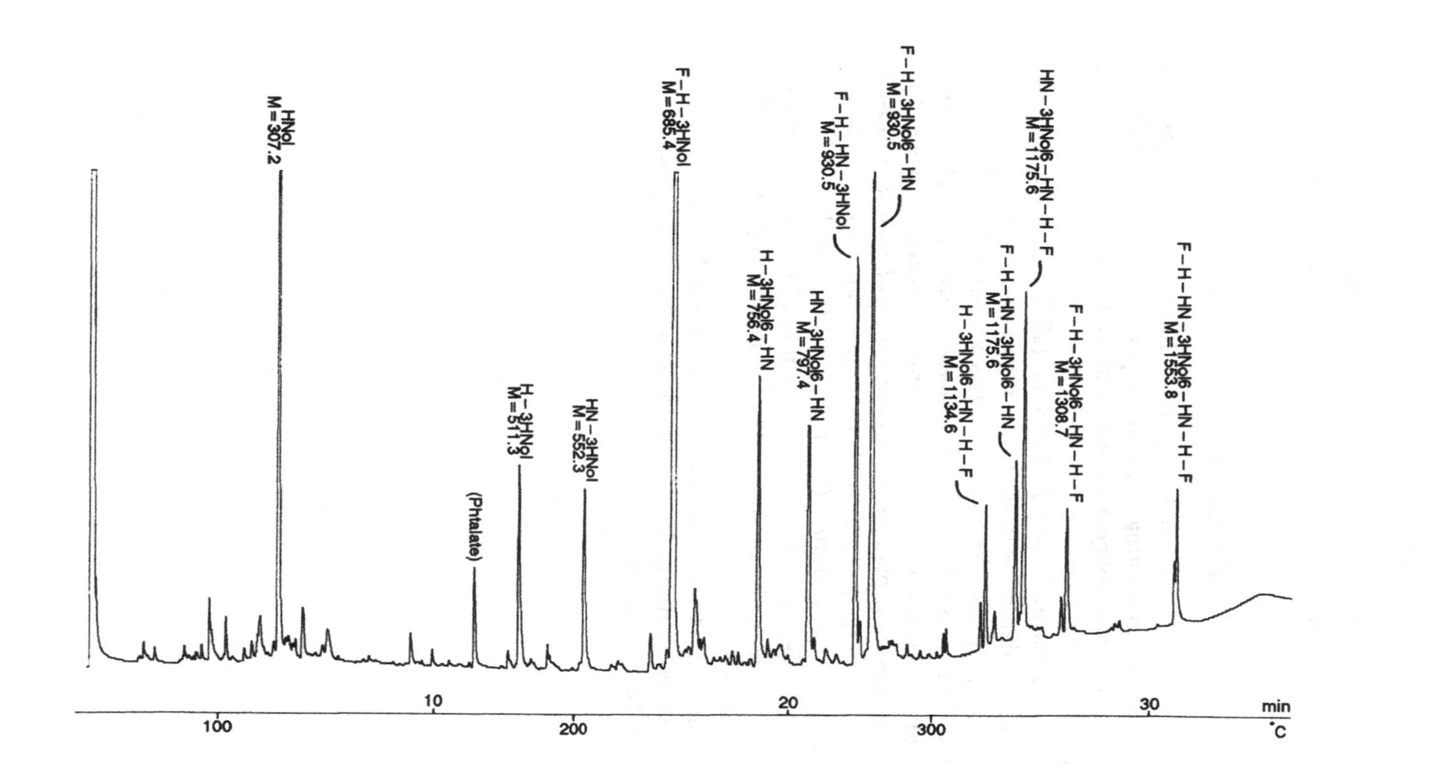

Fig. 1. Gas chromatogram of permethylated neutral oligosaccharide alditols released from rat small intestine mucin glycopeptides. Interpreted components from GC–MS are marked. Abbreviations are H for hexose, HN for *N*-acetylhexosamine, F for fucose, and HNol for *N*-acetylhexosaminitol (GalNAc linked to the peptide). A fused silica column (10 m × 0.25 mm id) coated with 0.04 µm of crosslinked PS 264 was used with a temperature program from 70°C (1 min) to 390°C (10°C/min). Hydrogen was used as carrier gas at a head pressure of 0.7 bar, giving a linear velocity of 90 cm/s at 70°C.

Acknowledgments

This chapter is based on work financed by grants from the Swedish Medical Research Council (No. 7461) and from the Swedish Board for Technical Development.

References

1. Karlsson, H., Carlstedt, I., and Hansson, G. C. (1989) The use of gas chromatography and gas chromatography–mass spectrometry for the characterization of permethylated oligosaccharides with molecular mass up to 2300. *Anal. Biochem.* **182,** 438–446.
2. Hansson, G. C., Carlstedt, I., and Karlsson, H. (1989) Characterization of glycosphingolipid mixtures with up to ten sugars by gas chromatography and gas chromatography–mass spectrometry as permethylated oligosaccharides and ceramides released by ceramide glycanase. *Biochemistry* **28,** 6672–6678.
3. Carlson, D. M. (1968) Structures and immunochemical properties of oligosaccharides isolated from pig submaxillary mucins. *J. Biol. Chem.* **243,** 616–626.
4. Ciucanu, I. and Kerek, F. (1984) A simple and rapid method for the permethylation of carbohydrates. *Carbohydr. Res.* **131,** 209–217.
5. Hakomori, S.-I. (1964) A rapid permethylation of glycolipid and polysaccharide catalyzed by methylsulfinyl carbanion in dimethyl sulphoxide. *J. Biochem.* **55,** 205–208.
6. Larson, G., Karlsson, H., Hansson, G. C., and Pimlott, W. (1987) Application of a simple methylation procedure for the analysis of glycosphingolipids. *Carbohydr. Res.* **161,** 281–290.

CHAPTER 5

Microscale Sequencing of *N*-Linked Oligosaccharides of Glycoproteins Using Hydrazinolysis, Bio-Gel P-4, and Sequential Exoglycosidase Digestion

Tsuguo Mizuochi

1. Introduction

Hydrazinolysis has been applied to the quantitative liberation of intact *N*-linked oligosaccharides from small amounts of glycoproteins. The usefulness of this approach was originally described by Mizuochi et al. *(1)*. This method has been extensively utilized for preparation of *N*-linked oligosaccharides from a number of glycoproteins for structural analysis *(2–15)*. When glycoproteins with *N*-linked oligosaccharides are heated with anhydrous hydrazine at 100°C for 10 h, almost all peptide bonds in the polypeptide moiety are cleaved and the amino acids are converted to hydrazides, whereas the glycosidic bonds are stable. Simultaneously, there is quantitative cleavage of the GlcNAc-Asn linkage and release of acyl groups linked to the amino groups of amino sugars and sialic acids. Therefore, the procedure for preparing *N*-linked oligosaccharides from glycoproteins for structural analysis consists of three steps: hydrazinolysis, re-*N*-acetylation, and reduction with NaB^3H_4 for radiolabeling a small amount of liberated oligosaccharides.

Gel-filtration chromatography on a Bio-Gel P-4 column (extrafine) has been applied to the separation of oligosaccharides for preparative purposes. Using this method, glucose oligomers can be separated singly starting with the monomer to about the 24-mer. To date, about

From: *Methods in Molecular Biology, Vol. 14: Glycoprotein Analysis in Biomedicine*
Edited by: E. F. Hounsell

200 kinds of tritium-labeled oligosaccharides have been subjected to the method using a mixture of glucose oligomers added as internal standards, and their sizes have been expressed in glucose units *(16,17)*.

Bio-Gel P-4 column chromatography is a convenient method for microscale sequencing of oligosaccharides when combined with sequential exoglycosidase digestion. When the oligosaccharide chain is reduced with NaB^3H_4, tritium is incorporated into the reducing terminal of the chain. Upon incubation with a specific exoglycosidase to which a particular oligosaccharide is susceptible, nonradioactive monosaccharide is released from the nonreducing terminal. Comparison of the elution profiles on the Bio-Gel P-4 column of the radioactive oligosaccharides before and after digestion informs us of the number of monosaccharide residues released by the treatment. In addition, the radioactive product can be recovered after analysis by the Bio-Gel P-4 column and resubjected to the next exoglycosidase digestion. Therefore, by repeating this procedure using various exoglycosidases with distinct specificities for substrates, one can effectively determine the anomeric configuration and sequence of each monosaccharide in the oligosaccharide chain. The method has been extensively utilized for microscale sequencing of a small quantity (approx 5×10^4 cpm) of radioactive oligosaccharides *(1–15)*.

2. Materials

2.1. Hydrazinolysis and NaB^3H_4 Reduction

1. Anhydrous hydrazine is prepared essentially as described by Kusama *(18)*. A mixture of 80% hydrazine hydrate (50 g), toluene (500 g), and CaO (500 g) is allowed to stand overnight. The mixture is refluxed for 3 h using a cold condenser and an NaOH tube. The mixture is then subjected to azeotropic distillation with toluene at 93–94°C under anhydrous conditions. Anhydrous hydrazine is collected from the bottom layer and stored in an airtight screw cap tube with a Teflon™ disk seal under dry conditions at 4°C in the dark. Commercially available anhydrous hydrazine (such as that from Aldrich Chemical Co., Inc., Milwaukee, WI) can also be used. It is important to check the quality with a glycoprotein of which the oligosaccharide structure has already been established before using for analysis, because contamination by trace amounts of water in some lots could modify the reducing terminal *N*-acetylglucosamine of *N*-linked oligosaccharides.

Caution: Anhydrous hydrazine is a strong reducing agent, is highly toxic, corrosive, suspected to be carcinogenic, and is flammable. Therefore, great caution should be exercised during handling.

2. Toluene.
3. Saturated sodium bicarbonate solution prepared at room temperature.
4. Acetic anhydride.
5. 1-Octanol.
6. Lactose.
7. Acetic acid (1*N*).
8. NaOH (1*N*).
9. Methanol.
10. NaOH (0.05*N*) freshly prepared from 1*N* NaOH just before use.
11. Sodium borotritide (NaB^3H_4, approx 22 GBq/mmol) and approx 40 m*M* in dimethylformamide (silylation grade #20672, Pierce Chemical Co., Rockford, IL). When stored in a small airtight tube with a Teflon™ disk seal at –18°C, this NaB^3H_4 solution is stable for at least 1 yr. Dimethylformamide should be stored with molecular sieves in a small screw cap bottle with a Teflon™ disk seal under dry conditions.
12. 1-Butanol:ethanol:water (4:1:1, v/v).
13. Ethyl acetate:pyridine:acetic acid:water (5:5:1:3, v/v).
14. Dowex 50W-X12 (H^+ form, 50–100 mesh).
15. Whatman 3MM chromatography paper.
16. Airtight screw cap tube with a Teflon™ disk seal.
17. Dry heat block capable of maintaining 100°C.
18. Vacuum desiccator.
19. High-vacuum oil pump.
20. Descending paper chromatography tank.

2.2. Bio-Gel P-4 Column Chromatography and Sequential Exoglycosidase Digestion

1. Standard glucose oligomer mixture: One gram of high-mol-wt dextran (170–200 kDa) is dissolved in 10 mL of 0.1*N* HCl and then heated in a screw cap tube with a Teflon™ disk seal at 100°C for 4 h. The solution is then passed through a small column (1 mL) of Bio-Rad AG 3-X4A (OH^- form, 100–200 mesh) to remove HCl, and the column is then washed with 5 mL of distilled water. The pooled effluent should be neutral. If it is not, the procedure is repeated again with a new column. Twenty-five milligrams of the dextran are dissolved in the solution, and the volume adjusted to 50 mL. The solution of glucose oligomers (about 20 mg/mL) thus obtained may be stored frozen.

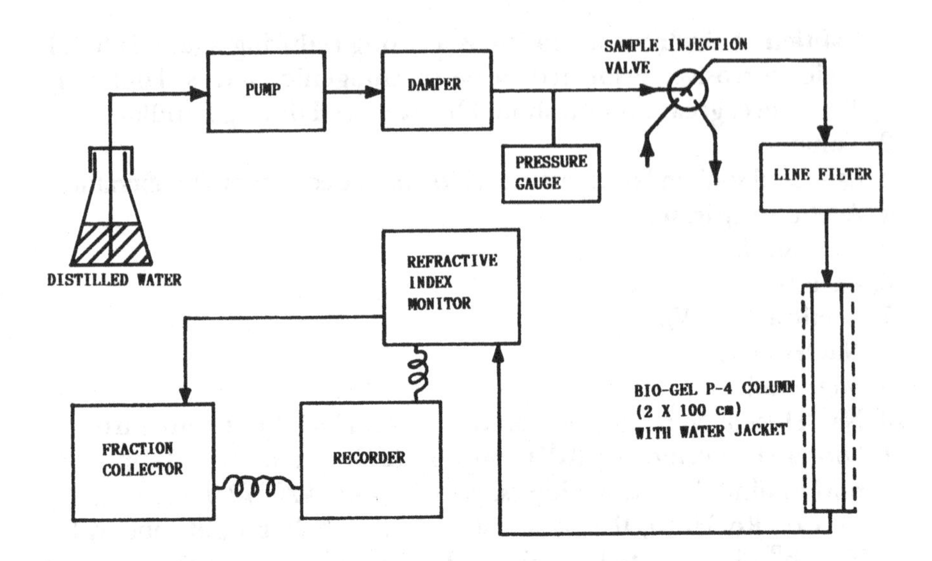

Fig. 1. Diagram of a liquid chromatograph equipped with a Bio-Gel P-4 column (extrafine, –400 mesh).

2. Oligosaccharide samples: When oligosaccharides are acidic, they should be neutralized before subjection to chromatography, because acidic oligosaccharides are eluted in the void volume of the column. For microsequencing, oligosaccharides should be labeled with tritium by reduction with NaB^3H_4.
3. Liquid chromatograph employing a Bio-Gel P-4 column (*see* Fig. 1): Distilled water is used as the mobile phase. An HPLC pump is recommended to give a constant flow rate. The usual flow rate for separation of oligosaccharides is 0.4 mL/min (*see* Note 1). The volume of the sample loop in the injection valve should be 1 mL. A line filter protects the column from impurities in the sample solution. When the pressure gage shows a high reading, the filter is replaced. Bio-Gel P-4 (extrafine, –400 mesh) is packed in a water-jacketed thick-wall glass column (100 × 2 cm id) and kept at 55°C by circulating warm water. The recorder should be equipped with two pens or with one pen and an event marker to record the refractive index of effluent and fractionation by the fraction collector. The column end fittings are equipped with a Teflon™ filter. Teflon™ tubing (0.5 mm id) is recommended.
4. Preparation of the Bio-Gel P-4 column: Suspend Bio-Gel P-4 (extrafine, –400 mesh) in distilled water, and remove very small particles and impurities by repeating decantation. The suspension is then deaerated under reduced pressure at 60°C. The warm slurry is packed at

constant pressure (about 5 Kg/cm^2) in the water-jacketed glass column prewarmed to 55°C by circulating water, and the column is maintained at this temperature during the analysis.
5. Exoglycosidases used for sequential digestion of oligosaccharides should be highly purified (*see* Note 2). One unit of exoglycosidase is defined here as the amount of enzyme required to hydrolyze 1 µmol of *p*-nitrophenyl glycosides/min at 37°C.
6. Buffers for exoglycosidase digestions may be stored frozen. The buffers and enzymes are described in Sections 3.2, steps 2,9, and 12.
7. Toluene.
8. pH test paper.
9. Small airtight tube.
10. Incubator to maintain the temperature at 37°C during enzyme digestion.
11. Dowex 50W-X12 (H^+ form, 50–100 mesh).
12. Bio-Rad AG 3-X4A (OH^- form, 100–200 mesh).

3. Methods

3.1. Hydrazinolysis and NaB^3H_4 Reduction

1. Add the glycoprotein (0.1–100 mg) to an airtight screw cap tube with a Teflon™ disk seal, and dry *in vacuo* overnight in a desiccator over P_2O_5 and NaOH.
2. Add anhydrous hydrazine (0.2–1.0 mL) with a glass pipet. The pipet must be dried to avoid introducing moisture into the anhydrous hydrazine (*see* Note 3).
   Caution: Anhydrous hydrazine is strong reducing, highly toxic, flammable, corrosive, and a suspected carcinogen. Caution should be exercised during handling.
3. Heat at 100°C for 10 h using a dry heat block. Glycoprotein is readily dissolved at 100°C.
4. Remove hydrazine by evaporation *in vacuo* in a desiccator. To protect the vacuum oil pump from hydrazine, connect traps between the desiccator and the pump in the following order: cold trap with dry ice and methanol, conc. H_2SO_4-trap, and NaOH-trap on the desiccator side. Remove the last trace of hydrazine by coevaporation with several drops of toluene.
5. To re-*N*-acetylate, dissolve the residue in ice-cold saturated $NaHCO_3$ solution (1 mL/mg of protein). Add 10 µL of acetic anhydride, mix, and incubate for 10 min at room temperature. Re-*N*-acetylation is continued at room temperature by further addition of 10 µL (three times) and then 20 µL (three times) of acetic anhydride at 10-min intervals. The total volume of acetic anhydride is 100 µL/1 mL of

saturated $NaHCO_3$ solution. Keep the solution on ice until the addition of acetic anhydride to avoid epimerization of the reducing terminal sugar.

6. To desalt, pass the reaction mixture through a column (1 mL for 1 mL of the $NaHCO_3$ solution) of Dowex 50W-X12, and wash with 5 column bed vol of distilled water. Evaporate the effluent to dryness under reduced pressure at a temperature below 30°C. Addition of a drop of 1-octanol is effective in preventing bubbling over.
7. Dissolve the residue in a small amount of distilled water, and spot on a sheet of Whatman 3MM paper. Perform paper chromatography overnight using 1-butanol:ethanol:water (4:1:1, v/v) as developing solvent. This procedure is indispensable for the next tritium-labeling step of the liberated oligosaccharides. This is because oligosaccharides larger than trisaccharides remain very close to the origin, whereas the degradation products derived from the peptide moiety, which react with NaB^3H_4, move a significant distance on the paper.
8. Cut the area 0–4 cm from the origin, recover the oligosaccharides by elution with distilled water, and then evaporate to dryness under reduced pressure. On this chromatogram, lactose migrates <4 cm from the origin.
9. To label oligosaccharides with tritium, dissolve the oligosaccharide fraction thus obtained in 100 μL of ice-cold 0.05*N* NaOH (freshly prepared from 1*N* NaOH). Verify that the pH of the oligosaccharide solution is above 11 with pH test paper using a <1-μL aliquot. If not, adjust the pH of the solution with 1*N* NaOH to that which gives the same color on pH paper as 0.05*N* NaOH. After addition of NaOH solution, keep the oligosaccharide solution on ice; otherwise, part of the reducing terminal *N*-acetylglucosamine may be converted to *N*-acetylmannosamine by epimerization.
10. Add a 20-molar excess of NaB^3H_4 solution to the oligosaccharide solution, mix, and incubate at 30°C for 4 h to reduce the oligosaccharides. Then, add an equal weight of $NaBH_4$ (20 mg/mL of 0.05*N* NaOH, freshly prepared) as the original glycoprotein, and continue the incubation for an additional 1 h to reduce the oligosaccharides completely. Stop the reaction by acidifying the mixture with 1*N* acetic acid. During the reduction and addition of acetic acid, keep the reaction mixture in a draft chamber, since tritium gas is generated (*see* Note 4).
11. To desalt, apply the reaction mixture to a small Dowex 50W-X12 column, wash with 5 column bed vol of distilled water, and then evaporate the effluent under reduced pressure below 30°C. The volume of

the column should be calculated based on the amount of NaOH and $NaBH_4$, and the capacity of the resin. Then, remove the boric acid by repeated (three to five times) evaporation with methanol under reduced pressure. Dimethyl formamide used to dissolve NaB^3H_4 is usually coevaporated during the repeated evaporation.

12. Dissolve the residue in a small amount of distilled water, spot on a sheet of Whatman 3MM paper, and perform paper chromatography overnight using ethyl acetate:pyridine:acetic acid:water (5:5:1:3, v/v) as developing solvent. This procedure is effective in removing the radioactive components originating from NaB^3H_4, which migrate a significant distance on the chromatogram.
13. Recover radioactive oligosaccharides, which migrate slower than lactitol (about 20 cm from origin), from the paper by elution with distilled water, and evaporate to dryness under reduced pressure.
14. Finally, subject the radioactive *N*-linked oligosaccharides thus obtained to high voltage paper electrophoresis at pH 5.4 or HPLC to separate oligosaccharides by charge.

3.2. Bio-Gel P-4 Column Chromatography and Sequential Exoglycosidase Digestion

Bio-Gel P-4 column chromatography of radioactive *N*-linked oligosaccharides derived from human fibrinogen by hydrazinolysis followed by NaB^3H_4 reduction and microscale sequencing of the oligosaccharides by sequential exoglycosidase digestion will be described here as an example.

1. Radioactive *N*-linked oligosaccharides of human fibrinogen obtained by hydrazinolysis followed by NaB^3H_4 reduction are composed of neutral, monosialo, and disialo oligosaccharides as determined by high-voltage paper electrophoresis at pH 5.4 *(6)*. Since acidic oligosaccharides are eluted in the void volume of a Bio-Gel P-4 column in this system, the radioactive oligosaccharide mixture is subjected to sialidase digestion to obtain a neutral oligosaccharide mixture for the microsequencing.
2. For sialidase digestion, evaporate the radioactive *N*-linked oligosaccharide mixture (2–100 × 10^4 cpm) in a small airtight tube, dissolve it in 50 µL of 0.1*M* sodium acetate buffer, pH 5.0, and add 50 mU of *Arthrobacter ureafaciens* sialidase dissolved in 10 µL of 0.1*M* sodium acetate buffer, pH 5.0. Ensure that the pH of the reaction mixture is 5.0, which is the optimal pH for the enzyme, by comparing colors on pH test paper between the buffer (pH 5.0) and <1 µL aliquot of the

reaction mixture. If not the same, adjust the pH with 0.5*M* buffer while comparing the colors. Pipet a very small amount of toluene onto the tube wall to maintain a toluene atmosphere in the tube during enzyme digestion, and then incubate at 37°C for 16–18 h.

3. After incubation, subject the reaction mixture to paper electrophoresis at pH 5.4, and recover the neutral oligosaccharide component as already described. To desalt, pass the neutral oligosaccharide fraction through a small column that contains approx 0.5 mL each of Bio-Rad AG 3-X4A (lower layer) and Dowex 50W-X12 (upper layer), wash the column with 5 bed vol of distilled water, and then evaporate the effluent to dryness. The electrophoresis can be also replaced by HPLC.
4. For chromatography on the Bio-Gel P-4 column, dissolve $1–10 \times 10^4$ cpm of the neutral oligosaccharide mixture in 0.2 mL of a standard glucose oligomer mixture added as internal standards and distilled water in a total vol of 0.8 mL (*see* Note 5).
5. Apply the sample solution to the column via a sample injection valve, and start collecting 1.6-mL fractions of the effluent just before the void volume is eluted. It is advisable to obtain an elution profile of the standard glucose oligomer mixture beforehand. A refractive index monitor detects glucose oligomers in the effluent, but not the radioactive oligosaccharides because of their very low concentration in comparison to glucose oligomers added as internal standards. Determine the radioactivities of the effluent in each tube by liquid scintillation counting of appropriate aliquots.
6. Plot the radioactivity data on a sheet of paper where the refractive index of the effluent is recorded, after compensating for the time-lag between detection of the refractive index and fraction collection. Based on the comparison between elution positions of the radioactive oligosaccharide peaks and the glucose oligomers added as internal standards, the size of the radioactive oligosaccharide is expressed in glucose units (*see* Fig. 2). Neutral oligosaccharides derived from human fibrinogen behave on the P-4 column, as does 13.5 glucose units (Figs. 2 and 3a).
7. For exoglycosidase digestion, pool and evaporate the effluent, which contains the radioactive oligosaccharide peak.
8. Desalt on a small column (approx 0.5 mL each of Bio-Rad AG 3-X4A and Dowex 50W-X12), as described in step 3, in case of contamination by charged materials, which may inhibit exoglycosidases.
9. To subject the oligosaccharide to jack bean β-galactosidase digestion, a step in the sequential exoglycosidase digestion, dissolve the radioactive oligosaccharide in a small airtight tube with 20 µL of 0.4*M* sodium

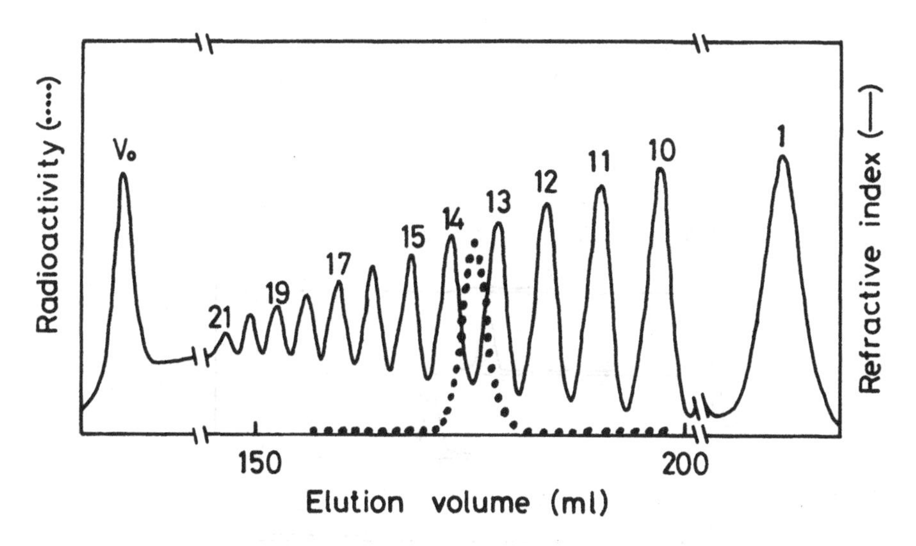

Fig. 2. Analysis of an oligosaccharide on a Bio-Gel P-4 column. A radioactive oligosaccharide mixture obtained from human fibrinogen by hydrazinolysis followed by NaB^3H_4 reduction is desialylated by sialidase treatment. The resultant radioactive neutral oligosaccharide mixture is dissolved in the standard glucose oligomer mixture and then analyzed on a Bio-Gel P-4 column (extrafine, –400 mesh) as described in the text.

citrate-phosphate buffer, pH 3.5, add 0.5 U (10 µL) of jack bean β-galactosidase and 20 µL of distilled water, and mix. Ensure that the pH of the reaction mixture is 3.5, which is the optimal pH for the enzyme, by comparing colors on pH test paper of the buffer (pH 3.5) and a <1-µL aliquot of the mixture. If not the same, adjust the pH with 0.5*M* buffer while comparing colors. Pipet a very small amount of toluene on the tube wall to maintain a toluene atmosphere in the tube during enzyme digestion, and then incubate at 37°C for 16–18 h.

10. After incubation, heat for 3 min in boiling water to stop the reaction, and then remove salts and heat-inactivated enzyme in the mixture with a small column containing Bio-Rad AG 3-X4A and Dowex 50W-X12, as described in step 3.
11. For analysis of the reaction product on a Bio-Gel P-4 column, dissolve the product in standard glucose oligomer mixture, as described in step 4. Then, carry out chromatography as described in steps 5 and 6. As shown in Fig. 3b, upon incubation with β-galactosidase, there is a shift of approx 2 glucose units on the P-4 column corresponding to the release of two galactose residues (compare with Fig. 3a).

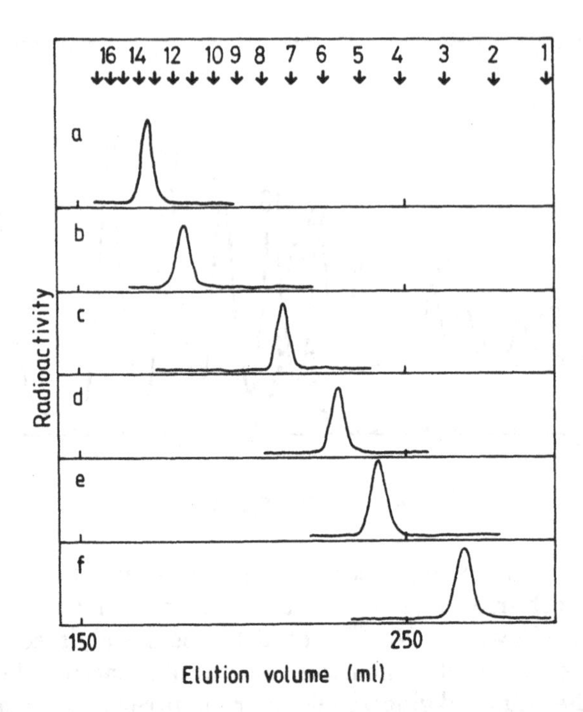

Fig. 3. Sequential exoglycosidase digestion of the radioactive neutral oligosaccharide derived from human fibrinogen. Radioactive products are analyzed on a Bio-Gel P-4 column at each step of exoglycosidase digestion as described in the text. The arrows indicate the elution positions of glucose oligomers added as internal standards (numbers indicate the glucose units) and void volume (Vo). (a) The radioactive neutral oligosaccharide mixture obtained from human fibrinogen, as shown in Fig. 2 and the text (same sample as that in Fig. 2); (b) the radioactive oligosaccharide in (a) after incubation with jack bean β-galactosidase; (c) the radioactive peak in (b) after incubation with jack bean β-*N*-acetylhexosaminidase; (d) the radioactive peak in (c) after incubation with jack bean α-mannosidase; (e) the radioactive peak in (d) after incubation with snail β-mannosidase; (f) the radioactive peak in (e) after incubation with jack bean β-*N*-acetylhexosaminidase.

12. For the next step in sequential exoglycosidase digestion and analysis of the reaction product, repeat steps 7, 8, 9, 10, and 11, altering the reaction conditions for exoglycosidase digestion of radioactive oligosaccharides according to the enzymes used. The conditions used for sequential exoglycosidase digestion of the radioactive neutral oligosaccharides derived from human fibrinogen (Figs. 2 and 3a) are as

follows: jack bean β-galactosidase digestion, 0.5 U of enzyme in 0.16*M* sodium citrate-phosphate buffer, pH 3.5 (50 μL); jack bean β-*N*-acetylhexosaminidase digestion, 0.5 U of enzyme in 0.16*M* sodium citrate-phosphate buffer, pH 5.0 (50 μL); jack bean α-mannosidase digestion, 1 U of enzyme in 0.1*M* sodium acetate buffer, pH 4.5 (50 μL); snail β-mannosidase digestion, 20 mU of enzyme in 0.1*M* sodium acetate buffer, pH 4.5 (50 μL); diplococcal β-galactosidase digestion, 5 mU of enzyme in 0.2*M* sodium acetate buffer, pH 6.0 (50 μL); diplococcal β-*N*-acetylglucosaminidase digestion, 5 mU of enzyme in 0.2*M* sodium citrate-phosphate buffer, pH 5.0 (50 μL).

Elution profiles on the Bio-Gel P-4 column of the reaction products at each step of exoglycosidase digestion are shown in Fig. 3a–f. The radioactive neutral oligosaccharides derived from human fibrinogen shown in Fig. 3a is resistant to digestion by β-*N*-acetylhexosaminidase, α-mannosidase, and β-mannosidase (data not shown). As already described, the radioactive oligosaccharides in Fig. 3a release two β-galactose residues upon incubation with jack bean β-galactosidase (Fig. 3b). The radioactive product in Fig. 3b is susceptible to jack bean β-*N*-acetylhexosaminidase digestion (Fig. 3c), and there is a shift of approx 4 glucose units on the P-4 column corresponding to the release of two β-*N*-acetylglucosamine residues *(16,17)*. The radioactive product in Fig. 3c sequentially releases two α-mannose residues (Fig. 3d), one β-mannose residue (Fig. 3e), and then one β-*N*-acetylhexosamine residue (Fig. 3f) by sequential exoglycosidase digestion using jack bean α-mannosidase, snail β-mannosidase, and then jack bean β-*N*-acetylhexosaminidase, respectively, finally resulting in [^{3}H]-*N*-acetylglucosaminitol (Fig. 3f). From these results, it was confirmed that the radioactive oligosaccharide in Fig. 3a has the sequence (Galβ1→HexNAcβ1→Manα1→)$_2$Manβ1→HexNAcβ1→[^{3}H]-*N*-acetylglucosaminitol. In separate experiments, it was shown that when jack bean β-galactosidase and jack bean β-*N*-acetylhexosaminidase were replaced with the diplococcal enzymes, the reaction products yielded the same elution profiles on the P-4 column as those obtained upon incubation with the jack bean enzymes (data not shown). Since diplococcal β-galactosidase cleaves the Galβ1→4GlcNAc linkage, but not Galβ1→3GlcNAc and Galβ1→6GlcNAc linkages *(19)* and since diplococcal β-*N*-acetylglucosaminidase hydrolyzes the GlcNAcβ1→2Man linkage, but not GlcNAcβ1→4Man and GlcNAcβ1→6Man linkages *(20)*, it was confirmed that the oligosaccharide has the sequence [Galβ1→4GlcNAcβ1→2(or 3)Manα1→]$_2$Manβ1→GlcNAcβ1→[^{3}H]-*N*-acetylglucosaminitol.

4. Notes

1. Connection of two to four Bio-Gel P-4 columns in series with Teflon™ tubing (0.5 mm id) is effective for separation of oligosaccharides with close retention times. When a flow rate of 0.4 mL/min cannot be obtained because of a rapid increase in column pressure caused by P-4 gel quality depending on lot number, decrease the flow rate to obtain a column pressure below approx 15 kg/cm^2. Repacking the column is also recommended.
2. Exoglycosidases utilized for sequential digestion of oligosaccharides should be highly purified. Commercially available exoglycosidases are also useful. It is essential that they are not contaminated with other glycosidases; otherwise results of the enzyme digestion will be misleading. Several commercially available glycosidases are frequently contaminated. Some of these preparations do not hydrolyze synthetic substrates, such as *p*-nitrophenyl glycosides, but cleave natural oligosaccharides. Therefore, exoglycosidases utilized for microsequencing of oligosaccharides should be certified to be uncontaminated with other glycosidase activities, not only with synthetic substrates (e.g., *p*-nitrophenyl glycosides), but also with radioactive oligosaccharides as natural substrates of which the structures have already been established.
3. For quantitative liberation of intact *N*-linked oligosaccharides from glycoproteins by hydrazinolysis, great care should be taken to maintain anhydrous conditions until the re-*N*-acetylation step. Introduction of moisture into glycoprotein samples or anhydrous hydrazine results in diverse modifications of reducing terminal *N*-acetylglucosamine residues, especially when unsubstituted with an Fucα1→6 group, and causes the release of *O*-linked oligosaccharides accompanied with various degradations of the reducing end.
4. When oligosaccharides are reduced by NaB^3H_4 with high specific activity (e.g., 555 GBq/mmol), the sensitivity of detection of oligosaccharides increases about 20-fold. To label oligosaccharides with tritium at high efficiency, it is recommended to keep the concentration of NaB^3H_4 high in the incubation mixture by reducing the vol of 0.05*N* NaOH (e.g., to the same vol as the NaB^3H_4 solution). A 20-molar excess of NaB^3H_4 solution is required for complete reduction of *N*-linked oligosaccharides, whereas a five-molar excess of NaB^3H_4 solution is sufficient for that of monosaccharides. The amount of NaB^3H_4 solution required for complete reduction of *N*-linked oligosaccharides derived from glycoprotein samples is roughly estimated

from data on the carbohydrate content or amino acid sequence. If generation of tritium gas is to be avoided, continue the incubation for an additional 1 h with large amounts of glucose to absorb excess NaB^3H_4 before acidifying the mixture.

5. The amount of standard glucose oligomer mixture added as an internal standard for analysis of radioactive oligosaccharides using the P-4 column depends on the sensitivity of the refractive index monitor.

References

1. Mizuochi, T., Yonemasu, K., Yamashita, K., and Kobata, A. (1978) The asparagine-linked sugar chains of subcomponent C1q of the first component of human complement. *J. Biol. Chem.* **253,** 7404–7409.
2. Mizuochi, T., Yamashita, K., Fujikawa, K., Kisiel, W., and Kobata, A. (1979) The carbohydrate of bovine prothrombin. *J. Biol. Chem.* **254,** 6419–6425.
3. Mizuochi, T., Yamashita, K., Fujikawa, K., Titani, K., and Kobata, A. (1980) The structures of the carbohydrate moieties of bovine blood coagulation factor X. *J. Biol. Chem.* **255,** 3526–3531.
4. Mizuochi, T. and Kobata, A. (1980) Different asparagine-linked sugar chains on the two polypeptide chains of human chorionic gonadotropin. *Biochem. Biophys. Res. Commun.* **97,** 772–778.
5. Yoshima, H., Matsumoto, A., Mizuochi, T., Kawasaki, T., and Kobata, A. (1981) Comparative study of the carbohydrate moieties of rat and human plasma α_1-acid glycoproteins. *J. Biol. Chem.* **256,** 8476–8484.
6. Mizuochi, T., Taniguchi, T., Asami, Y., Takamatsu, J., Okude, M., Iwanaga, S., and Kobata, A. (1982) Comparative studies on the structures of the carbohydrate moieties of human fibrinogen and abnormal fibrinogen Nagoya. *J. Biochem.* **92,** 283–293.
7. Mizuochi, T., Taniguchi, T., Shimizu, A., and Kobata, A. (1982) Structural and numerical variations of the carbohydrate moiety of immunoglubulin G. *J. Immunol.* **129,** 2016–2020.
8. Mizuochi, T., Taniguchi, T., Fujikawa, K., Titani, K., and Kobata, A. (1983) The structures of the carbohydrate moieties of bovine blood coagulation factor IX (christmas factor). Occurrence of penta- and tetrasialyl triantennary sugar chains in the asparagine-linked sugar chains. *J. Biol. Chem.* **258,** 6020–6024.
9. Mizuochi, T., Nishimura, R., Derappe, C., Taniguchi, T., Hamamoto, T., Mochizuki, M., and Kobata, A. (1983) Structures of the asparagine-linked sugar chains of human chorionic gonadotropin produced in choriocarcinoma. Appearance of triantennary sugar chains and unique biantennary sugar chains. *J. Biol. Chem.* **258,** 14,126–14,129.
10. Taniguchi, T., Mizuochi, T., Banno, Y., Nozawa, Y., and Kobata, A. (1985) Carbohydrates of lysosomal enzymes secreted by *Tetrahymena pyriformis*. *J. Biol. Chem.* **260,** 13,941–13,946.

11. Taniguchi, T., Adler, A. J., Mizuochi, T., Kochibe, N., and Kobata, A. (1986) The structures of the asparagine-linked sugar chains of bovine interphotoreceptor retinol-binding protein. Occurrence of fucosylated hybrid-type oligosaccharides. *J. Biol. Chem.* **261,** 1730–1736.
12. Mizuochi, T., Hamako, J., and Titani, K. (1987) Structures of the sugar chains of mouse immunoglobulin G. *Arch. Biochem. Biophys.* **257,** 387–394.
13. Mizuochi, T., Spellman, M. W., Larkin, M., Solomon, J., Basa, L. J., and Feizi, T. (1988) Carbohydrate structures of the human immunodeficiency-virus (HIV) recombinant envelope glycoprotein gp120 produced in Chinese-hamster overy cells. *Biochem. J.* **254,** 599–603.
14. Mizuochi, T., Matthews, T. J., Kato, M., Hamako, J., Titani, K., Solomon, J., and Feizi T. (1990) Diversity of oligosaccharide structures on the envelope glycoprotein gpl20 of HIV-1 from the lymphoblastoid cell line H9: presence of complex type oligosaccharides with bisecting *N*-acetylglucosamine residues. *J. Biol. Chem.* **265,** 8519–8524.
15. Mizuochi, T., Hamako, J., Nose, M., and Titani, K. (1990) Structural changes in the oligosaccharide chains of immunoglobulin G in autoimmune MRL/Mp-lpr/lpr mice. *J. Immunol.* **145,** 1794–1798.
16. Yamashita, K., Mizuochi, T., and Kobata, A. (1982) Analysis of oligosaccharides by gel filtration. *Methods Enzymol.* **83,** 105–126.
17. Kobata, A., Yamashita, K., and Takasaki, S. (1987) Bio-Gel P-4 column chromatography of oligosaccharides: effective size of oligosaccharides expressed in glucose units. *Methods Enzymol.* **138,** 84–94.
18. Kusama, K. (1957) On the species difference in carboxyterminal amino acids of serum albumins from various animals. *J. Biochem.* **44,** 375–381.
19. Paulson, J. C., Prieels, J. P., Glasgow, L. R., and Hill, R. L. (1978) Sialyl- and fucosyltransferases in the biosynthesis of asparagine-linked oligosaccharides in glycoproteins. *J. Biol. Chem.* **253,** 5617–5624.
20. Yamashita, K., Ohkura, T., Yoshima, H., and Kobata, A. (1981) Substrate specificity of diplococcal β-*N*-acetylhexosaminidase. A useful enzyme for the structural studies of complex type asparagine-linked sugar chains. *Biochem. Biophys. Res. Commun.* **100,** 226–232.

Chapter 6

Analysis of Sugar Chains by Pyridylamination

Sumihiro Hase

1. Introduction

For analyses of structures of sugar chains with high sensitivity, reducing ends of sugar chains are tagged with 2-aminopyridine as shown in Fig. 1 *(1)*. A pyridylamino (PA) derivative of a sugar chain having fluorescence and a positive charge has the following advantages:

1. Sensitive detection is possible; 0.03 pmol of a PA-sugar chain can be detected by a commercially available HPLC apparatus;
2. Excellent separation is achieved by reversed-phase HPLC;
3. The pyridylamino group is stable; fluorescence does not disappear under the conditions of procedures used for structure elucidation, such as light, acid hydrolysis, alkaline hydrolysis, hydrazinolysis, partial acetolysis, methylation analysis, or Smith periodate degradation; and
4. A positive charge is useful for separation of PA-sugar chains by ion exchangers or electrophoresis.

For pyridylamination of sugar chains from glycoconjugates, sugar chains are first liberated by hydrazinolysis-*N*-acetylation *(2)*, a glycopeptidase *(3,4)*, or an endoglycoceramidase *(5,6)*. The reaction mixture is directly pyridylaminated without purification of sugar chains. PA-sugar chains are purified by three kinds of HPLC with different separation principles:

1. Anion-exchange HPLC;
2. Size-fractionation HPLC; and
3. Reversed-phase HPLC.

From: *Methods in Molecular Biology, Vol. 14: Glycoprotein Analysis in Biomedicine*
Edited by: E. F. Hounsell

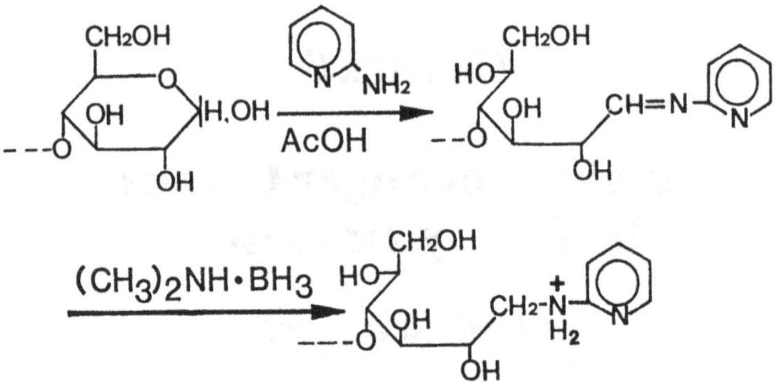

Fig. 1. Pyridylamination of sugar chains.

Table 1
Fluorescence Intensities of PA-*N*-Linked Sugar Chains

PA-GlcNAc$_2$	1.05
PA-Xylomannose	0.99
M6B	1.00
PA-Biantenna	1.07
PA-Biantenna-NeuAc$_2$	0.95
PA-Triantenna	0.95
PA-Tetraantenna	0.95

Figures indicate peak areas per mole when PA-sugar chains were analyzed under Section 3.2.3. M6B (*see* Fig. 6 later in this chapter) is taken as unity. Amounts of PA-sugar chains were determined by gas-liquid chromatography after methanolysis.

The additivity rule, which correlates elution times and chemical structures for reversed-phase chromatography, will support the analyses of PA-sugar chains as described in the following. Since the fluorescence intensities of PA-derivatives of *N*-linked sugar chains are almost the same (Table 1, Hase, S. et al., unpublished data), peak-area ratios are considered as molar ratios. Sensitivity of structure analyses, such as partial acetolysis *(7)*, Smith degradation *(8)*, component sugar analysis *(9)*, and exoglycosidase digestion studies *(10)*, are raised owing to the fluorescence and excellent separation by reversed-phase HPLC. NMR *(11)*, methylation analysis, and mass spectrometry can be performed as before. Two-dimensional HPLC sugar mapping *(12)* is useful for analyses of sugar chains from tissues *(13)*. This chapter will try to help the researcher prepare and purify PA-sugar chains.

2. Materials

1. Glass test tubes (10 × 100 mm) tapered at the bottom or Reacti-Vials (0.3 mL; Pierce, Rockford, IL).
2. Distilled water is used throughout the experiment.
3. 2-Aminopyridine: Colorless leaflet crystals stored in a desiccator. The commercial reagent is recrystallized from hexane. **Caution:** This compound is toxic, and an irritant to skin, eye, and mucosa. Avoid inhalation.
4. Coupling reagent: Dissolve 552 mg of 2-aminopyridine in 200 µL of distilled acetic acid. (When the reagent is diluted with 9 vol of water, the pH of the solution should be 6.8.) The reagent is stored at –20°C in a tube sealed with Parafilm.
5. Reducing reagent: Prepare just before use by dissolving 200 mg of borane-dimethylamine complex ($[CH_3]_2NH\text{-}BH_3$: Aldrich Chemical Company, Inc., Milwaukee, WI) in a mixture of 50 µL of water and 80 µL of distilled acetic acid. **Caution:** This compound is corrosive to eye, skin, and mucosa. Avoid inhalation.
6. A Tosoh TSKgel HW-40F column (1.5 × 45 cm, Tosohaas, Philadelphia, PA) washed well with 0.01*M* ammonium acetate, pH 6.0 (0.01*M* acetic acid is titrated to pH 6.0 with 1.5*M* aqueous ammonia). Sephadex G-15, Sephadex G-25, and Bio-Gel P-2 can also be used. PA-*N*-linked sugar chains appear between 25 and 55 mL of elution volume (25–43 mL for PA-sialosugar chains and 43–55 mL for PA-asialosugar chains).

   Excess 2-aminopyridine appears from 95 mL. Contaminating peaks appear between 60 and 90 mL (Fig. 2). Take out the used gel and wash the gel with the same solvent before the next use.
7. Mono Q HR 5/5 (0.5 × 5.0 cm, Pharmacia, Uppsala). Tosoh TSKgel DEAE-5PW (0.75 × 7.5 cm) can also be used. Solvent A: 0.7 m*M* aqueous ammonia (pH 9). Solvent B: 0.5*M* ammonium acetate, pH 9.0. (The pH of 0.5*M* acetic acid is adjusted to 9.0 with 4*M* aqueous ammonia.)
8. MicroPak AX-5 (0.46 × 15 cm, Varian Aerograph, Walnut Creek, CA). Tosoh TSKgel Amido-80 and YMC PA-03 (YMC Inc., Morris Plains, NJ) can also be used. A precolumn (silica gel, 0.75 × 7.5 cm) is placed between an injector and a pump to prevent damage to the MicroPak AX-5. Solvent A: 3% acetic acid in acetonitrile-water (4:1) is titrated to pH 7.3 with triethylamine. Solvent B: 3% acetic acid is titrated to pH 7.3 with triethylamine. The conditions are a modification of those reported *(14)*.

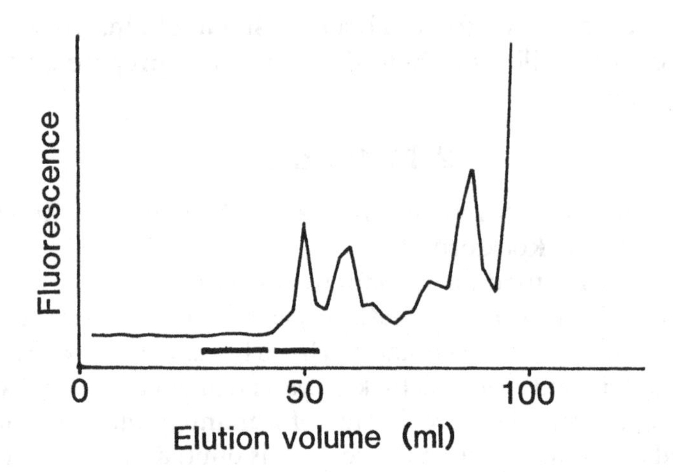

Fig. 2. Purification of PA-sugar chains by gel chromatography. Sugar chains liberated from 13 nmol of Taka-amylase A by hydrazinolysis-*N*-acetylation were pyridylaminated. The reaction products were subjected to a TSKgel HW-40F column.

9. Nacalai Cosmosil 5C18-P (0.46 × 15 cm, JM Science Inc., Buffalo, NY) or other C18 reversed-phase columns. Solvent A: 0.1*M* ammonium acetate, pH 4.0 (0.1*M* acetic acid is titrated to pH 4.0 with 4*M* aqueous ammonia). Solvent B: 0.1*M* ammonium acetate, pH 4.0–0.5% 1-butanol.
10. An HPLC apparatus and a fluorescence spectrophotometer equipped with a 1-cm cuvet, a flow cell (8–16 µL), a 150-W xenon lamp, and two monochrometers.
11. A water bath (90 and 80°C), small centrifuge, and lyophilizer.
12. Standard PA-sugar chains available from Takara Biochemical Inc., Berkeley, CA.

3. Method

3.1. Procedure for Pyridylamination (15)

1. Lyophilize 0.05–50 nmol of a sugar chain or sugar chains liberated from a glycoconjugate (<1.5 mg; *see* Note 1) in a glass test tube tapered at the bottom or in a Reacti-Vial.
2. Add to the residue 20 µL of coupling reagent (the reagent and pipets are warmed before use), and mix well. Seal the tube, and spin to concentrate the reaction mixture at the bottom of the tube.
3. For Schiff base formation, heat the reaction mixture at 90°C for 60 min (care should be taken to heat the *whole* tube). Cool the tube to room temperature and open.

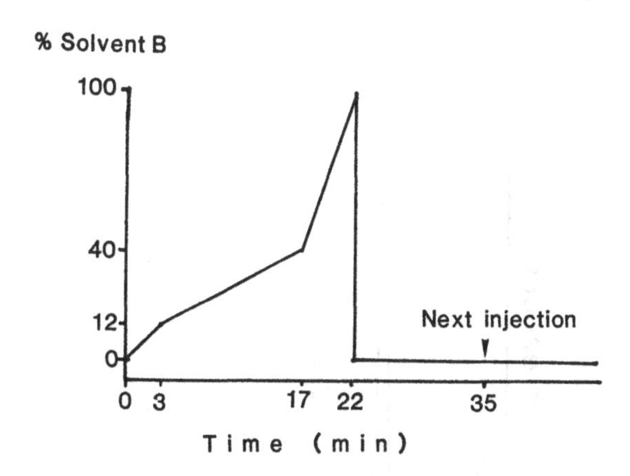

Fig. 3. An elution procedure for anion-exchange chromatography.

4. Add reducing reagent (70 µL), and mix well.
5. For reduction of the Schiff base, reseal the tube, spin, and heat at 80°C for 35 min.
6. Remove excess reagents by either method A or method B. Method A: Open the tube, add 0.5 mL of water, and subject to gel filtration on a column of TSKgel HW-40F (Fig. 2) eluted with 0.01*M* ammonium acetate, pH 6.0. Collect the PA-sugar chain fraction, and freeze-dry repeatedly for removal of ammonium acetate.

   Method B: After opening the tube, add 40 µL of a mixture of methanol and triethylamine (3:1, v:v) to the reaction mixture, and mix well. Then add 40 µL of toluene. Remove excess reagents by flashing the nitrogen (300 mL/min) at 50°C for 12 min under a vacuum at about 150 mmHg (*see* Note 2). Repeat the evaporation with 60 µL 2:1 toluene:methanol and 50 µL toluene. To remove a minute amount of reagents still remaining, carry out gel filtration using a column (0.8 × 20 cm) smaller than that described under method A.

3.2. Separation of PA-Sugar Chains by HPLC

3.2.1. Anion-Exchange Chromatography (16)

1. Before injection of approx 10 samples, wash Mono Q HR 5/5 with 6% acetic acid for 10 min and 0.4*M* aqueous ammonia for 10 min at a flow rate of 1 mL/min.
2. Equilibrate the column with solvent A.
3. Inject a sample made up to pH 9.0, and carry out gradient elution as described in Fig. 3 at a flow rate of 1.0 mL/min, with the column temperature being 25°C.

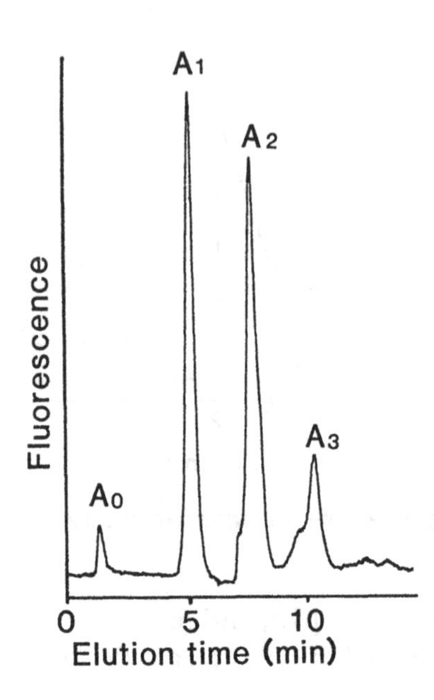

Fig. 4. Anion-exchange chromatography of PA-sugar chains from fetuin. A_0, PA-asialosugar chains; A_1 ~ A_3, PA-monosialosugar chains ~PA-trisialosugar chains, respectively.

4. Detect using an excitation wavelength at 310 nm and emission wavelength at 380 nm. Under these conditions, PA-sugar chains are separated according to the number of sialic acid residues as shown in Fig. 4.

3.2.2. Size-Fractionation HPLC (10)

1. Wash MicroPak AX-5 with methanol, and equilibrate the column with solvent A.
2. Inject a sample (<10 µL), and chromatograph using the gradient described in Fig. 5 at a flow rate of 1.0 mL/min, with column temperature of 25°C.
3. Detect using the same wavelengths as in Section 3.2.1., step 4.

The elution time of a PA-sugar chain is increased by addition of a sugar residue. When PA-isomaltooligosaccharides are used as a scale, addition of a mannose, galactose, or fucose residue causes 1.0 glucose unit; an *N*-acetylhexosamine residue, 0.6 glucose unit; "bisecting" *N*-acetylglucosamine residue, 0.2 glucose unit; fucosylα1-6 residue,

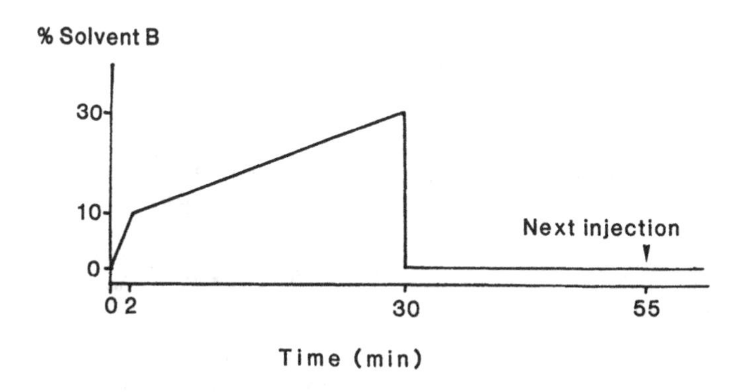

Fig. 5. An elution procedure for size-fractionation HPLC.

0.5 glucose unit; PA-glycose residue, 1.0 glucose unit. The values are almost independent of their linkage points and anomeric configurations. Separation of a mixture of 12 PA-oligomannose type sugar chains (Fig. 6) is shown in Fig. 7.

3.2.3. Reversed-Phase HPLC (10; see *Note 3)*

1. Wash Cosmosil 5C18-P with methanol, and equilibrate the column with the starting solvent.
2. Inject a sample, and carry out the gradient elution as shown in Fig. 8 at a flow rate of 1.5 mL/min and column temperature of 25°C. Separation of M_5 to M_9 (Fig. 7) is shown in Fig. 9 *(12; see* Note 4).
3. Detect using fluorescence wavelength at 315 nm and excitation wavelength at 400 nm.

3.2.4. Reversed-Phase HPLC for PA-Sialosugar Chains (16)

1. Prepare Cosmosil 5C18-P as described in 3.2.3., step 1.
2. Inject a sample, and carry out the gradient as shown in Fig. 8 using as solvent A: 0.1*M* AcOH and solvent B: 0.1*M* AcOH-0.5% butanol at a flow rate of 1.5 mL/min and the column temperature of 25°C.
3. Detect using the same wavelengths as in 3.2.3., step 3.

Separation of PA-sialylbiantennary sugar chains is described in ref. *18.* This HPLC is suited for purification of PA-sialosugar chains for better separation from PA-asialosugar chains, though the conditions cited in Section 3.2.3. can also be used.

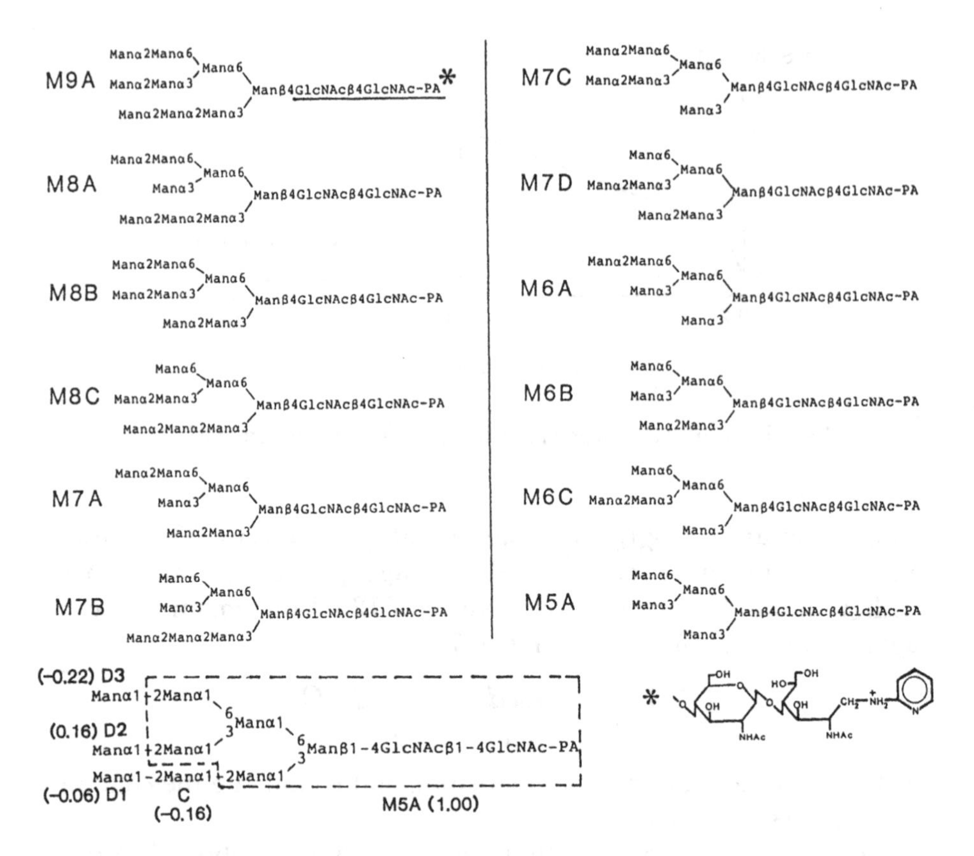

Fig. 6. Abbreviations and structures of 12 PA-oligomannose-type sugar chains. Coding of sugar residues *(17)*. Figures in the parentheses are examples of partial elution times relative to M5A obtained in Fig. 9. M5A is taken as a base PA-sugar chain (E_0 = 1.00).

4. Notes

1. For tagging sugar chains obtained from glycoconjugates, sugar chains are first liberated by hydrazinolysis-*N*-acetylation, glycopeptidases, or endoglycoceramidases. The reaction mixture can be directly pyridylaminated without purification of sugar chains. In this case, however, the amount of the glycoconjugate must be <1.5 mg if the scale of pyridylamination is done as cited.
2. An apparatus for this purpose, Palstation, is made by Takara Schuzo Co., Ltd., Kyoto.
3. Additivity rule for reversed-phase HPLC *(12,19)*: A sugar residue in a PA-sugar chain has an intrinsic contribution to the (relative) elution

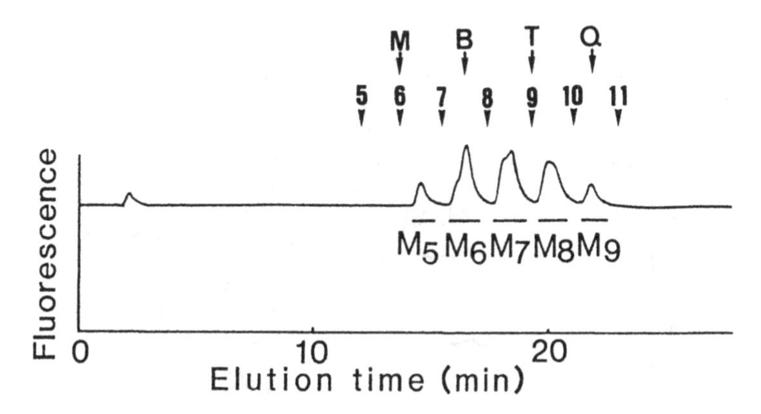

Fig. 7. Size-fractionation of an equimolar mixture of 12 PA-sugar chains shown in Fig. 6. Arrowheads 5–11 indicate the elution positions of PA-isomaltopentaose to PA-isomaltoundecaose, respectively. An arrow M indicates the elution position of PA-monoantennary sugar chain; B, PA-biantennary sugar chain; T, PA-triantennary sugar chain; Q, PA-tetraantennary sugar chain.

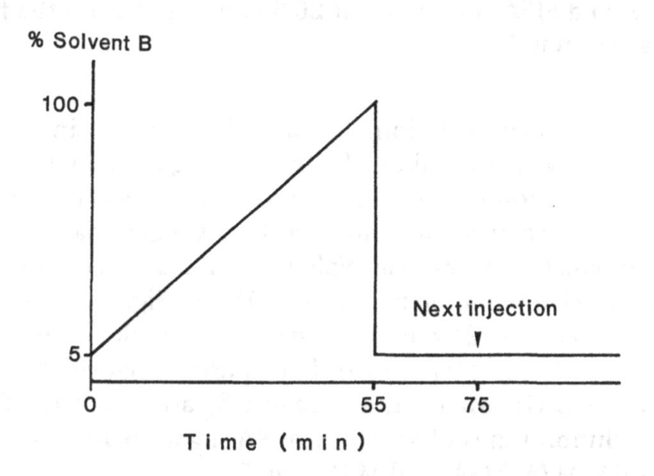

Fig. 8. An elution procedure for reversed-phase HPLC.

time of the PA-sugar chain. The contribution is referred to as "partial (relative) elution time, Ei" *(19)*. To compensate the deviation owing to column aging, buffer solutions, and so on, relative elution times are used (M5A is taken as unity in this chapter). The additivity rule is expressed as below:

$$E = E_0 + \sum_{i=1}^{n} Ei \qquad (1)$$

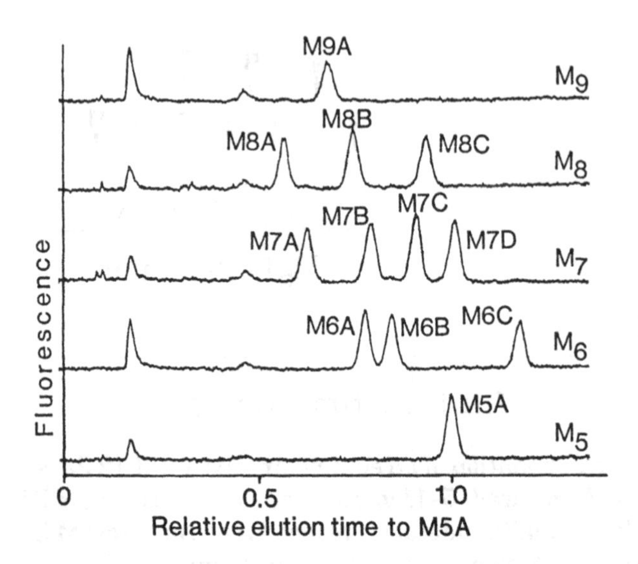

Fig. 9. Reversed-phase HPLC of PA-oligomannose-type sugar chains (two-dimensional HPLC mapping). The gradient procedure used was from reference *12*, and M5A was eluted at 20.6 min. M_5 –M_9 are the fractions collected as shown in Fig. 7.

where E is the relative elution time of the PA-sugar chain in question, E_0 is the relative elution time of a base PA-sugar chain, Ei is the partial relative elution time of sugar residue *i*, and *n* is the number of sugar residues required to construct the PA-sugar chain in question from the base PA-sugar chain. Values <1 or 2 are recommended.

Partial relative elution time of Man D3 (*see* Fig. 6), for example, is calculated to be –0.22 by subtracting the relative elution time of M5A (1.00) from that of M6A (0.78). Four partial relative elution times (Man C, Man D1, Man D2, and Man D3) are calculated from the relative elution times of at least five PA-sugar chains, for example, M5A, M6A, M7A, M8A, and M9A (Fig. 6).

Relative elution times of other seven PA-sugar chains are calculated using the four partial relative elution times, even if PA-sugar chains are not available. A relative elution time of M6B, for example, is calculated to be 0.84 by adding the relative elution time of M5A (1.00) and partial relative elution time of Man C (–0.16). The same rule is applicable for the sugar residue in PA-derivatives of oligomannose type, *N*-acetyllactosamine-type, xylomannose-type *(13)*, and sialylsugar chains *(18)*. The error in the elution times calculated was less than a few percent among the 70 PA-sugar chains tested. Partial relative elution times are dependent on the column, the column temperature, the method of gradient elution, and the HPLC apparatus used, so these conditions must be kept constant.

4. Two-dimensional HPLC mapping is useful for detection of sugar chains of tissues *(13)* analyzed in the way shown in Fig. 9. The additivity rule indicates, on the two-dimensional HPLC map shown in Fig. 9, that the vectors caused by addition of the same sugar residue to PA-sugar chains are parallel (e.g., M5A to M6A, M6B to M7A, and so on, by addition of Man D3).

References

1. Hase, S., Ikenaka, T., and Matsushima, Y. (1978) Structure analyses of oligosaccharides by tagging of the reducing end sugars with a fluorescent compound. *Biochem. Biophys. Res. Commun.* **85,** 257–263.
2. Yosizawa, Z., Sato, T., and Schmid, K. (1966) Hydrazinolysis of α_1-acid glycoprotein. *Biochim. Biophys. Acta* **121,** 417–420.
3. Takahashi, N. (1977) Demonstration of a new amidase acting on glycopeptides. *Biochem. Biophys. Res. Commun.* **76,** 1194–1201.
4. Plummer, T. H., Jr., Elder, J. H., Alexander, S., Phelan, A. W., and Tarentino, A. L. (1984) Demonstration of peptide: *N*-Glycosidase F activity in Endo-β-*N*-acetylglucosaminidase F preparations. *J. Biol. Chem.* **259,** 10,700–10,704.
5. Ito, M. and Yamagata, T. (1986) A novel glycosphingolipid-degrading enzyme cleaves of the linkage between the oligosaccharide and ceramide of neutral and acidic glycosphingolipids. *J. Biol. Chem.* **261,** 14,278–14,282.
6. Li, S. C., DeGasperi, R., Muldrey, J. E., and Li, Y. T. (1986) A unique glycosphingolipid-splitting enzyme (ceramide-glycanase from leech) cleaves the linkage between the oligosaccharide and the ceramide. *Biochem. Biophys. Res. Commun.* **141,** 346–352.
7. Natsuka, S., Hase, S., and Ikenaka, T. (1987) Fluorescence method for the structural analysis of oligomannose-type sugar chains by partial acetolysis. *Anal. Biochem.* **167,** 154–159.
8. Hase, S., Kikuchi, N., Ikenaka, T., and Inoue, K. (1985) Structures of sugar chains of the third component of human complement. *J. Biochem.* **98,** 863–874.
9. Suzuki, J., Kondo, A., Kato. I., Hase, S., and Ikenaka, T. (1991) Analysis by high-performance anion-exchange chromatography of component sugars as their fluorescent pyridylamino derivatives. *Agric. Biol. Chem.* **55,** 283–284.
10. Hase, S., Koyama, S., Daiyasu, H., Takemoto, H., Hara, S., Kobayashi, Y., Kyogoku, Y., and Ikenaka, T. (1986) Structure of a sugar chain of a protease inhibitor isolated from barbados pride seeds. *J. Biochem.* **100,** 1–10.
11. Koyama, S., Daiyasu, H., Hase, S., Kobayashi, Y., Kyogoku, Y., and Ikenaka, T. (1986) ^{1}H-NMR analysis of the sugar structures of glycoproteins as their pyridylamino derivatives. *FEBS Lett.* **209,** 265–268.

12. Hase, S., Natsuka, S., Oku, H., and Ikenaka, T. (1987) Identification method for twelve oligomannose-type sugar chains thought to be processing intermediates of glycoproteins. *Anal. Biochem.* **167,** 321–326.
13. Hase, S., Ikenaka, K., Mikoshiba, K., and Ikenaka, T. (1988) Analysis of tissue glycoprotein sugar chains by two-dimensional high-performance liquid chromatographic mapping. *J. Chromatogr. (Biomedical Applications)* **434,** 51–60.
14. Mellis, S. J. and Baenziger, J. U. (1983) Size fractionation of anionic oligosaccharides and glycopeptides by high-performance liquid chromatography. *Anal. Biochem.* **134,** 442–449.
15. Kuraya, N. and Hase, S. (1992) Release of *O*-linked sugar chains from glycoproteins with anhydrous hydrazine and pyridlamination of the sugar chains with improved reaction conditions. *J. Biochem.* **112,** in press.
16. Yamamoto, S., Hase, S., Fukuda, S., Sano, O., and Ikenaka, T. (1989) Structures of the sugar chains of interferon-γ produced by human myelomonocyte cell line HBL-38. *J. Biochem.* **105,** 547–555.
17. Vliegenthart, J. F. G., Dorland, L., and van Halbeek, H. (1983) High-resolution, ^{1}H-nuclear magnetic resonance spectroscopy as a tool in the structural analysis of carbohydrates related to glycoproteins. *Adv. Carbohydr. Chem. Biochem.* **41,** 209–374.
18. Yamamoto, S., Hase, S., Yamauchi, H., Tanimoto, T., and Ikenaka, T. (1989) Studies on the sugar chains of interferon-γ from human peripheral-blood lymphocytes. *J. Biochem.* **105,** 1034–1039.
19. Hase, S. and Ikenaka, T. (1990) Estimation of elution times on reverse-phase HPLC of pyridylamino derivatives of sugar chains from glycoproteins. *Anal. Biochem.* 184, 135–138.

CHAPTER 7

Analysis of Carbohydrates in Glycoproteins by High-Performance Liquid Chromatography and High-Performance Capillary Electrophoresis

Kazuaki Kakehi and Susumu Honda

1. Introduction

1.1. Analysis of Oligosaccharide Chains in Glycoproteins

Analysis of the diverse species of carbohydrate chains released from glycoproteins inevitably requires high-resolution methods. In addition, because of the relative nonvolatility even after derivatization, high-performance liquid chromatography (HPLC), rather than gas chromatography, becomes the technique of choice. Various separation modes are employed for HPLC, such as normal phase partition (on amine-bonded silica) and ligand exchange (on sulfonated styrene-divinylbenzene copolymers). Reversed-phase partition (RP-HPLC) as reductively pyridyl aminated derivatives (*see* Chapter 6, this vol.) or condensation products with 1-phenyl-3-methyl-5-pyrazolone *(1)* is also useful.

Another convenient procedure has been developed for direct analysis of carbohydrate chains by anion-exchange mode of HPLC (AE-HPLC) with pulsed amperometric detection. We present here some applications of HPLC based on this mode to a few *N*- and *O*-glycosidically bound carbohydrates, using thyroglobulin and bovine submaxillary mucin as glycoprotein models *(2)*.

From: *Methods in Molecular Biology, Vol. 14: Glycoprotein Analysis in Biomedicine*
Edited by: E. F. Hounsell

High-performance capillary electrophoresis (HPCE) is a merging method for separation and commercial automated apparatus have just become available. This chapter also describes a series of HPCE procedures for analysis of oligosaccharide derivatives using ovalbumin as a model glycoprotein.

1.2. Analysis of Constituent Monosaccharides

Analysis of constituent monosaccharides in glycoproteins offers fundamental information on the type of carbohydrate chain and the number of chains per molecule. Prior to analysis of constituent monosaccharides, they must be released by acid hydrolysis or enzyme digestion. Excellent reviews have been recently published by two groups *(3,4)*. Procedures for acid hydrolysis to release neutral monosaccharides and amino sugars described in this chapter are the standard procedures employed in our laboratory, and are similar to those described by Hardy *(3)*.

Neutral monosaccharides, including galactose, mannose, and fucose, which are ubiquitous in glycoproteins, can be quantitatively released with 2*M* trifluoroacetic acid in nitrogen atmosphere at 100°C. On the other hand, amino sugars should be hydrolyzed in 4*M* hydrochloric acid in nitrogen atmosphere. The *N*-acetyl group is incidentally removed under these conditions; hence, re-*N*-acetylation is necessary prior to analysis. This can be easily accomplished by adding acetic anhydride in an aqueous sodium bicarbonate solution. Hydrolysis in 4*M* hydrochloric acid causes serious degradation of neutral monosaccharides. For these reasons, a sample has to be hydrolyzed under two different conditions mentioned earlier, one for neutral sugars and the other for amino sugars. Sialic acids must be hydrolyzed under much milder conditions using dilute acetic or sulfuric acid at a lower temperature, since they are labile to acid. Determination of sialic acid content is usually performed separately, for instance, by colorimetry *(5)*; direct HPLC analysis of neuraminic acids by using malononitrile as a postcolumn labeling reagent is an alternative method *(6)*.

Various methods for analysis of released neutral monosaccharides and amino sugars have been reported by many groups. We describe herein an example based on RP-HPLC with UV detection after precolumn labeling with 1-phenyl-3-methyl-5-pyrazolone (PMP, ref. *1)* developed in our laboratory.

2. Materials

2.1. Analysis of Oligosaccharide Chains in Glycoproteins

2.1.1. AE-HPLC

1. *N*-glycosidically bound oligosaccharides (*see* Note 1):
   a. Oligosaccharide mixture from porcine thyroglobulin (5 mg): prepared from porcine thyroids according to the method of Ui and Tarutani *(7)*.
   b. Sodium hydroxide (analytical grade) (*see* Note 2).
   c. Sodium acetate.
   d. Water: double distilled and filtered through a membrane filter (0.2 µm).
   e. The HPLC apparatus with gradient elution: set up from a Dionex pump capable of programmable gradient elution, a Rheodyne 7125 sample injector equipped with a Tefzole seal with a 20-µL loop, a Dionex HPIC-AS-6 analytical column (4 mm id, 25 cm) with an HPIC-AS-6 guard column (4 mm id, 5 cm), and a Dionex triple-pulsed amperometric detector equipped with a gold electrode (PAD II).
2. *O*-gylcosidically bound oligosaccharides:
   a. Bovine submaxillary mucin (1 mg): obtained from fresh bovine submaxillary glands by the method of Tettamanti and Pigman *(8)*.
   b. Sodium hydroxide (0.05*M*) containing sodium borohyride to a concentration of 1*M*.
   c. Acetic acid.
   d. A column of Amberlite CG-120 (H^+ form, 10 mL).
   e. The HPLC system with isocratic elution: built up from a Hitachi 655 dual-plunger pump. Other equipment is the same as that used in HPLC with gradient elution described earlier.

2.1.2. Analysis by HPCE as Reductively Pyridylaminated Derivatives

1. Direct electrophoresis (CZE-HPCE) in an acidic carrier.
   a. The HPCE system: a Bio-Rad HPE 100 apparatus.
   b. Carrier: 100 m*M* phosphate buffer (pH 2.5) (*see* Note 3).
   c. A capillary tube: a cassette-mounting polyacrylamide-coated tube (25 µm id, 20 cm) designed for this apparatus; available from Bio-Rad (Hercules, CA).
2. Zone electrophoresis as borate complexes (ZEBC-HPCE).
   a. The HPCE system *(9)*: a handmade apparatus constructed from a high-voltage power supply (Matsusada Precision Devices, Model

HEL-30-13), a pair of PTFE electrode vessels (1-mL vol), a Hitachi 650-10 LC fluoromonitor, and an SIC Chromatocorder 12 (*see* Note 4).

b. A fused silica capillary tube (Scientific Glass Engineering, Melbourne, Australia; 50 µm id, 95 cm) (*see* Note 5).
c. Borate buffer as carrier: prepared by adding sodium hydroxide pellets to a solution (200 m*M*) of boric acid, pH being adjusted to 10.5.

2.2. Analysis of Constituent Monosaccharides

2.2.1. Hydrolysis

1. Trifluoroaacetic acid.
2. Hydrochloric acid.
3. Amberlite CG-120 (H^+ form, 3 mL).
4. Acetic anhydride.
5. A centrifugal concentrator.
6. A block heater (70°C, 100°C).
7. A polypropylene tube with a screw cap (1.5 mL).
8. A glass tube (7 mm id, 12 cm).

2.2.2. Precolumn Conversion to PMP Derivatives

1. 1-Phenyl-3-methyl-5-pyrazolone: available from Kishida (Doshomachi, Osaka). The same reagent is also available from Aldrich (Milwaukee, WI) in the name of 3-methyl-1-phenyl-5-pyrazolone. (Note 6).
2. An aqueous sodium hydroxide solution: prepared by diluting standardized 1.0*N* sodium hydroxide to 0.3*M* concentration.
3. Diluted hydrochloric acid: prepared by dilution of standardized 1.0*N* hydrochloric acid to 0.3*M* concentration.
4. Chloroform.
5 A centrifugal concentrator.
6. A polypropylene tube with a screw cap (1.5 mL).

2.2.3. Analysis of PMP Derivatives by RP-HPLC

1. The HPLC apparatus: set up from a Hitachi 655A-12 dual-plunger pump, a Rheodyne 7125 injector with a 20-µL loop, a Shimadzu SPD-6A-UV detector, and a Hitachi D-2000 data processor. Volume of detector cell, 8 µL; wavelength for detection, 245 nm; column (4.6 mm id, 25 cm), Capcell Pak C-18 (Shiseido, Ginza, Chuo-ku, Tokyo); eluent, phosphate buffer (70 m*M*, pH 6.8) containing acetonitrile to a concentration of 18%; flow rate, 1.0 mL/min.
2. Acetonitrile (HPLC grade).

3. Phosphate buffer (70 m*M*, pH 6.8): Prepare the buffer by dissolution of KH_2PO_4 (18.15 g) and Na_2HPO_4. $12H_2O$ (47.75 g) in 4.0 L of double-distilled water followed by filtration through a membrane filter (0.2 μm).

3. Methods

3.1. Analysis of Oligosaccharide Chains in Glycoproteins

3.1.1. AE-HPLC

1. *N*-glycosidically bound oligosaccharides:
   a. Dissolve a sample prepared by hydrazinolysis, re-*N*-acetylation, and borohydride reduction (*see* Chapter 5) of thryoglobulin (5 mg) in distilled water (1 mL).
   b. Inject an aliquot (20 μL) to the HPLC column.
   c. Elute the column with eluent prepared by continuously adding 0.1*M* sodium hydroxide containing sodium acetate (0.5*M*) into a solution of 0.1*M* sodium hydroxide in linear gradient mode for 200 min at a flow rate of 1.0 mL/min. Set the applied potentials and pulse durations of the detector as follows: E_1 = +0.045 V (0.6 s), E_2 = +0.60 V (0.12 s), E_3 = –0.80 V (0.42 s). The elution profile of oligosaccharide mixtures from porcine thyroglobulin is shown in Fig. 1. Each peak is assigned to the structure given in Scheme 1 by comparing the elution volume to that of the authentic specimen. It is indicated that high mannose-type oligosaccharides are eluted with low alkali concentrations. Increase of the mannose residue results in more retardation. Sialooligosaccharides have longer retention times, and the retardation becomes larger as the number of the sialic acid residue increased (*see* Note 7).
2. *O*-glycosidically bound oligosaccharides:
   a. Release the *O*-glycosidically linked oligogaccharides by incubation of bovine submaxillary mucin (1 mg) in 0.05*M* sodium hydroxide (1 mL) containing sodium borohydride to a concentration of 1*M*, for 24 h at 45°C (*see* Note 8).
   b. Add acetic acid into the mixture carefully to decompose excess borohydride.
   c. Pass the reaction mixture through a column of Amberlite CG-120 (H^+ form, 10 mL), and wash the column with water (50 mL). Evaporate the combined eluate and washing fluids. Dissolve the residue in a small volume of methanol, and evaporate the solution. Repeat the procedure several times to remove boric acid as the volatile borate ester.

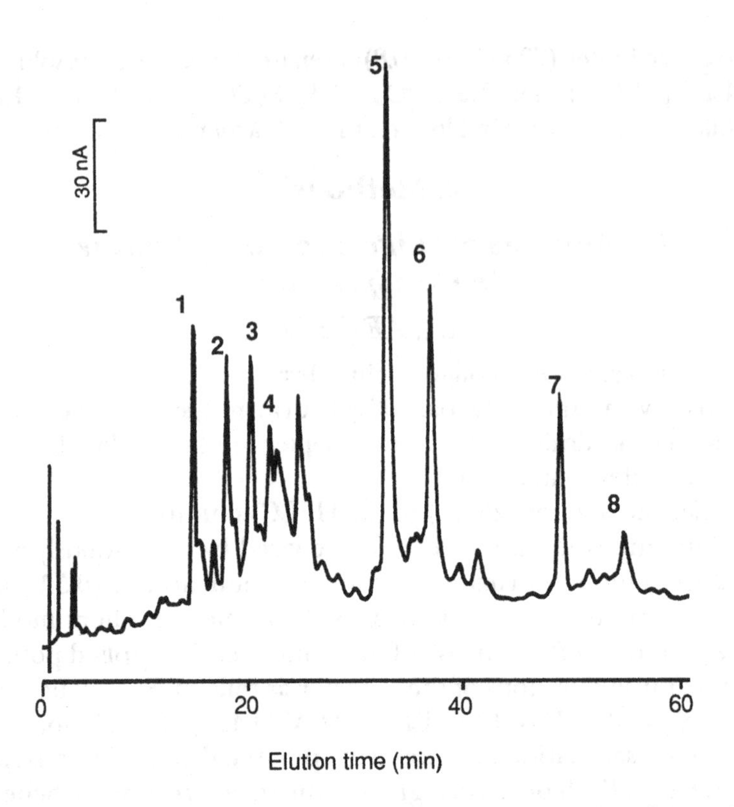

Fig. 1. Analysis of borohydride-reduced oligosaccharides obtained from porcine thyroglobulin. Column, Dionex HPIC As-6 (4 mm id, 25 cm); flow rate, 1.0 mL/min; detection, pulse amperometry on a gold electrode (E_1 = + 0.045 V, 0.6 s; E_2 = + 0.60 V, 0.12 s; E_3 = - 0.80 V, 0.42 s); elution, linear gradient (0.1*M* sodium hydroxide–0.1*M* sodium hydroxide containing sodium acetate, 0.5*M*, in 200 min); sample scale, 50 µg as glycoprotein. Peak numbers are corresponding to the compound numbers in Scheme 1. (Reproduced from ref. 2 with permission.)

d. Dissolve the final residue in water (1 mL), and inject an aliquot (20 µL) into the HPLC column.
e. Elute the column with 0.3*M* sodium hydroxide using a Hitachi 655 dual-plunger pump for isocratic elution under similar conditions to those described in analysis of *N*-glycosidically bound oligosaccharides (*see* Note 9).

Figure 2 shows the profile of isocratic elution of the mixture of *O*-glycosidically bound oligosaccharides obtained from bovine submaxillary mucin. The peaks are assigned to the structures in

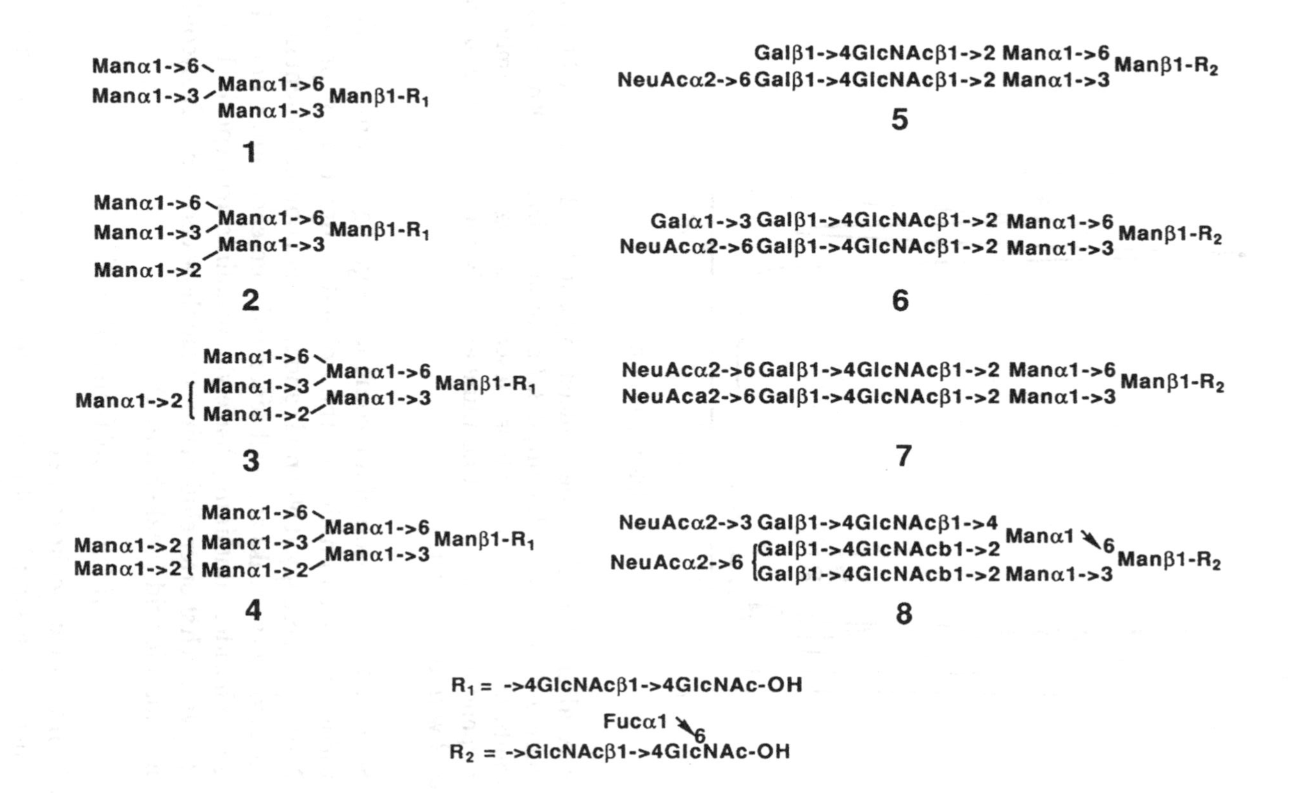

Scheme 1. Borohydride-reduced oligosaccharides from porcine thryoglobulin.

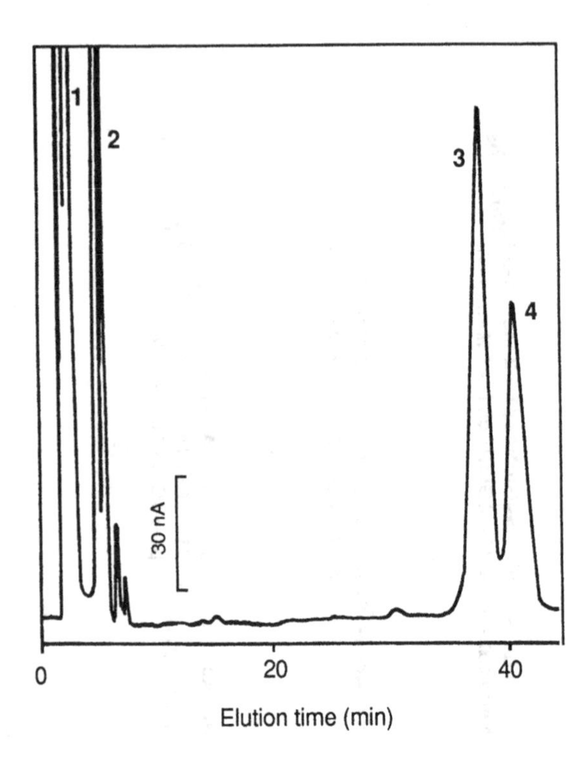

Fig. 2. Analysis of borohydride-reduced oligosaccharides obtained from bovine submaxillary mucin. Eluent, 0.3*M* sodium hydroxide. Other analytical conditions are the same as those described in Fig. 1. Peak numbers are corresponding to the compound numbers in Scheme 2. (Reproduced from ref. *2* with permission.)

Scheme 2 by isolation of the compounds giving individual peaks, followed by examination of the isolated compounds by proton NMR spectroscopy at 500 mHz and fast atom bombardment mass spectrometry. Peaks are well separated fom each other, and oligosaccharides containing *N*-acetylneuraminic acid (peaks 1 and 2) give much shorter elution times than those containing *N*-glycolylneuraminic acid (peaks 3 and 4).

3.1.2. HPCE (see *Note 10)*

1. CZE-HPCE in an acidic carrier:
   a. Introduce a sample solution (10 μL) to the inlet cavity with a microsyringe (*see* Note 11).
   b. Transfer pyridylaminated derivaties of carbohydrates (*see* Chapter 5) in the cavity to the capillary tube by the electromigration method

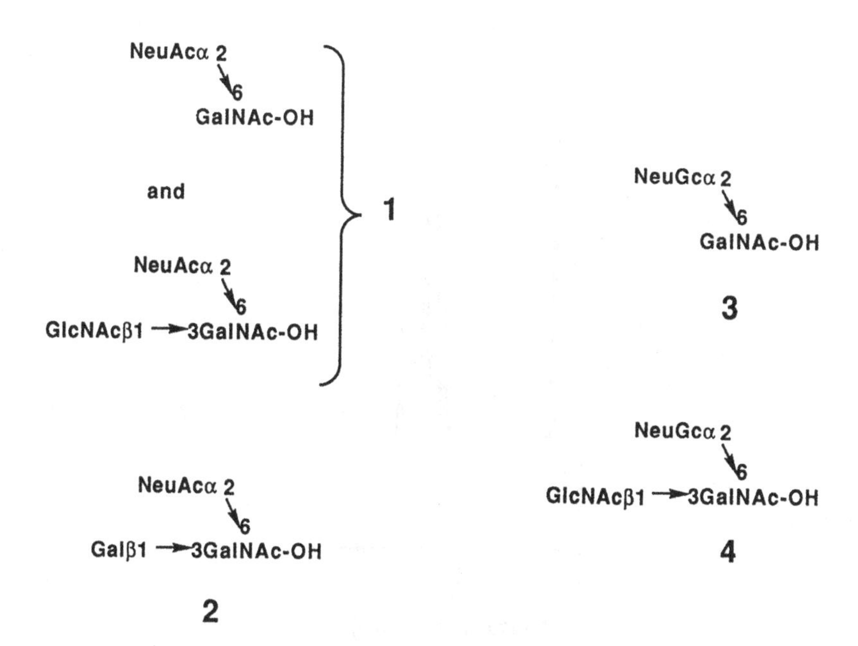

Scheme 2. Borohydride-reduced oligosaccharides from bovine submaxillary mucin.

with application of a potential of 8 kV for 30 s.

c. Perform HPCE at the potential of 15 kV (*see* Note 12).

Figure 3 shows separation of *N*-glycosidically bound oligosaccharides based on their molecular sizes. Five major peaks are assignable to hepta-, octa-, nona-, deca-, and undecasaccharide derivatives, respectively, by comparison with isomaltooligosaccharide derivatives. Oligosaccharides having the identical degree of polymerizations (d.p.s) cannot be separated from each other by this mode.

2. ZEBC-HPCE:
   a. Introduce a sample solution to a capillary tube by the hydrodynamic method under conditions of a 10-cm rise for 10 s.
   b. The electrophoretogram is recorded at 395 nm (emission) with irradiation at 316 nm (excitation). Figure 4 shows an electrophoretogram of oligosaccharide derivatives. The first (11.7 min) and the last (24.3 min) peaks are owing to the remaining reagent (AP) and reductively pyridylaminated glucose (G-AP), respectively, added as an internal reference. The nine peaks 1–9 in the range of 14–17 min are assigned to the oligosaccharide deriva-

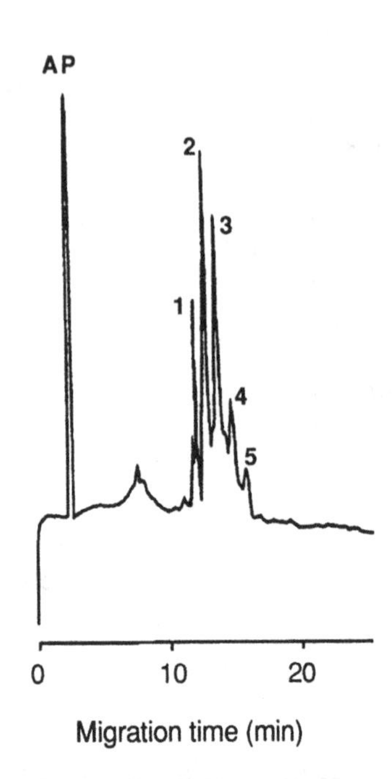

Fig. 3. Analysis of reductively pyridylaminated oligosaccharides derived from ovalbumin by ZE-HPCE. Capillary, fused silica coated with polyacrylamide (Bio-Rad, 25 μm id, 20 cm); carrier, 100 m*M* phosphate buffer (pH 2.5); applied voltage, 8 kV; detection, UV absorption at 240 nm. AP, 2-aminopyridine (excess reagent). Peaks 1, 2, 3, 4, and 5 are assignable to the derivatives of hepta-, octa-, nona-, deca-, and undecasaccharides, respectively, in Scheme 3. (Reproduced from ref. *9* with permission.)

tives. Scheme 3 gives proposed assignment based on comparison of their relative mobilities to the G-AP with those of authentic specimens.

3.2. Analysis of Constituent Monosaccharides

3.2.1. Hydrolysis of Glycoprotein Samples

1. Neutral monosaccharides:
   a. Add 2*M* trifluoroacetic acid (200 μL) to a glycoprotein sample (100–500 μg) in a glass tube (7 mm id, 120 mm).
   b. Flush the solution with nitrogen, and seal the tube.
   c. Keep the tube in the block heater at 100°C for 4 h.
   d. Cool the tube, and open it.

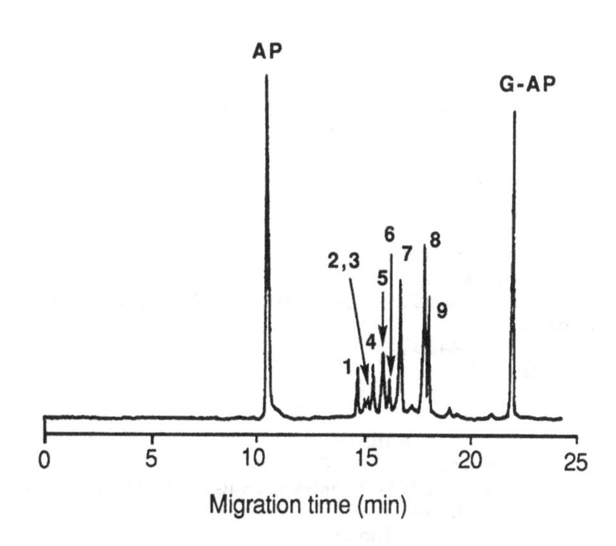

Fig. 4. Analysis of reductively pyridlyaminated oligosaccharides derived from ovalbumin by ZEBC-HPCE. Capillary, fused silica (Scientific Glass Engineering, 50 µm id, 95 cm); carrier, 200 m*M* borate buffer (pH 10.5); applied voltage, 20 kV; detection, fluorescence at 395 nm (irradiated al 316 nm). AP, 2-aminopyridine; G-AP, reductively pyridylaminated glucose (internal standard). Peak numbers are corresponding to the compound numbers in Scheme 3, though the assignment is tentative. Peaks 2 and 3 could not be assigned. (Reproduced from ref. *9* with permission.)

   e. Transfer the solution and the washing fluid (300 µL) to a polypropylene tube.
   f. Evaporate the solution to dryness (*see* Note 13).
2. Amino sugars:
   a. Add 4*M* hydrochloric acid (200 µL) to a glycoprotein sample (100–500 µg) in a glass tube (7 mm id, 120 cm).
   b. Flush the solution with nitrogen, and seal the tube.
   c. Keep the tube in the block heater at 100°C for 4 h.
   d. Cool the tube, and open it.
   e. Transfer the solution and the washing fluid (300 µL) to a polypropylene tube.
   f. Evaporate the solution to dryness.
   g. Add an aqueous saturated solution of sodium bicarbonate (500 µL) and acetic anhydride (20 µL).
   h. Keep the solution overnight in the refrigerator.

```
                 Manα1↘
          Manα1→3↗       6
                 GlcNAcß1→4Manß1→4GlcNAcß1→4GlcNAc-R
1                        3
                 Manα1↗
        GlcNAcß1→2


                 Manα1↘
          Manα1→3        6
                 GlcNAcß1→4Manß1→4GlcNAcß1→4GlcNAc-R
4       GlcNAcß1→4       3
                 Manα1↗
        GlcNAcß1→2


                      Manα1↘
                Manα1→3        6
5                     GlcNAcß1→4Manß1→4GlcNAcß1→4GlcNAc-R
     Galß1→4GlcNAcß1→4         3
                      Manα1↗
             GlcNAcß1→2

           Manα1→6
                 Manα1↘
           Manα1→3       6
6                GlcNAcß1→4Manß1→4GlcNAcß1→4GlcNAc-R
        GlcNAcß1→4       3
                 Manα1↗
        GlcNAcß1→2
```

R = —NH (2-pyridyl, N)

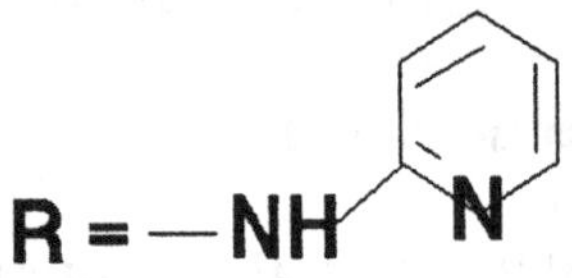

Scheme 3. Oligosaccharide derivatives from ovalbumin.

i. Pass the solution through a column of Amberlite CG-120 (H^+ form, 3 mL), and wash the column with water (20 mL).
j. Evaporate the combined eluate and the washing fluid to dryness.
k. Add methanol (2 mL) to the residue, and evaporate the mixture to dryness. Repeat the procedure several times.
l. Transfer the residue to a polypropylene tube with a small vol of water.
m. Evaporate the solution to dryness.

```
Manα1→6
        Manα1↘
Manα1→3       6
     GlcNAcß1→4Manß1→4GlcNAcß1→4GlcNAc-R          ┐
              3                                    │
        Manα1↗                                     │
GlcNAcß1→2                                         ├        7
                                                   │
        Manα1→6                                    │
                Manα1↘                             ┘
        Manα1→3       6
             GlcNAcß1→4Manß1→4GlcNAcß1→4GlcNAc-R
Galß1→4GlcNAcß1→4    3
                Manα1↗
        GlcNAcß1→2

        Manα1→6
                Manα1↘
        Manα1→3       6
                      Manß1→4GlcNAcß1→4GlcNAc-R       ┐
                      3                                │
        Manα1→2Manα1↗                                  │
                                                       ├    8
        Manα1→2Manα1→6                                 │
                      Manα1↘                           │
                Manα1→3       6                        ┘
                              Manß1→4GlcNAcß1→4GlcNAc-R
                              3
                Manα1→2Manα1↗


                Manα1→6
                        Manα1↘
                Manα1→3       6
                              Manß1→4GlcNAcß1→4GlcNAc-R     9
                              3
                        Manα1↗
```

Scheme 3: *continued.*

3.2.2. Precolumn Conversion to PMP Derivatives and Analysis of the Derivatives

1. Into a hydrolyzate obtained by one of the procedures described in Section 3.2.1. or a mixture of standard aldoses (*see* Note 14), add a 0.3*M* aqueous solution of sodium hydroxide (50 µL) and a 0.5*M* methanolic solution (50 µL) of PMP.
2. Stand the mixture for 30 min at 70°C (*see* Note 15).
3. Cool the mixture to room temperature, and add an equivalent vol of 0.3*M* hydrochloric acid (50 µL) for neutralization.
4. Evaporate the solution to dryness. Add water (200 µL) and chloroform (200 µL) to the residue, and shake the mixture vigorously.
5. Discard the chloroform layer, and repeat extraction with chloroform to remove the excess reagent from the aqueous layer (*see* Note 16).

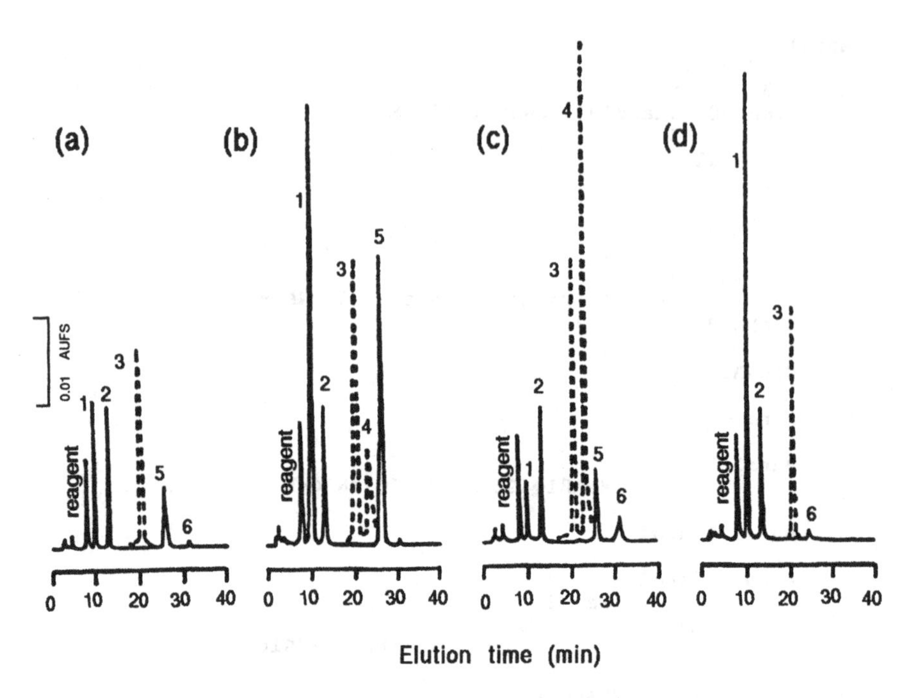

Fig. 5. Analysis of constituent monosaccharides of human serum transferrin (a), calf serum fetuin (b), bovine submaxillary mucin (c), and hen egg ovalbumin (d). Sample amount injected, 20 µg each as protein. Solid and dotted lines represent the results obtained with the trifluoroacetic acid and hydrochloric acid hydrolysates, respectively. Column, Capcell Pak C-18 (4.6 mm, id, 25 cm); eluent, 0.1*M* phosphate buffer (pH 7.0) containing acetonitrile (18%); flow rate, 1.0 mL/min; sample amount, 20 µg each as protein. Peak assignment: 1, mannose; 2, rhamnose (internal standard); 3, *N*-acetylglucosamine; 4, *N*-acetylgalactosamine; 5, galactose; 6, fucose. (Reproduced from ref. *1* with permission.)

6. Evaporate the aqueous layer to dryness, dissolve the residue in a small vol (200 µL) of eluent for HPLC, and inject an aliquot (20 µL) onto the HPLC column.

   Elution profiles obtained from the hydrolysates of some glycoprotein samples are shown in Fig. 5, and the monosaccharide contents are summarized in Table 1. The accuracy and precision of this method are sufficiently high when rhamnose is used as the internal standard. For example, the calibration curve of glucose shows excellent linearity in the range of 5–1000 pmol. The lower limit of detection at the signal-to-noise ratio of 5 was ca. 1 pmol. Repeated determination ($n = 7$) of glucose-PMP gave SD of 1.9, 1.1, and 2.3% at the 10-, 200- and 1000-pmol levels, respectively.

Table 1
Determination of Component Monosaccharides in Glycoproteins[a]
(Reproduced from ref. *1* with permission.)

Glycoprotein	Content, w/w%				
	Fucose	Mannose	Galactose	Glucosamine	Galactosamine
Transferrin[b]	0.040	1.08	1.00	2.52	0.053
(human serum)	(0.031)	(1.11)	(0.86)	(1.79)	(–)
Fetuin[b]	0.033	2.45	3.49	2.62	-
(calf)	(0.027)	(2.73)	(4.59)	(5.30)	(0.674)
Thyroglobulin[c]	0.38	2.26	1.49	5.55	1.38
(bovine thyroid)	(0.37)	(2.09)	(1.20)	(2.95)	(1.38)
Albumin[b]	-	2.80	0.12	2.73	-
(hen egg)	(–)	(2.40)	(0.15)	(1.28)	(–)
Mucin[b]	0.53	0.16	1.24	6.22	14.68
(bovine - submaxillary)	(0.95)	(0.21)	(1.52)	(6.92)	(16.80)

[a]The numbers in parentheses are reported values.
[b]Ref. *10*.
[c]Ref. *11*.

4. Notes

1. Readers should refer to Chapter 5 for release of *N*-glycosidically bound oligosaccharides by a combination of hydrazinolysis, re-*N*-acetylation, and borohydride reduction.
2. Make aqueous 50% sodium hydroxide as the stock solution. The eluent for AE-HPLC is prepared by appropriate dilution of the stock solution. Surface and bottom layers of the stock solution should not be used to avoid contamination with sodium carbonate.
3. An aqueous solution of sodium hydroxide (100 m*M*) was added to 100 m*M* phosphoric acid, and pH was adjusted to 2.5.
4. The fluoromonitor is slightly modified. A quartz convex lens is placed between the light source and the cell holder, so as to focus the irradiation light on a capillary tube.
5. A 5-mm portion of the polyimide coating is removed by burining at a distance of 30 cm from the outlet of the tube, where the excitation light irradiates.
6. The reagent is crystallized from hot methanol before use. A 0.5 *M* solution is prepared by dissolution of the reagent (79 mg) in methanol (1.0 mL). The solution stored in a refrigerator for several months can be used.
7. This system employing a pulsed amperometric detector on a gold electrode is sensitive to not only carbohydrates, but also amino acids.

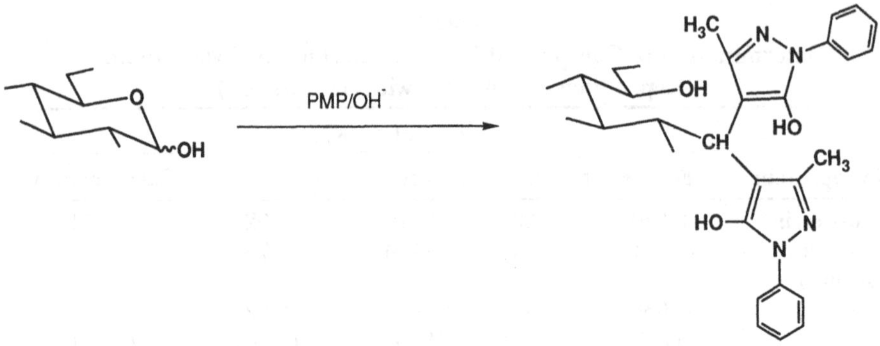

Scheme 4. Conversion of monosaccharides to PMP derivatives.

It is weakly positive to proteins, such as albumin. The problem of interference by these substances can be simply solved by clean up of the sample solutions on a small column of Sephadex G-25 column (1.0 cm id, 30 cm) with water as eluent.

8. Under these conditions, removal of the *N*-acyl groups on the sialic acid and hexosamine residues is negligible, whereas *O*-acyl groups are completely removed.
9. A progammable gradient pump may also be used for separation of *O*-glycosidically linked oligosaccharides.
10. Precolumn labeling of reducing oligosaccharides with 2-aminopyridine by reductive amination should be referred to Chapter 6 in this vol.
11. Oligosaccharide mixtures are prepared from ovalbumin (10 mg). Hydrazinolysis, re-*N*-acetylation, and pyridylamination (*see* Chapters 5 and 6 in this vol.) affords a pyridylaminated-oligosaccharide mixture. The final residue is dissolved in water (200 µL).
12. This mode allows separation based on molecular size, because all oligosaccharide derivatives have commonly one imino group and, accordingly, the same electric charge. The smaller species migrate faster, giving earlier peaks.
13. The dry-up procedure following hydrolysis seems to be important for reproducible determination and should be completed in a short period by using an efficient evaporator, such as a centrifugal concentrator.
14. Each monosaccharide in a range of 0.2 nmol–1 µmol can be determined.
15. Under these conditions, each monosaccharide molecule is condensed with two molecules of PMP, as indicated in Scheme 4.
16. Further repetition of extraction causes loss of 6-deoxyhexose derivatives.

References

1. Honda, S., Akao, E., Suzuki, S., Okuda, M., Kakehi, K., and Nakamura, J. (1989) High-performance liquid chromatography of reducing carbohydrates as strongly ultraviolet-absorbing and electrochemically sensitive 1-phenyl-3-methyl-5-pyrazolone derivatives. *Anal. Biochem.* **180,** 351–357.
2. Honda, S., Suzuki, S., Zaiki, S., and Kakehi, K. (1990) Analysis of *N*- and *O*-glycosidically bound sialooligosaccharides in glyoproteins by high performance liquid chromatography with pulsed amperometric detection. *J. Chromatogr.* **523,** 189–200.
3. Hardy, M. R. (1989) Monosaccharide analysis of glycoconjugates by high performance anion-exchange chromatography with pulsed amperometric detection, in *Methods in Enzymology* vol. 179 (Ginsburg, V., ed.), Academic, New York, pp. 76–82.
4. Biermann, C. J. (1988) Hydrolysis and other cleavages of glycosidic linkages in polysaccharides. *Adv. Carbohydr. Chem.* **46,** 251–271.
5. Uchida, Y., Tsukada, Y., and Sugimori, T. (1977) Distribution of neuraminidase in *Arthrobacter* and its purification by affinity chromatography. *J. Biochem.* **82,** 1425–1433.
6. Honda, S., Iwase, S., Suzuki, S., and Kakehi, K. (1987) Fluorometric determination of sialic acids in weakly alkaline media and its application to postcolumn labeling in high-performance liquid chromatography. *Anal. Biochem.* **160,** 455–461.
7. Ui, N. and Tarutani, O. (1961) Purification of hog thyroglobulin. *J. Biochem. (Tokyo)* **50,** 508–518.
8. Tettamanti, G. and Pigman, W. (1968) Purification and characterization of bovine and ovine submaxillary mucin. *Arch. Biochem. Biophys.* **124,** 41–50.
9. Honda, S., Makino, A., Suzuki, S., and Kakehi, K. (1990) Analysis of the oligosaccharides in ovalbumin by high performance capillary electrophoresis. *Anal. Biochem.* **191,** 222–234.
10. Honda, S. and Suzuki, S. (1984) Common conditions for high performance liquid chromatographic microdetermination of aldoses, hexosamines, and sialic acids in glycoproteins. *Anal. Biochem.* **142,** 167–174.
11. Spiro, R. G. and Spiro, M. J. (1965) The cabohydrate composition of the thyroglobulin from several species. *J. Biol. Chem.* **240,** 997–1001.

CHAPTER 8

The Microanalysis of Glycosyl-Phosphatidylinositol Glycans

M. Lucia Güther and Michael A. J. Ferguson

1. Introduction

The name glycosyl-phosphatidylinositol (GPI) is a trivial name for a family of structures that contains the structural motif: Manα1-4GlcNH$_2$α1-6*myo*-Inositol. This common substructure suggests that this family of molecules is biosynthetically related and differentiates it from other glycosylated phosphoinositides, such as the glycosylated phosphatidylinositols of mycobacteria and inositol phosphoceramides of yeasts and plants. The GPI family (for recent comprehensive reviews, *see 1–3)* can be conveniently divided into two groups based on structural homology and function. The first group is composed of the membrane protein anchors (Fig. 1), which are found covalently linked to the *C*-termini of a wide variety of externally disposed plasma membrane proteins throughout the eukaryotes. These GPI anchors afford a stable attachment of proteins to the membrane and can be viewed as an alternative mechanism of membrane attachment to a single-pass hydrophobic transmembrane peptide domain. The second group has been found so far only in the parasitic kinetoplastid protozoans *Leishmania major, L. donovani L. mexicana,* and the South American trypanosome *Trypanosoma cruzi.* These molecules exist as free glycophospholipids, such as the "glycosyl inositol phospholipids" (GIPLs) of the *Leishmania (4,5),* or attached to phosphorylated repeating units known as the lipophosphoglycans (LPGs) of the *Leishmania (6,7).* In this chapter, we will consider only the protein-linked GPI anchors.

From: *Methods in Molecular Biology, Vol. 14: Glycoprotein Analysis in Biomedicine*
Edited by: E. F. Hounsell

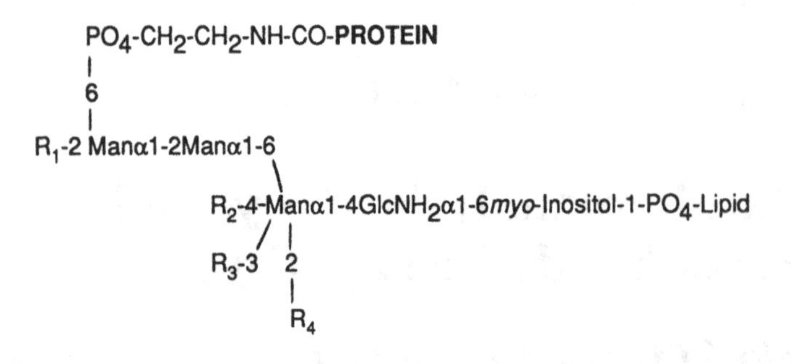

Example	R_1	R_2	R_3	R_4	Lipid
1. *T.brucei* VSG	OH	OH	α-Gal_{2-4}	OH	diacyl-glycerol
2. *L.major* PSP	OH	OH	OH	OH	alkyl/acyl-glycerol
3. Rat Thy-1	α-Man	β-GalNAc	OH	EtNP	?
4. Human AChE	OH	OH	OH	EtNP	alkyl/acyl-glycerol

Fig. 1. Consensus structure of GPI anchors. The consensus structure shown is based on refs. *8–11*. EtNP indicates ethanolamine phosphate. OH indicates no substitution.

The complete structures of several members of the GPI anchor family are now available (Fig. 1). These structures have been determined by one- and two-dimensional proton NMR, FAB-MS, GC-MS, and chemical and enzymatic modifications *(8–11)*. During the course of these studies, a strategy has been developed to isolate and radiolabel chemically GPI glycan moieties from small quantities (0.5–5 nmol) of GPI containing molecules for chromatographic and chemical characterization. It is this microsequencing strategy (Fig. 2) that is described in this chapter.

The chromatographic properties on Bio-Gel P4 gel-filtration columns and Dionex anion-exchange carbohydrate HPLC of a variety of deaminated and reduced GPI glycans are described. Each of the 16 structures has unique chromatographic properties when P4 and Dionex HPLC elution positions are collated. These tabulated values should aid the rapid identification of subnanomolar quantities of GPI glycans. In addition exoglycosidase and chemical cleavage protocols, as well as a high-sensitivity GC–MS methylation analysis protocol, are presented.

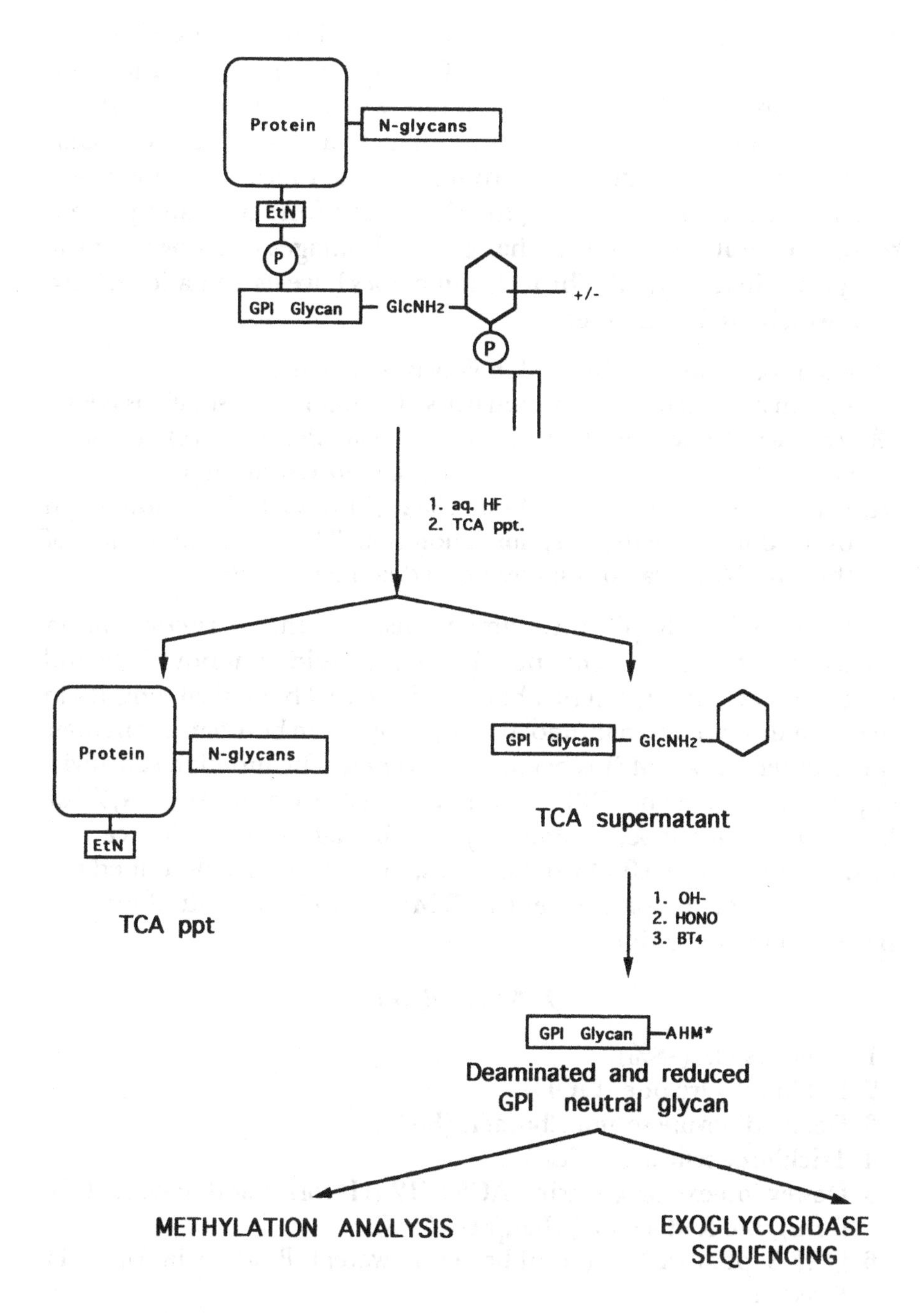

Fig. 2. Microsequencing strategy for GPI glycans. A scheme to isolate GPI glycans directly from whole glycoproteins for sequence analysis.

As a working example of the methods, the solution of a GPI glycan structure, from 100 μg (1.8 nmol) of a *Trypanosoma brucei* variant surface glycoprotein (VSG variant MIT at 1.5), is presented (*see* Note 20).

The protocols described in this chapter are designed to obtain structural details of GPI glycans from small quantities of starting material. The approach of releasing the GPI glycan from lipid and protein by aq. HF dephosphorylation has great advantages in recovering the GPI glycan in high yield. The technique does have its limitations, however, which are listed here:

1. No information on the lipid moiety is obtained.
2. Information on the position and types of phosphate substitutions is lost.
3. Very acid-labile substituents, such as galactofuranose *(6)* and sialic acid *(12)*, can be lost during the dephosphorylation step.
4. Partial de-*N*-acetylation (about 15–20%) of GalNAc residues can occur during the dephosphorylation step. This results in the loss of these de-*N*-acetylated residues upon deamination *(9)*.

Nevertheless, despite these limitations, substantial structural information on the glycan sequence, including residue anomericity and linkage composition, can be obtained from real biological samples in the subnanomolar to nanomolar range. Fast atom bombardment mass spectrometry (FAB-MS) is becoming a very useful and high-sensitivity approach to studying GPI anchors and related molecules *(4,7,12)*; however, residue type, anomericity, and linkage composition cannot be determined by FAB-MS alone. Thus, the strategy just described can be used either to complement FAB-MS data or to stand alone as a partial characterization.

2. Materials

1. Aqueous HF (48%).
2. Lithium hydroxide (LiOH).
3. Defatted bovine serum albumen (BSA).
4. Trichloroacetic acid (TCA).
5. Dowex ion-exchange resins: AG50X12 (H^+ form) and AG3X4 (OH^- form), Bio-Rad, Hemel Hempstead, UK.
6. QAE-Sephadex A25 (equilibrated in water), Pharmacia, Uppsala, Sweden.
7. Methanol, HPLC grade.
8. Concentrated ammonia solution (35% ammonia).
9. 1*M* Acetic acid.

10. 300 m*M* Sodium acetate.
11. 1*M* Sodium nitrite.
12. 400 m*M* Boric acid.
13. Sodium hydroxide.
14. Sodium borotritiide (NaB^3H_4), 10–15 Ci/mmol, New England Nuclear, Hertfordshire, UK.
15. Sodium borodeuteride, Fluka puriss, Fluka, Glossop, UK.
16. Whatman 3MM chromatography paper.
17. 1-Butanol and ethanol.
18. Pyridine.
19. Chelex 100 (Na^+ form) ion-exchange resin, Bio-Rad, Hemel Hempstead, UK.
20. Anotop 0.2-µm filters, BDH, Poole, UK.
21. Bio-Gel P4 (–400 mesh), Bio-Rad, swollen at 55°C and fines removed three times. The column was calibrated with a hydrolysate of dextran detected by a refractive index monitor. Radioactive GPI were detected using an on-line radioactivity monitor (*see* Section 3.4.3.).
22. HPLC apparatus for high-pH anion-exchange chromatography equipped for pulsed electrochemical detection, e.g., Dionex Bio-LC with a CarboPac PA1 column.
23. Hydrochloric acid, Pierce constant boiling HCl, Pierce Warriner, Chester, UK.
24. Trifluoroacetic acid (TFA), Pierce.
25. Xylitol, galacitol, glucitol, mannitol, and 2,5-anhydromannitol.
26. Acetic anhydride.
27. Sulfuric acid.
28. Jack bean α-mannosidase (JBAM), Boehringer, Lewes, UK, dialyzed into buffer.
29. Coffee bean α-galactosidase (CBAG), Boehringer, dialyzed into buffer.
30. Jack bean β-hexosaminidase (JBBH), Sigma, Poole, UK, dialyzed into buffer.
31. *Aspergillus phoenicis* α-mannosidase (APAM), Oxford Glycosystems, Oxford, UK.
32. Mannoic acid-γ-lactone, Genzyme, Cambride, MA.
33. *Scyllo*-Inositol, Calbiochem, Nottingham, UK.
34. TMS reagent: hexamethyldisalazane, trimethylchlorosilane, pyridine (3:1:10), prepared weekly. Reagents from Fluka.
35. Dimethylsulfoxide (DMSO).
36. Methyl iodide (iodomethane), Aldrich gold grade, Aldrich, Gillingham, UK.
37. Sodium thiosulfate.

38. Acetonitrile, HPLC grade.
39. Toluene.
40. Dichloromethane.
41. Equipment and columns for GC–MS and RP-HPLC (*see* Sections 3.8.1. and 3.8.2.).

3. Methods

3.1. Isolation of Protein Anchor GPI Glycans (see *Note 1*)

1. Place freeze-dried protein (0.01–4 mg) in a polypropylene or Teflon™ tube.
2. Add 50 µL ice-cold 50% aq. HF, briefly sonicate in a chilled sonicating water bath, and, together with several identical blank tubes, incubate on ice-water for 48–60 h.
3. Use the blank tubes to assess how much freshly prepared saturated LiOH is needed to neutralize the HF (typically 250–300 µL).
4. Freeze a vol of saturated LiOH sufficient to adjust the pH of the aq. HF to between 3 and 5 on dry ice in an Eppendorf tube.
5. Transfer the sample aq. HF digest to the frozen LiOH, and vortex (the frozen LiOH solution prevents significant warming of the neutralization mixture).
6. Remove the LiF precipitate by centrifugation for a few seconds in a microfuge, and transfer the supernatant back to the original sample tube. Wash the LiF precipitate twice with 50 µL of water, and combine all the supernatants (*see* Note 2.).
7. Add defatted BSA carrier protein to a final concentration of 1 mg/mL, and precipitate the proteins at 4°C for 2 h by adding 50 µL 50% TCA. Centrifuge to remove the protein precipitate.
8. Remove the TCA and the residual F^- ions by applying the supernatant to a 0.6-mL column of AG3X4(OH^-), layered over 0.2 mL of QAE-Sepharose A25. Elute the GPI glycan and remaining Li^+ ions with four column vol of water. At this stage, take a small aliquot (1%) of the sample for inositol quantitation by selected ion monitoring GC–MS (*see* Section 3.8.1.).
9. Neutralize the residual LiOH with 0.04 meq of HF (40 µL of a 1*M* solution), and freeze-dry.
10. Extract the water-soluble GPI glycan from the residue with water (50 µL followed by 20 µL). Combine the supernatants, and dry in an Eppendorf tube.

11. Suspend the sample in 40 µL of methanol, 35% aq. NH_3 (1:1), and incubate for 1 h at 40°C (*see* Note 3). Remove the reagents under vacuum in a Speedvac concentrator.

3.2. Deamination and NaB^3H_4 Reduction

1. Dissolve the isolated GPI glycan in 15 µL 300 m*M* sodium acetate buffer, and deaminate by adding 5 µL freshly prepared 1*M* sodium nitrite, 2.5 h at room temperature (*see* Note 4).
2. Adjust the pH of the deamination mixture to approx 10.5 by adding 5 µL 400 m*M* boric acid followed by 10 µL of 2*M* NaOH (the pH is checked by spotting 0.5 µL on a pH paper).
3. Reduce the glycan by the rapid addition of 5 µL 36 m*M* NaB^3H_4 dissolved in 0.1*M* NaOH (*see* Note 5).
4. Incubate in an efficient fume hood for 1.5 h and for a further 2 h after adding 20 µL 1*M* $NaBD_4$.
5. Destroy the excess reductant by carefully adding 1*M* acetic acid (*see* Note 6).
6. Desalt the acidified solution by passage through a small column (0.3 mL) of AG50X12(H^+), to remove sodium ions, and elute with 1.5 mL water. Dry, and evaporate twice with 0.25 mL 5% acetic acid in methanol and twice with methanol to remove boric acid.

3.3. Removal of Radiochemical Contaminants

1. Dry the deaminated and reduced neutral glycan (NG) twice from water (0.5 mL followed by 0.1 mL). Redissolve in 20 µL water, and apply to a strip (3 × 45 cm) of Whatman 3MM paper. Perform downward chromatography for 48–60 h in 1-butanol:ethanol:water (4:1:0.6) (*see* Note 7.)
2. Cut out the region containing the GPI glycan, roll it up, and place in the barrel of a 2-mL plastic syringe, which is suspended in a 15-mL glass tube. Soak the paper with water (20 µL cm^2), and leave for 5 min.
3. Recover the water extract by centrifugation, and repeat the extraction four more times.
4. Dry the recovered neutral glycan fraction, and redissolve in 20 µL of water.
5. Transfer to a sheet of Whatman 3MM paper, and subject to high-voltage paper electrophoresis (Camag HVE, BDH Ltd.) in pyridine:acetic acid:water (3:1:387), pH 5.2 for 60 min at 80 V/cm (*see* Note 8).
6. Recover the neutral glycan as described in steps 2 and 3.

7. Pass the products through a column of 100 µL each Chelex100(Na^+) over AG50X12(H^+) over AG3X4(OH^-) over QAE-Sephadex A25, elute with 1.5 mL water, and filter through a 0.2-µm membrane (*see* Note 9).

3.4. Bio-Gel P4 Chromatography

1. The use of Bio-Gel P4 chromatography follows the methodology of Kobata and colleagues (*see* Chapter 5). Mix the labeled neutral glycans (typically aliquots of 1×10^4 to 1×10^5 cpm) with 750 µg of mixed β-glucan oligomers (*see* Note 10).
2. Separate according to hydrodynamic volume on a 1m × 1.5 cm column of Bio-Gel P4 (–400 mesh, jacketed at 55°C) eluted with HPLC-grade water at 0.2 mL/min.
3. Detect the glycan standards with a refractive index monitor (Erma 7512, ACS Ltd., Macclesfield, UK) and the radioactive GPI glycans using a Raytest Ramona radioactivity flow monitor equipped with a 200-µL X-cell solid scintillator flow cell (Raytest Instruments, Sheffield, UK). Collect the eluate in 1-mL (5-min) fractions (*see* Note 11).

3.5. Dionex Anion-Exchange Carbohydrate HPLC

1. Mix labeled neutral glycans (typically aliqouts of 1×10^4 to 1×10^5 cpm) with 150 µg of β-glucan oligomer standards (*see* Note 10) in 25 µL water.
2. Separate on a Dionex Carbopac PA1 column using a Dionex Bio-LC chromatograph (*see* Note 12).

3.6. Sequencing Reactions

3.6.1. Acid Hydrolysis (see Note 13)

1. To check for the presence or absence of radiolabeled 2,5-anhydromannitol (AHM), dry aliquots (about $1–2 \times 10^4$ cpm) in screw-top Eppendorf tubes, and hydrolyze in 200 µL 4*M* TFA, 2 h, 100°C.
2. Dry in a Speedvac, and redissolve the samples in 25 µL water containing 0.1 m*M* xylitol, galacitol, glucitol, mannitol, and 2,5-anhydromannitol (Sigma, Poole, UK) standards.
3. Analyze by Dionex HPLC (*see* Note 14).

3.6.2. Acetolysis (see Note 15)

1. Dry the labeled glycan and peracetylate in 20 µL acetic anhydride, pyridine (1:1) for 30 min at 100°C.
2. Remove the reagents under vacuum in a Speedvac.
3. Perform the acetolysis in 30 µL acetic anhydride:acetic acid:concentrated sulfuric acid (10:10:1) for 6 h at 37°C.
4. Quench the reaction by adding 20 µL pyridine and 0.5 mL water.

5. Recover the peracetylated products by extracting into 0.25 mL chloroform and wash three times with 0.5 mL water.
6. Dry the chloroform phase, and de-*O*-acetylate the products with 100 µL methanol:35% aq. NH_3 (1:1) for 16 h at 37°C.
7. Remove the base in a Speedvac, and redissolve the acetolysis cleaved glycans in water for Bio-Gel P4 and/or Dionex HPLC analysis.

3.7. Exoglycosidase Digestion

1. Dissolve samples in enzyme containing buffers (at a concentration of at least 10 µ*M*), and digest for 2–3 h at room temperature, followed by 16 h at 37°C.
2. For jack bean α-mannosidase (JBAM), use 30 µL 25U/mL JBAM in 0.1*M* sodium acetate, pH 5.0.
3. For coffee bean α-galactosidase (CBAG), use 30 µL 25U/mL CBAG in 0.1*M* sodium acetate, pH 6.0.
4. For jack bean β-hexosaminidase (JBBH), use 30 µL 4U/mL JBBH in 0.1*M* citrate-phosphate, pH 4.2, containing 10 m*M* D-mannoic-γ-lactone as a mannosidase inhibitor.
5. For *Aspergillus phoenicis* Manα1-2Man specific α-mannosidase (APAM), use 10 µL 1mU/mL APAM in 0.1*M* sodium acetate, pH 5.0.

3.8. Analyses by Gas Chromatography–Mass Spectrometry

3.8.1. Inositol Analysis (see Note 16)

1. Mix a small amount of the sample, usually 1%, with 10 pmol of *scyllo*-inositol internal standard, and hydrolyze in 50 µL 6*M* HCl, 110°C, 16–24 h. Hydrolyze a mixture of 10 pmol each *myo*- and *scyllo*-inositol standards in parallel.
2. Dry the hydrolysate in a Speedvac, and redry from 20 µL methanol.
3. Trimethylsilylate the products with 15 µL TMS reagent for at least 30 min prior to analysis of 1 µL by GC–MS using a Hewlett-Packard 5890-MSD system (*see* Note 17).

3.8.2. Methylation Linkage Analysis (see Note 18)

1. Dry neutral glycan samples in a 2-mL Pierce reactivial, with a Teflon™-covered stirring vane, and redry from 20 µL methanol.
2. Dissolve in 50 µL DMSO with constant stirring for 20 min.
3. Add 50 µL of a fine suspension of NaOH in DMSO (120 mg/mL), freshly prepared in a glass pestle and mortar, and stir for 20 min.
4. Make three additions of methyl iodide (10,10, and 20 µL) at 10-min intervals with continuous stirring.

5. Quench the reaction by adding 1 mL 100 mg/mL sodium thiosulfate, and extract the permethylated glycans by vortexing with 0.25 mL chloroform.
6. Wash the chloroform phase three times with 1 mL water, and dry it under a stream of N_2.
7. Dissolve the products in 100 µL 25% acetonitrile, and separate on a Hypersil 5-µm C18 reversed-phase column (25 × 0.4 cm, Hichrom, Reading, UK) eluted with a linear gradient of 20–80% acetonitrile in water over 60 min at 1 mL/min. Monitor the eluate for radioactive permethylated glycans using a Raytest Ramona monitor as described earlier.
8. Pool the pooled peak fractions, dry, and hydrolyze in 100 µL 0.25*M* H_2SO_4 in 93% acetic acid and 7% water, 80°C, 2.5 h.
9. Neutralize the sulfuric acid with 70 µL 1*M* NaOH, and add an internal standard of 20 µL 0.1 m*M* *scyllo*-inositol.
10. Transfer the mixture with 300 µL 50% methanol to a 15-mL glass tube, and dry using an Eyela S10 (Jencons, Hemel Hempstead, UK) multiport rotary evaporator.
11. Remove residual acetic acid on a high-vacuum line by coevaporation with two additions of 25 µL toluene.
12. Dissolve the products (partially methylated monosaccharides and sodium acetate) in 100 µL 10 m*M* NaOH and deuterium reduce by the addition of 100 µL 1*M* $NaBD_4$ (3 h at room temperature or overnight at 4°C).
13. Destroy excess reductant with acetic acid, dry, and remove boric acid, by evaporation with 0.25 mL 5% acetic acid in methanol (two times) and 0.25 mL of methanol (two times).
14. Dry the resulting partially methylated alditols, and acetylate them with 250 µL acetic anhydride (100°C, 3 h).
15. Remove the anhydride under vacuum, add 2 mL water, and vortex with 0.5 mL dichloromethane. Wash the dichloromethane phase twice with 2 mL water.
16. Recover the organic phase containing the partially methylated alditol acetates (PMAAs), and concentrate under a stream of N_2 to 10–20 µL.
17. Analyze aliquots of 1 µL by GC–MS (*see* Note 19).

4. Notes

1. The isolation of whole GPI anchors, following the proteolytic cleavage of most of the protein, has been described in detail for the anchors of *Trypanosoma brucei* variant surface glycoprotein (VSG), rat brain and thymocyte Thy-1 antigen, human erythrocyte acetylcholinesterase (AChE) and *Leishmania major* promastigote surface protease

(PSP) *(8–11)*. Here we describe the isolation of the GPI glycan moiety free from lipid and/or protein in a single step via dephosphorylation with cold 50% aqueous HF, and its subsequent radiolabeling by nitrous acid deamination and NaB^3H_4 reduction (Fig. 2). The success of the procedure relies on obtaining purified GPI-anchored protein substantially free of salts and detergent. In this respect, GPI-anchored proteins solubilized from membranes by the action of bacterial PI-PLC or *T. brucei* GPI-PLC present a relatively simple case, since they can be purified by conventional or affinity chromatography in the absence of detergent and finally dialyzed against water to remove salts. The isolation of soluble-form VSG (sVSG) from osmotically lysed *T. brucei* is a good example. The sVSG is produced by the artifactual cleavage of the GPI dimyristyl-glycerol moiety by an endogenous GPI-PLC *(13)* and purified by ion-exchange chromatography on DEAE-cellulose. Other examples include the preparation of placental alkaline phosphatase after the removal of phosphatidic acid by the action of serum-derived GPI-PLD *(14,15)*. In other cases this approach may not be feasible; for example, many GPI-anchored proteins are resistant to GPI-PLC and PI-PLC digestion *(1–3,10)* and can therefore only be solubilized by detergent containing buffers. This obstacle has been overcome in one case where a GPI-anchored surface antigen of *Trypanosoma cruzi* was purified from a whole cell detergent lysate by affinity chromatography on an immobilized monoclonal antibody matrix *(16)*. After washing the matrix with detergent buffers, the column was washed with 3 column vol of water to remove salts and detergent, and the absorbed antigen was eluted with 1*M* propionic acid. Propionic acid is an effective low-pH eluant with apparently good solvation properties for GPI-anchored proteins. The eluate can then be freeze-dried to yield pure buffer-free glycoprotein. If eluate volumes are large, we recommend the addition of deffated BSA to 0.1 mg/mL to act as a glycan-free carrier protein.

2. This fraction contains the free dephosphorylated GPI glycan and the residual (mostly undegraded) protein in approx 400 µL of saturated LiF (approx 100 m*M*). Importantly, the other common classes of glycoprotein glycans (*N*-linked and *O*-linked carbohydrate) remain attached to the protein backbone.
3. This step results in deacylation of the sample if it contains a palmitoylated inositol ring *(10)* and appears to improve the subsequent deamination efficiency of all small samples.
4. The deamination of the $GlcNH_2$ residue causes its conversion to 2,5-anhydromannose with simultaneous release of the *myo*-inositol resi-

due. The deaminated GPI glycan is not isolated before reduction, since the aldehyde group of 2,5-anhydromannose is highly reactive.

5. The NaB^3H_4 solution can be stored in sealed containers at –70°C for up to 2 mo. However, best results are obtained with freshly dissolved reductant.
6. There is a considerable evolution of tritium gas and tritiated water at this stage. The sample must be handled in a fume hood at all stages until it has been dried repeatedly from water.
7. Carbohydrates larger than disaccharides remain within 3 cm of the origin and can be detected using a Raytest Rita linear analyzer (Raytest Instruments, Sheffield). Substantial amounts of radiochemical contaminants and exchangeable tritium are removed in the paper chromatography step.
8. This procedure removes most of the remaining radiochemical contaminants. The neutral labeled GPI glycan remains at the origin and can be located by linear analyzer.
9. The resulting solution contains the highly purified GPI glycan bearing a labeled 2,5-anhydromannitol (AHM) at its reducing terminus. The specific activity of the introduced label is usually between 0.5 and 3 Ci/mmol. Thus, as little as 1 nmol of starting glycoprotein can yield of the order of 0.5×10^6 to 3×10^6 cpm of pure labeled glycan. With quantities of 1 nmol or more, approx 75% of the sample is saved for GC–MS methylation analysis, and the remaining material is used for chromatographic analysis before and after chemical and enzymatic sequencing reactions.
10. The β-glucan oligomers (Glc_1–Glc_{30}) are prepared by partial acid hydrolysis (0.1*M* HCl, 3 h, 100°C) of dextran (BDH, grade C). Hydrolysis is performed at 100 mg/mL, and the acid is removed by passage of the hydrolysate through a fivefold excess of Dowex AG3X4(OH^-).
11. The refractive index monitor output is digitized by the Ramona radioactivity detector, and the data collected using Raytest Ramona chromatography software. The two chromatograms are offset to correct for delay between the detectors. The effective size of the labeled glycan is expressed in relative glucose units (Gu) by interpolation of the radioactive glycan peaks between adjacent β-glucan standards.
12. The following program is routinely used for the resolution of GPI glycans: buffer A = 0.15*M* NaOH, buffer B = 0.15*M* NaOH, 0.25*M* sodium acetate, starting conditions 95%A, 5%B followed by a linear gradient to 70%A, 30%B over 75 min at 0.6 mL/min. A wash cycle of 100% B for 10 min is followed by reequilibration in 95%A, 5%B for at least 15 min. The β-glucan standards are detected by pulsed-

amperiometric detection (PAD) (Dionex), and the pH of the eluate lowered by passage through a DIONEX AMMS anion suppressor prior to detection of the labeled GPI glycans by the Ramona detector. All data are collected and analyzed using the Raytest Ramona data system. The absolute retention times of glycans can vary substantially on this HPLC system from day to day. However, the elution positions of the GPI glycans relative to the set of β-glucan internal standards are almost constant. The elution position is expressed in so-called "Dionex units" (Du) by linear interpolation of the radioactive peak between adjacent β-glucan peaks. The Du value of a glycan has no specific meaning other than as a fixed chromatographic property.

13. Before sequencing any of the labeled glycan species resolved by Bio-Gel P4 and/or Dionex HPLC, it is worth checking that they terminate in a labeled 2,5-anhydromannitol residue. The NaB^3H_4 reduction step will, of course, label any reducible impurity.
14. Under isocratic conditions (0.15*M* NaOH) (0.6 mL/min), the xylitol, glucitol, galacitol, mannitol, and 2,5-anhydromannitol standards elute at 3.3, 4.2, 4.2, 4.7, and 7.4 min, respectively. The unlabeled standards are detected by PAD detector, and the labeled alditols by Ramona radioactivity monitor (*see* Note 12).
15. Acetolysis is relatively selective for the cleavage of Manα1-6Man glycosidic linkages *(17)* and is therefore useful in establishing the position of this linkage in GPI glycan cores. Reaction conditions are chosen that will cleave approx 70% of Manα1-6Man linkages, leaving a small amount of starting material. These conditions are used for two reasons: (1) in order to be able to see any partial products that would arise if a structure contained two or more Manα1-6Man linkages, and (2) because the typical Manα1-4AHM linkage is itself relatively labile to acetolysis *(18)*. Some typical acetolysis data demonstrating these points can be seen in Fig. 3B,C.
16. The *myo*-inositol content of the isolated GPI glycan is usually measured prior to deamination/reduction to assess yield and to establish the inositol isomer present in the structure.
17. The samples are separated on a Econocap SE-54 column (30 m × 0.25 mm, Alltech Associates, Carnforth, UK) using a direct on-column injector with a head pressure of 5 psi helium. The column is held at 140°C (1 min) followed by a linear temperature gradient to 250°C at 15°C/min. Inositol-TMS_6 derivatives elute well within a time window of 6–16 min. During this time, the inositol-TMS-specific ions *m/z* 305 and *m/z* 318 are selectively monitored with a dwell time of 100 ms each; ionization is by electron impact at 70 eV. Inositol

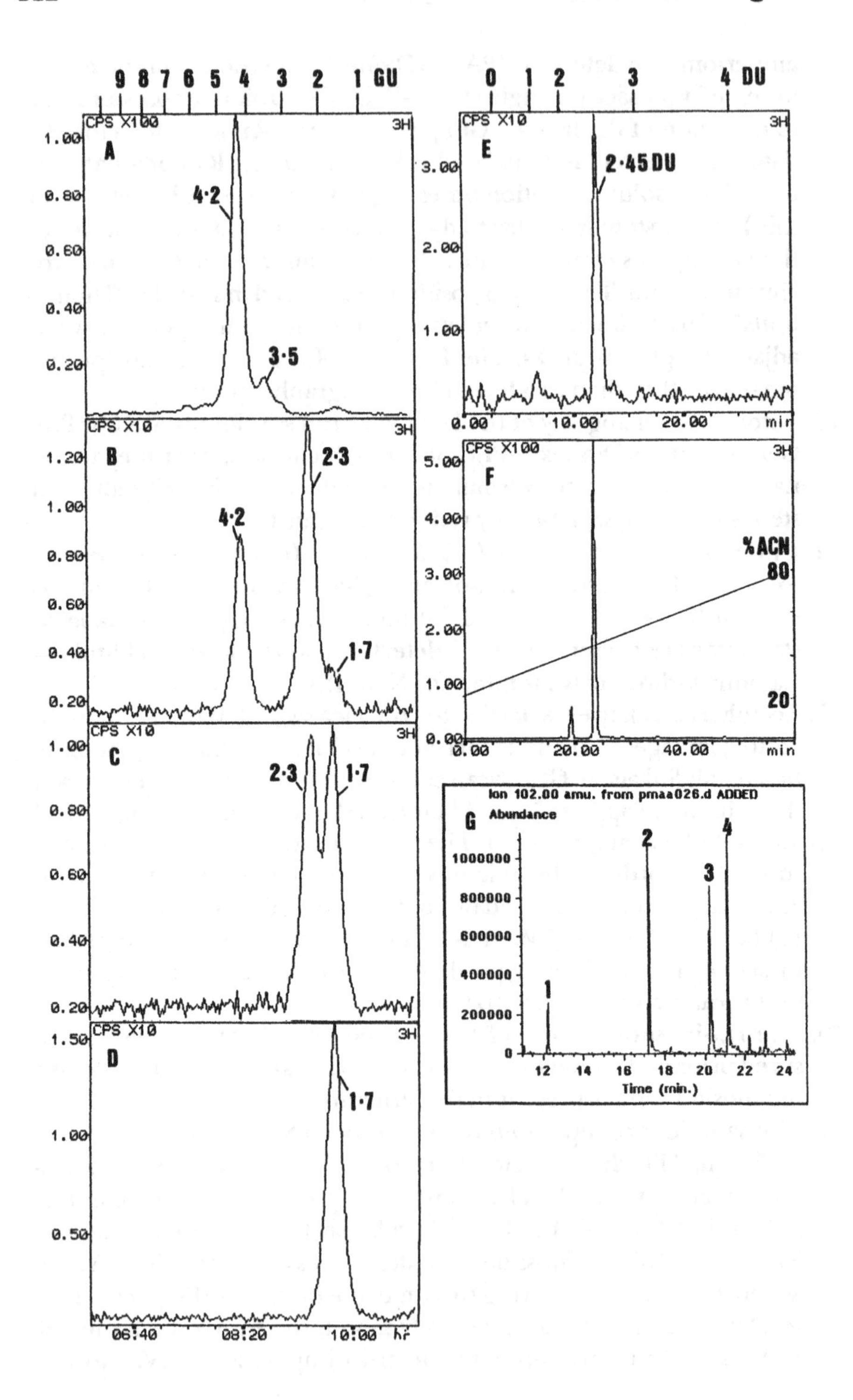
9 8 7 6 5 4 3 2 1 GU
0 1 2 3 4 DU
CPS X100
CPS X10
3H
A
B
C
D
E
F
G
4·2
3·5
2·3
1·7
2·45 DU
%ACN
80
20
06:40
08:20
10:00
hr
0.00
10.00
20.00
40.00
min
Ion 102.00 amu. from pmaa026.d ADDED
Abundance
Time (min.)

isomer identification is based on characteristic retention time and *m/z* 305:*m/z* 318 ratio. Quantitation is made by total ion current integration. Relative response factors for the isomers are made by analyzing a set of standards in parallel with each set of samples.

18. The linkage composition of the labeled GPI glycan by GC–MS can be made on as little as 500 pmol of material. The success of this high-sensitivity analysis relies in the presence of the tritium label in the glycan for repurification of the permethylated glycan by reverse-phase HPLC and the use of extracted ion chromatograms in the subsequent GC–MS analysis.
19. PMAA derivatives are analyzed on an SE-54 column (*see* Note 17) and on an SP2380 bonded-phase column (30 m × 0.25 mm, Supelco, Saffron Walden, UK). Direct on-column injection is used at a column head pressure of 5 psi helium with a temperature program of: 80°C (1 min) rising to 140°C at 15°C/min and then to 250°C at 5°C/min, held for 20 min. Electron impact spectra are recorded at 70 eV scanning between 40 and 350 mass units. The peaks owing to PMAA derivatives in the total ion chromatogram are easily identified by comparison with extracted ion profiles at *m/z* 102, 118, 129, 130, 161, 162, 189, 205, 233, and 234 (characteristic of pentose- and hexose-derived deutero-reduced PMAAs) and at *m/z* 117, 159, 144, 168, and 210

Fig. 3 (opposite page). Working example: the microsequencing of an authentic GPI glycan. The GPI neutral glycan was prepared from 1.8 nmol of a variant surface glycoprotein (VSG variant MIT at 1.5) of *Trypanosoma brucei* and processed as described in Note 20. **A:** The column is calibrated with glucan oligomers (Gu) Bio-Gel P4 chromatography of the untreated radiolabeled neutral glycan. **B** and **C:** Bio-Gel P4 chromatography of the 4.2-Gu peak after acetolysis for 6 h and 24 h, respectively. Note that quantitative cleavage of the Manα1-6Man linkage (panel **C**) also results in substantial cleavage of the Manα1-4AHM linkage. **D:** Bio-Gel P4 chromatography of the 4.2-Gu peak after digestion with jack bean α-mannosidase (JBAM). **E:** Dionex HPLC of the 4.2-Gu peak from panel A. **F:** Reverse-phase HPLC of the permethylated neutral glycan fraction. **G:** Methylation linkage analysis of the major permethylated glycan. The chromatogram is an extracted ion chromatogram of ions at *m/z* 102, 118, 129, 130, 161, 162, 189, 205, 233, and 234 only. Peak 1, 4-acetyl-1,3,6-trimethyl-1-deutero-2,5-anhydromannitol; peak 2, 1,5-diacetyl-2,3,4,6-tetramethyl-1-deutero-mannitol; peak 3, 1,2,5-triacetyl-3,4,6-trimethyl-1-deutero-mannitol; peak 4, 1,5,6-triacetyl-2,3,4-trimethyl-1-deutero-mannitol. The derivatives were identified by their characteristic mass spectra and retention times.

(characteristic of deutero-reduced *N*-acetylhexosamine and inositol derivatives). The complete spectra of peaks presenting one or more of these ions are investigated using a background subtraction program. The PMAAs are identified by their mass-spectra and retention times relative to the late eluting *scyllo*-inositol hexaacetate internal standard peak.

20. Working example—the analysis of the GPI glycan of *T. brucei* VSG (variant MIT at 1.5): The compositional analysis of the GPI anchor of VSG variant MIT at 1.5 suggested that this particular variant contained no galactose in its GPI anchor *(19)*. To investigate the precise structure of its mannose-containing GPI anchor, 100 µg (1.8 nmol) were dephosphorylated, the GPI glycan isolated, deaminated, and reduced as described here. The purified tritium-labeled glycan (1.5 nmol of *myo*-inositol, measured by GC–MS, and 4×10^6 cpm) was split into aliquots for chromatographic analysis and the remaining 75% taken for methylation analysis. Figure 3A shows the chromatography of the native glycan on Bio-Gel P4. A major peak at 4.2 Gu and a minor peak at 3.5 Gu were observed. Acid hydrolysis and isocratic Dionex HPLC analysis revealed that only the 4.2 Gu peak contained labeled 2,5-anhydromannitol (data not shown). The 3.5-Gu contaminant peak was not analyzed further. The 4.2-Gu peak was rechromatographed by Dionex HPLC (Fig. 3E), and a peak at 2.45 Du was observed. Following permethylation, the remaining neutral glycan fraction was resolved into a major and a minor peak by reverse-phase HPLC (Fig. 3F). The methylation (linkage composition) analysis of the major HPLC-purified permethylated glycan is shown in Fig. 3G. The PMAA derivatives corresponding to terminal-Man, 2-*O*-substituted-Man, 6-*O*-substituted-Man and 4-*O*-substituted-2,5-anhydromannitol were observed, suggesting a linear glycan structure.

The glycan had a size of 4.2 Gu on Bio-Gel P4. Since authentic AHM elutes at 1.7 Gu, this size predicts a trisaccharide linked to AHM, consistent with the methylation analysis. Furthermore, the cochromatography of this glycan with authentic Manα1-2Manα1-6Manα1-4AHM on both Bio-Gel P4 and Dionex HPLC (4.2 Gu and 2.45 Du; *see* Table 1) strongly suggests that it may have the same structure. Treatment of the glycan with JBAM removed three α-Man residues to yield AHM eluting at 1.7 Gu on Bio-Gel P4 (Fig. 3D). The identity of this 1.7-Gu peak was confirmed by Dionex HPLC using the isocratic alditol analysis program (data not shown). Acetolysis of the glycan produced a fragment eluting at 2.3 Gu (Fig. 3B,C) corresponding to $Man\alpha_1$-AHM (*see* Table 1). Taken together, these data unambiguously

Table 1
Chromatographic Properties of GPI Neutral Glycans

Structure	Bio-Gel P4	Dionex HPLC
AHM	1.7 Gu	1.0 Du
Manα1-4AHM	2.3 Gu	1.1 Du
Manα1-6Manα1-4AHM	3.2 Gu	2.2 Du
Manα1-2Manα1-6Manα1-4AHM	4.2 Gu	2.5 Du
Manα1-2Manα1-6Manα1-4AHM Galα1-3	5.2 Gu	3.6 Du
Manα1-2Manα1-6Manα1-4AHM Galα1-6Galα1-3	6.1 Gu	3.8 Du
Manα1-2Manα1-6Manα1-4AHM Galα1-6Galα1-3 Galα1-2	6.8 Gu	4.4 Du
Manα1-2Manα1-6Manα1-4AHM Galα1-2Galα1-6Galα1-3	6.8 Gu	4.0 Du
Manα1-2Manα1-6Manα1-4AHM Galα1-2Galα1-6Galα1-3 Galα1-2	7.6 Gu	4.7 Du
Manα1-2Manα1-2Manα1-6Manα1-4AHM	5.2 Gu	3.0 Du
Manα1-2Manα1-6Manα1-4AHM GalNAcβ1-4	5.7 Gu	3.0 Du
Manα1-2Manα1-2Manα1-6Manα1-4AHM GalNAcβ1-4	6.5 Gu	3.5 Du

define the glycan as: Manα1-2Manα1-6Manα1-4AHM. Since AHM is an unnatural sugar, made by the deamination of $GlcNH_2$, the sequence may be further refined to: Manα1-2Manα1-6Manα1-4$GlcNH_2$.

Acknowledgments

We thank Dr. Malcolm McConville (Dundee) for helpful suggestions and Pascal Schneider (University of Lausanne) for his assistance. This work was supported by the EC (contract No. TS2* 0271-UK (SMA)) and the The Wellcome Trust. M. L. S. G. thanks Coordenacao de Aperfeicoamento de Pessoal de Nivel Superior (CAPES) for a Ph.D fellowship.

References

1. Thomas, J. R., Dwek, R. A., and Rademacher, T. W. (1990) Structure, biosynthesis, and function of glycosylphosphatidylinositols. *Biochemistry* **29,** 5413–5422.
2. Cross, G. A. M. (1990) Glycolipid anchoring of plasma membrane proteins. *Annu. Rev. Cell Biol.* **6,** 1–39.
3. Ferguson, M. A. J. (1991) Lipid anchors on membrane proteins. *Current Opinion in Structural Biology* **1,** 522–529.
4. McConville, M. J., Homans, S. W., Thomas-Oates, J. E., Dell, A., and Bacic, A. (1990) Structures of the glycoinositolphospholipids from *Leishmania major. J. Biol. Chem.* **265,** 7385–7394.
5. McConville, M. J. and Blackwell, J. (1991) Developmental changes in the glycosylated-phosphatidylinositols of *Leishmania donovani.* Characterization of the promastigote and amastigote glycolipids. *J. Biol. Chem.* **266,** 15,170–15,179.
6. Turco, S. J. and Orlandi, P. A., Jr. (1989) Structure of the phosphosaccharide-inositol core of the *Leishmania donovani* lipophosphoglycan. *J. Biol. Chem.* **264,** 6711–6715.
7. McConville, M. J., Thomas-Oates, J. E., Ferguson, M. A. J., and Homans, S. W. (1990) Structure of the lipophosphoglycan from *Leishmania major. J. Biol. Chem.* **265,** 19,611–19,623.
8. Ferguson, M. A. J., Homans, S. W., Dwek, R. A., and Rademacher, T. W. (1988) Glycosyl-phosphatidylinositol moiety that anchors *Trypanosoma brucei* variant surface glycoprotein to the membrane. *Science* **239,** 753–759.
9. Homans, S. W., Ferguson, M. A. J., Dwek, R. A., Rademacher, T. W., Anand, R., and Williams, A. F. (1988) Complete structure of the glycosyl phosphatidylinositol membrane anchor of rat brain Thy-1 glycoprotein. *Nature* **333,** 269–272.
10. Roberts, W. L., Santikarn, S., Reinhold, V. N., and Rosenberry, T. L. (1988) Structural characterization of the glycoinositol phospholipid membrane anchor of human erythrocyte acetylcholinesterase by fast atom bombardment mass spectrometry. *J. Biol. Chem.* **263,** 18,776–18,784.
11. Schneider, P., Ferguson, M. A. J., McConville, M. J., Mehlert, A., Homans, S. W., and Bordier, C. (1990) Structure of the glycosyl-phosphatidylinositol membrane anchor of the *Leishmania major* promatisgote surface protease. *J. Biol. Chem.* **265,** 16,955–16,964.
12. Baldwin, M. A., Stahl, N., Reinders, L. G., Gibson, B. W., Prusiner, S. B., and Burlingame, A. L. (1990) Permethylation and tandem mass spectrometry of oligosaccharides having free hexosamine: analysis of the glycoinositol phospholipid anchor glycan from the scrapie prion protein. *Anal. Biochem.* **191,** 174–182.
13. Ferguson, M. A. J., Haldar, K., and Cross, A.M. (1985) *Trypanosoma brucei* variant surface glycoprotein has a sn-1,2-dimyristyl glycerol membrane anchor at its COOH terminus. *J. Biol. Chem.* **260,** 4963–4968.

14. Ogata, S., Hayashi, Y., Yasutake, K., and Ikehara, Y. (1987) Chemical identification of lipid components in the membranous form of rat liver alkaline phosphatase. *J. Biochem.* **102,** 1609–1615.
15. Huang, K-S., Li, S., Fung, W-J. C., Hulmes, J. D., Reik, L., Pan, Y-C. E., and Low, M. G. (1990) Purification and characterizadon of glycosyl-phosphatidylinositol-specific phospholipase D. *J. Biol. Chem.* **265,** 17,738–17,745.
16. Guther, M. L. S., Cardoso de Almeida, M. L., Yoshida, N., and Ferguson, M. A. J. (1992) Structural studies on the glycosylphosphatidyl inositol membrane anchor of *Trypanosoma cruzi* 1G7-antigen: The strucutre of the glycan core. *J. Biol Chem.* **267,** 6820–6828.
17. Natsuka, S., Hase, S., and Ikenaka, T. (1987) Fluorescence method for the structural analysis of oligomannose-type sugar chains by partial acetolysis. *Anal. Biochem.* **167,** 154–159.
18. Mayor, S., Menon, A.K., Cross, G. A. M., Ferguson, M. A. J., Dwek, R. A., and Rademacher, T. W. (1990) Glycolipid precursors for the membrane anchor of *Trypanosoma brucei* variant surface glycoproteins. *J. Biol. Chem.* **265,** 6164–6173.
19. Holder, A. A. (1985) Glycosylation in the African trypanosome. *Curr. Top. Microbiol. Immunol.* **117,** 57–74.

CHAPTER 9

Analysis of Bacterial Glycoproteins

Johann Lechner and Felix Wieland

1. Introduction

Investigation of bacterial glycoproteins is a relatively new field. Although biochemical research of glycoproteins was and is being focused on mammalian systems, only a few protein-linked bacterial carbohydrates have been solved so far. Since the bacterial structures described are not related closely enough to established structures, every bacterial carbohydrate analysis presents an adventurous challenge *per se*. Nevertheless, the monosaccharide composition of bacterial glycoproteins is not different from that of mammalian glyocoproteins in principal, although a few rare sugars are found in addition, e.g., 3-*O*-methylgalactosuronic acid, 3-*O*-methylglucose, and furanosidic galactose *(1–3)*. These monosaccharides together with the more common ones are composed to yield various types of oligosaccharides, which are linked to the core protein via *O*- and *N*-glycosyl linkages. In Fig. 1, our present knowledge of the structure of the cell-surface glycoprotein from halobacteria is summarized. So far, two novel types of *N*-glycosyl linkages have been discovered in this bacterial cell-surface glycoprotein. Here we would like to outline a few selected methods that have made bacterial glycoconjugate analysis easier in our laboratory, and also describe a more general way that has led us to uncover the novel linkage units between carbohydrate and protein.

From: *Methods in Molecular Biology, Vol. 14: Glycoprotein Analysis in Biomedicine*
Edited by: E. F. Hounsell Copyright © 1993 Humana Press Inc., Totowa, NJ

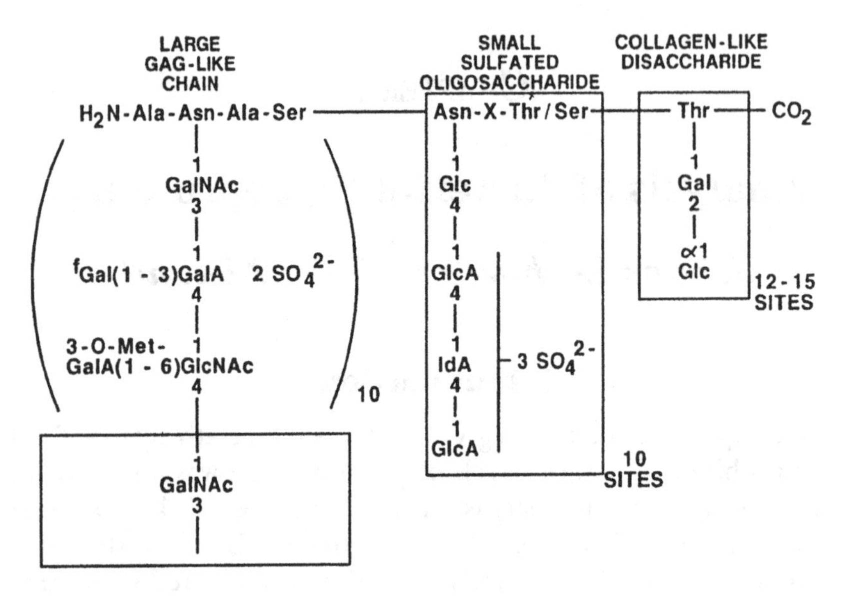

Fig. 1. Molecular structure of the glycoconjugates found in the cell-surface glycoprotein of Halobacteria. (For a review, *see 4*).

1.1. N-*Glycosyl Linkage Units*

A linkage unit can be regarded as established when it is isolated in pure form and analyzed either by composition analysis, NMR, or mass sprectroscopy. Alternatively, comigration with an authentic, synthesized compound in at least two chromatographic systems can be performed. To obtain the linkage unit, glycopeptides with the complete carbohydrate content are isolated by conventional techniques, mainly by reversed-phase HPLC. Interestingly, glycopeptides with a high carbohydrate content, which are therefore very soluble in water, are eluted from reversed-phase at higher organic solvent concentrations than their peptide cores.

An isolated and homogeneous glycopeptide obtained is then characterized by compositional amino acid and monosaccharide analysis, and submitted to alkaline conditions for β-elimination of the carbohydrate. Resistance to β-elimination indicates the presence of *N*-linked carbohydrate. Alternatively, the glycoconjugate could be linked to hydroxylysine or hydroxyproline, whose alcoholic groups are not in β-position and, therefore, cannot be eliminated by alkali. A technically easy method of discriminating between *O*- and *N*-glycosyl bonds is solvolysis of the carbohydrate with hydrogen fluoride (HF). HF splits all *O*-glycosyl bonds in a gentle way to yield the corresponding monosaccharide fluorides. After exposition to water, the fluorides are spontaneously converted into the free monosaccharides. *N*-glycosyl bonds and peptide bonds are stable against HF under the conditions given in Section 3. Thus, after HF treatment, the glycopeptide contains no carbohydrate but the single monosaccharide that is linked via nitrogen. Therefore, a composition analysis of the resulting peptide will reveal the monosaccharide part of the *N*-glycosyl unit. Two strategies can be applied to define the amino acid involved (which in all three *N*-linkage units established so far turned out to represent Asn).

The HF-treated glycopeptide may be further digested using pronase (a mixture of relatively unspecific proteases). This digestion has a high probability of releasing the aminoacylmonosaccharide, because removal of all carbohydrate but the monosaccharide by HF can decrease the steric hindrance for the attack of proteases. From the resulting mixture of small peptides and amino acids, the aminoacylmonosaccharide can be isolated either by ion-exchange chromatography *(4)* or by dansylation and subsequent HPLC *(2)*. Alternatively, the HF-treated glycopeptide can be submitted to Edman degradation *(1)*. Usually there is no influence on the yield of an Edman-step if an amino acid is substituted with a small saccharide. The phenylthiohydantoin of the aminoacylmonosaccharide can then be compared with the authentic, synthesized substance on various chromatographic systems. If authentic material is not available, appearance of an unknown product after an Edman step together with a lacking amino acid (when compared with the amino acid analysis) and the disappearance of the monosaccharide in the peptide remaining after the Edman cycle can be taken as evidence for the composition of the linkage unit. A better way to establish the nature of the unit, however, is to submit the unknown phenylthiohydantion derivative to fast atom bombardment mass spectroscopy (cf ref. *1*). Here we describe two simple methods to discriminate between *O*- and *N*-glycosyl bonds, a prerequisite for the characterization of a protein-carbohydrate linkage unit.

1.2. Identification of Uronic Acids, Amino Sugars, and Neutral Sugars from Bacterial Glycoproteins

A couple of problems arise when it comes to characterizing oligosaccharides that contain neutral sugars, amino sugars, and uronic acids at the same time. These problems are owing to chemical instability of some of the components under the conditions of hydrolysis, and to a high stability of the bonds formed by some uronic acids and amino sugars. For example, if *N*-deacylation of an acyl amino sugar occurs before its glycosyl bond is split, this bond becomes very resistant against the attack of protons. Therefore, mild but effective hydrolysis of these saccharides is desirable. For the subsequent characterization of carbohydrates, gas-liquid chromatography (GC) has turned out to be a powerful method. However, reduction and derivatization of uronic acids often is a tedious and time-consuming task, and GC analysis of amino sugars as their alditol acetates is hampered by destruction at the high temperature needed for separation of these derivatives on metal surfaces. Therefore, in the following, we describe methods that allow effective breakage of glycosyl bonds under very mild conditions and separation by GC of the resulting monosaccharides at relatively low temperature. With this method, all monosaccharides named here are separated in one GC run, including uronic acid derivatives and 2-deoxy-2-aminohexoses, which appear in good yields.

2. Materials

2.1. β-Elimination

1. 0.25*M* Sodium borohydride ($NaBH_4$) in 100 m*M* NaOH, prepared freshly.
2. AG50Wx8 (H^+-form) ion-exchange material, stored at 4°C.

2.2. Solvolysis with HF

1. HF in a lecture bottle equipped with outlet that allows connection of polyethylene tubing (diameter 4 mm).
2. 50 m*M* Sodium bicarbonate buffer.

2.3. Qualitative Analysis of Uronic Acids, Amino, and Neutral Sugars as Pentafluoropropionyl Derivatives by GC

1. Glass tubes used had an inner width of 0.5 cm, were long enough to be drawn out twice as described in Section 3. (10 cm), and thick enough (1 mm) to withstand the overpressure produced.
2. 0.5*N* HCl in methanol prepared fresh by adding acetyl chloride to absolute methanol (stored over a molecular sieve 4Å).
3. Pentafluoropropionic anhydride (Pierce, Ond Beijerland, The Netherlands) stored at 4°C, methylene chloride of highest quality, stored at room temperature, and sugar standards, stored as 1 mg/mL solutions at 20°C.

4. GC-MS apparatus equipped with a Durabond 1701 capillary column (30 m), carrier gas He at 1 mL/min, starting at 120°C with a temperature gradient of 2°C/min and mass detector, electron energy 70 eV. Six hundred and ninety atomic mass units were recorded per second between 100 and 700 amu and when pentafluoropropionyl derivatives were analyzed, and between 40 and 400 amu when partially methylated alditol acetates were analyzed.

2.4. Identification of Iduronic Acid as 6-D2-Hexaacetyl Iditol by GC-MS

1. Glass tubes, 0.5*N* HCl in methanol as in Section 2.3., step 1.
2. 2*N* HCl, 5*N* acetic acid, and 50 m*M* aqueous NaOH stored at room temperature.
3. Sodium borohydride and deuteride (Merck, Darmstadt, Germany) stored desiccated at room temperature.
4. 1*M* Sodium borohydride in 50 m*M* NaOH, prepared fresh.
5. AG50Wx8 (H^+form) ion-exchange material, stored at 4°C.
6. Pyridine (distilled twice), acetic anhydride, acetyl acetate of highest quality, stored at 4°C.
7. GC-MS apparatus as described earlier.

2.5. Identification of 3-O-Methylgalacturonic Acid

1. As under Section 2.4.
2. BBr_3 (Merck) stored at –20°C.

3. Methods

3.1. β-Elimination

1. Dissolve glycoprotein or glycopeptide (10 µg up to 10 mg/mL) in the appropriate vol of a solution of 250 m*M* $NaBH_4$ (or $NaBD_4$) in 100 m*M* NaOH, and incubate at 37°C for 14 h.
2. Stop the reaction by the dropwise addition of 2*M* acetic acid. **Caution:** Generation of hydrogen will cause strong foaming. After foaming has ceased, add one more drop of the acid.
3. Apply the sample onto a 1-mL column of AG50Wx8 (H^+form) in water, wash the colum with 5 mL of water, combine all eluents, and dry by evaporation (*see* Note 1).
4. Hydrolyze and analyze the eluants as in Section 3.3., steps 1–6.
5. For the analysis of amino group containing materials (*see* Note 2), recover those substances by passing 3 mL of NH_4OH (1 mL 25% NH_4OH, Merck, plus 2 mL H_2O) through the ion-exchange column used in step 3. Dry the eluants by rotary evaporation and analyze.

3.2. Solvolysis with HF (see *Note 3*)

1. Dry the sample (20 µg up to 2.5 mg) in an Eppendorf tube under vacuum.
2. Under a hood, condense hydrogen fluoride into a polyethylene tube cooled by liquid nitrogen (*see* Note 4).
3. Transfer tube to an ice bath, allow hydrogen fluoride to melt, and immediately (HF has a boiling temperature of 19°C!) pipet approx 50–500 µL to the sample. Incubate at 0°C for 3 h or at 23°C for 1 h.
4. Remove hydrogen fluoride by blowing a gentle stream of nitrogen onto the sample (hood!). If a larger amount of sample has been treated (≥0.2 mL), subsequently dry in the vacuum (*see* Note 5).
5. Dissolve in the buffer of choice (for example, in an $NaHCO_3$ solution) to neutralize any traces of acid left in the sample.

3.3. Qualitative Analysis of Uronic Acids, Amino, and Neutral Sugars as Pentafluoropropionyl Derivatives by GC

1. Vacuum dry the saccharide sample on the bottom of a glass tube for 3 h with P_2O_5 as dessicant (*see* Note 6).
2. In a hot flame, draw out the upper part of the glass tube leaving a small reservoir connected to the rest of the tube by a narrowing. Add 150 µL of 0.5*N* HCl in methanol into the reservoir, and suck the liquid into the tube by immersing the lower part of the tube in liquid nitrogen. Seal tube in a hot flame forming an ampule, and incubate at 80°C for 20 h.
3. Freeze sample in liquid nitrogen to reduce overpressure in the ampule, before opening the ampule with a glass cutter. After thawing, dry sample in a stream of nitrogen.
4. As in step 2, draw out upper part of the remaining glass tube, suck in 75 µL of methylene chloride and 75 µL of pentafluoropropropionic anhydride, seal the tube, and incubate at 100°C for 1 h. Methylene chloride and pentafluoropropionic anhydride form two layers at room temperature, but will mix at 100°C.
5. Freeze sample in liquid nitrogen as just described, open the ampule, and after thawing, concentrate sample to 2–5 µL in a stream of nitrogen. Avoid concentrating sample to complete dryness, since this will result in loss of the volatile pentafluoropropionyl derivatives.
6. Analyze 1–2 µL of the sample by GC as shown in Fig. 2.

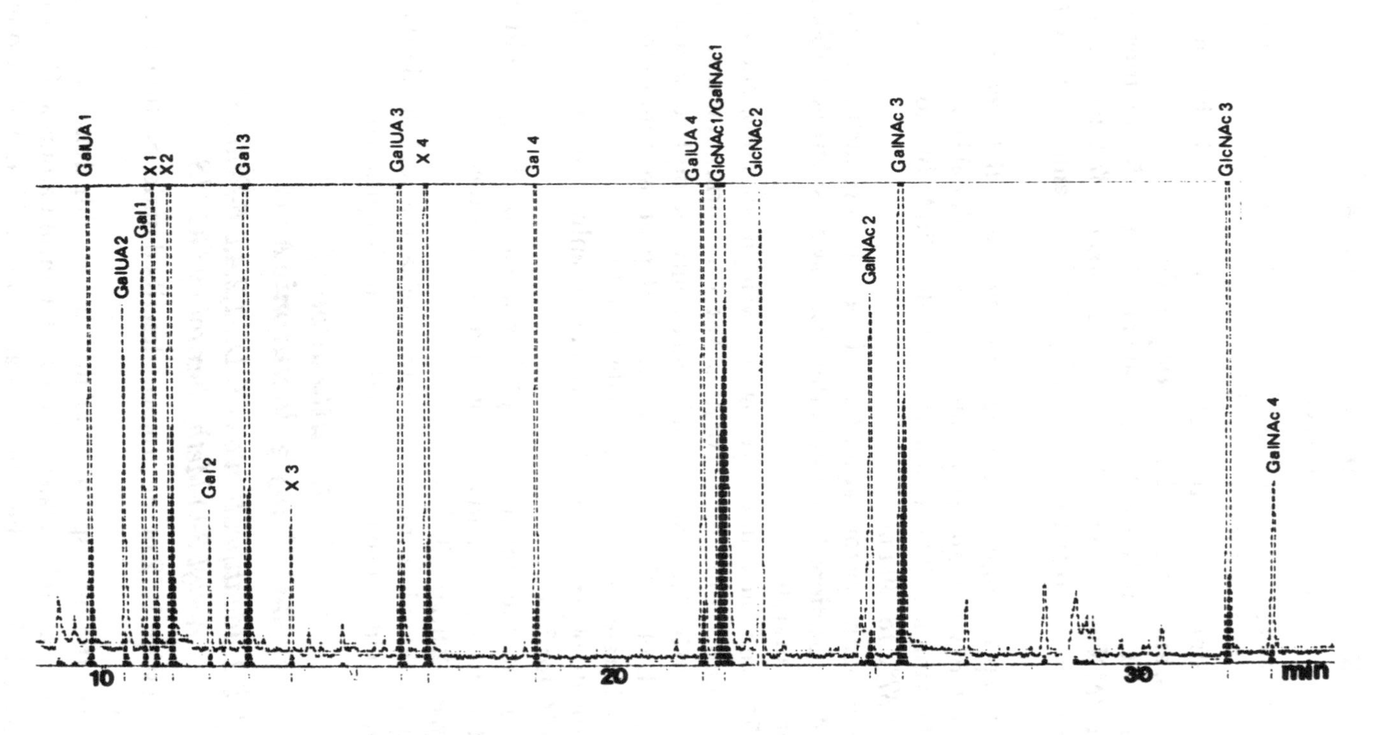

Fig. 2. GC/MS analysis of a mixture of uronic acids, neutral sugars, and amino sugars. The pentafluoropropionyl derivatives were separated as described under Section 2.3; step 4. X1-X4 represent the isomers of the 3-*O*-methyl galacturonic acid derivative.

3.4. Identification of Iduronic Acid as 6-D2-Hexaacetyl Iditol by GC-MS

1. Hydrolyze sample in 0.5*N* HCl in methanol, and remove solvent as described under Section 3., step 1 (*see* Note 7).
2. Incubate sample in 200 µL of 2*N* HCl at 100°C to hydrolyze methyl glycosides, and remove solvent by evaporation.
3. Add 100 mL 50 m*M* NaOH, incubate at 37°C for 15 min, then add 100 µL 2*M* $NaBH_4$ in 50 m*M* NaOH, and incubate at 37°C for 30 min.
4. To form an urono lactone, add 200 µL 2*N* HCl, and evaporate (*see* Note 8).
5. Repeat step 4 to improve lactonization (and therewith C_6 reduction), add 1 mL of water, and evaporate to remove residual HCl.
6. Dissolve sample in 300 µL of water, and add solid $NaBD_4$ to 1*M*. Incubate at 37°C for 30 min.
7. For quantitative lactonization (and therewith C_6 reduction), repeat steps 4–6. (**Caution:** Addition of the HCl may cause strong bubbling [second round step 4]).
8. Add 5*N* acetic acid until hydrogen development stops, and apply sample onto a 1-mL column of AG50Wx 8 (H^+ form) in water. Wash column with 5 mL of water, combine all eluents, and dry by evaporation.
9. Add 2 mL of methanol and evaporate. Repeat two times more to remove borate ions as volatile methyl borate.
10. Add 100 µL of pyridine and 100 µL of acetic anhydride, and incubate at 100°C for 30 min.
11. Concentrate the sample to 2–5 µL in a stream of nitrogen, add 200 µL of acetyl acetate, and concentrate as described earlier.
12. Analyze 1–2 µL by GC-MS.
13. Compare retention and mass spectrum of sample and standard prepared from idose using steps 3 and 9–12 of this protocol.

3.5. Identification of 3-O-Methylgalacturonic Acid

3.5.1. Identification as C6-D2-1,2,4,5,6-Penta-O -Acetyl-3-O-Methylhexitol by GC-MS

1. Perform steps 1–12 of the protocol described under Section 3.4. (*see* Note 9).
2. Compare the mass spectrum of the sample with those of partially methylated alditol acetates. The two deuterium atoms at C6 of the compound identify the sample as an uronic acid and allow one to distinguish between a 3- or 4-methyl hexuronic acid.

3.5.2. Demethylation with BBr3 and Identification as Hexa-O-Acetyl-Galactitol by GC-MS

1. Perform steps 1–11 of the protocol described under Section 3.4.
2. Add 200 µL of methylene chloride and 1 mL of BBr_3. Incubate at room temperature overnight.
3. Evaporate solvent, and dissolve sample in 1 mL of water. Purify sample by passage through a 1-mL column of AG50Wx8 ion-exchange material.
4. Reacetylate and analyze sample as in steps 12–13 of the protocol described under Section 3.4.
5. The disappearance of the peak corresponding to the 3-*O*-methyl-hexuronic acid and the appearance of a peak corresponding to galactose reveal the sample to be a 3-*O*-methylgalacturonic acid.

4. Notes

1. Neutral sugar-uronic acid- and *N*-acylaminosugar-containing saccharides will be found in the flow-through of this column. The oligosaccharides now contain their reducing end sugar in a reduced form. Any compound with a dischargeable positive charge (all amino acids, proteins, most peptides, and free amino sugars) will strongly bind to the column.
2. Proteins and many peptides will be degraded by the alkaline conditions needed for β-elimination; however, in some cases intact peptide cores can be recovered and used for further characterization. Even degraded material may well be worth recovering, because reductive β-elimination as described here will convert serine residues to alanine and threonine residues to α-aminobutyric acid. Quantitative analyses of these amino acids will allow discrimination between seryl- and threonyl-linkage units, and help determine the stoichometry between *O*-linked carbohydrate and glycoprotein. Favorably, elution with NH_4OH after rotary evaporation yields a completely salt-free sample.
3. Solvolysis with HF was first described in *(5)* for the analysis of Elastins.
4. Hydrogen fluoride is very hazardous to the respiratory system and the skin. Therefore, all manipulations should be performed under a fume hood and with strong rubber gloves. Any skin areas of accidental spillage should be washed under cold running water for 5 min and then calcium gluconate gel applied and medical help sought.
5. No special precautions have to be taken to exclude humidity during hydrogen fluoride condensation. However, excessive condensation of water during the removal of hydrogen fluoride should be avoided (by keeping the stream of nitrogen gentle, but effective). The aqueous

hydrogen fluoride produced this way might result in the unwanted hydrolysis of peptide- and *N*-glycosyl bonds.

6. The method requires at least 1 µg of individual sugar. It is adapted from a procedure describing analysis of sugars as trifluoroacetyl derivatives *(6)*, and allows analysis of uronic acids, amino and neutral sugars in a single experiment (Fig. 2). The use of pentafluoropropionic anhydride instead of trifluoroacetic anhydride allows the most prominent derivatives of glucose and glucuronic acid to be separated on the Durabond 1701 column. Alternative sugar analysis that involves derivatization to alditol acetates requires additional steps for uronic acid reduction *(7)* and amino sugar acetylation. Also, since uronic acids are reduced to the same alditol acetates as the corresponding neutral sugars, a control experiment that reveals the composition of neutral sugars only may be necessary for analysis by GLC. In the method described, hydrolysis of saccharides results in the formation of methylglycosides of monosaccharides. In addition, the carboxyl group of uronic acids is transformed to the the corresponding methylester. Derivatizing hexoses and pentoses by this method theoretically can yield four different configuration isomers: α- or β-methylglycopyranoside or furanoside. This can be observed in some cases, but more often two or three isomers are formed. The various isomers obtained need not be characterized individually, but are used as a set to identify the sample. Although this is a very unambiguous way of identifying a sugar, the occurrence of more than one signal for one compound makes this method less accurate for quantitation. It therefore is particularly recommended as a qualitative analysis.

   Care has to be taken that the sample is dry before treatment with HCl in methanol to prevent hydrolysis to free sugars. Also, sealing the glass tube is a step where possible failure may occur. Indicative for a leaky ampule is loss of solvent and—in Section 3.3., step 4—loss of sample during the incubations. It is therefore recommended that the intactness of the ampule be secured before incubation by simply checking for solvent smell. The distinct smell of pentafluoropropionic anhydride is especially revealing.
7. The method described requires that at least 2 µg of iduronic acid be present. This is similar to published protocols *(7,8)* but uses hydrolysis of saccharides by HCl in methanol, which is less destructive for uronic acids than the commonly used hydrolysis in 4*N* trifluoroacetic acid. Alternatively, hydrolysis according to ref. *9* can be applied if iduronic acid is suspected to be a constituent of the sample.

Iditol acetate derived from iduronic acid by this method contains two deuterium atoms introduced at C6 by the reduction of idurono lactone with sodium borodeuteride. Therefore, if GC-MS equipment is available, the reduced derivative can readily be distinguished from iditol acetate derived from idose. If only GC equipment is available, a control experiment analyzing the idose content of the sample (Section 3.4., steps 1–3 and 9–12) has to be performed.

8. Care has to be taken that HCl used to form uronic acid lactones in steps 4 and 5 is completely removed before adding $NaBD_4$. Checking the pH before $NaBD_4$ addition is therefore advisable. Also, there should be some $NaBD_4$ left after the incubation; that is, there should be D_2 development at the subsequent acid addition.
9. The method requires at least 5 µg of methyl galacturonic acid. The demethylation procedure was adapted from ref. *10.* Otherwise, the information given under Section 3.4. applies.

References

1. Paul, G., Lottspeich, F., and Wieland, F. (1986) Asparaginyl-*N*-acetylgalactosamine: Linkage unit of halobacterial glycosaminoglycan. *J. Biol. Chem.* **261,** 1020–1024.
2. Lechner, J., Wieland, F., and Sumper, M. (1985) Transient methylation of dolichyl oligosaccharides is an obligatory step in halobacterial sulfated glycoprotein synthesis. *J. Biol. Chem.* **260,** 8694–8699.
3. Paul, G. and Wieland, F. (1987) Sequence of the halobacterial glycosaminoglycan. *J. Biol. Chem.* **262,** 9587–9593.
4. Lechner, J. and Wieland, F. (1989) Structure and biosynthesis of prokaryotic glycoproteins. *Annu. Rev. Biochem.* **58,** 173–194.
5. Wieland, F., Heitzer, R., and Schäfer, W. (1983) Asparaginyl-glucose: Novel type of carbohydrate linkage. *Proc. Natl. Acad. Sci. USA* **80,** 5470–5474.
6. Mort, A. G. and Lamport, D. T. A. (1977) Anhydrous hydrogen fluoride deglycosylates glycoprotein. *Anal. Biochem.* **82,** 289–309.
7. Zannetta, J. B., Breckenridge, W. C., and Vincendon, G. (1972) Study of the carbohydrates of glycoproteins. *J. Chromatogr.* **69,** 291–304.
8. Lehrfeld, J. (1981) Differential gas-liquid chromatography method for determination of uronic acids in carbohydrate mixtures. *Anal. Biochem.* **115,** 410–418.
9. Spiro, R. G. (1972) Study of the carbohydrates of glycoproteins. *Methods Enzymol.* **28,** 3–43.
10. Lindhal, U. and Axelsson, O. (1971) Identification of iduronic acid as the major sulfated uronic acid in heparin. *J. Biol. Chem.* **246,** 74–82.
11. Hough, L. and Theobald, R. S. (1963) De-*O*-methylation with hydrobronic acid and boron tribromide. *Methods Carbohydr. Res.* **2,** 203–206.

CHAPTER 10

Isolation and Fractionation of Glycoprotein Glycans in Small Amounts

Rudolf Geyer and Hildegard Geyer

1. Introduction

The separation of highly diverse mixtures of glycoprotein glycans is often one of the major problems arising during carbohydrate structure analysis of glycoproteins. In the course of our studies on viral glycoproteins, we have established a fractionation procedure allowing the isolation of individual oligosaccharides at microscale. In order to facilitate the detection of glycoproteins and corresponding glycans during isolation and purification, they were biosynthetically labeled in their carbohydrate moieties by cultivation of virus-infected cells in the presence of [2-^{3}H]Man or [6-^{3}H]GlcN. The separation procedure used is based on sequential liberation of the different types of sugar chains present and their subsequent fractionation according to charge, size, and/or isomeric structure by a combination of different chromatographic techniques.

The purified radiolabeled glycoprotein is digested with a specific protease, such as trypsin. Resulting glycopeptides are desalted by gel filtration and subsequently treated with endo-β-*N*-acetylglucosaminidase H (endo H). The high-mannose and/or hybrid-type *N*-glycans released are separated from endo-H-resistant glycopeptides by reversed-phase HPLC (step 1 in Scheme 1), reduced with sodium borohydride, fractionated by high-pH anion-exchange (HPAE) chromatography (step 2), and eventually subfractionated by HPLC on a LiChrosorb-Diol

From: *Methods in Molecular Biology, Vol. 14: Glycoprotein Analysis in Biomedicine*
Edited by: E. F. Hounsell

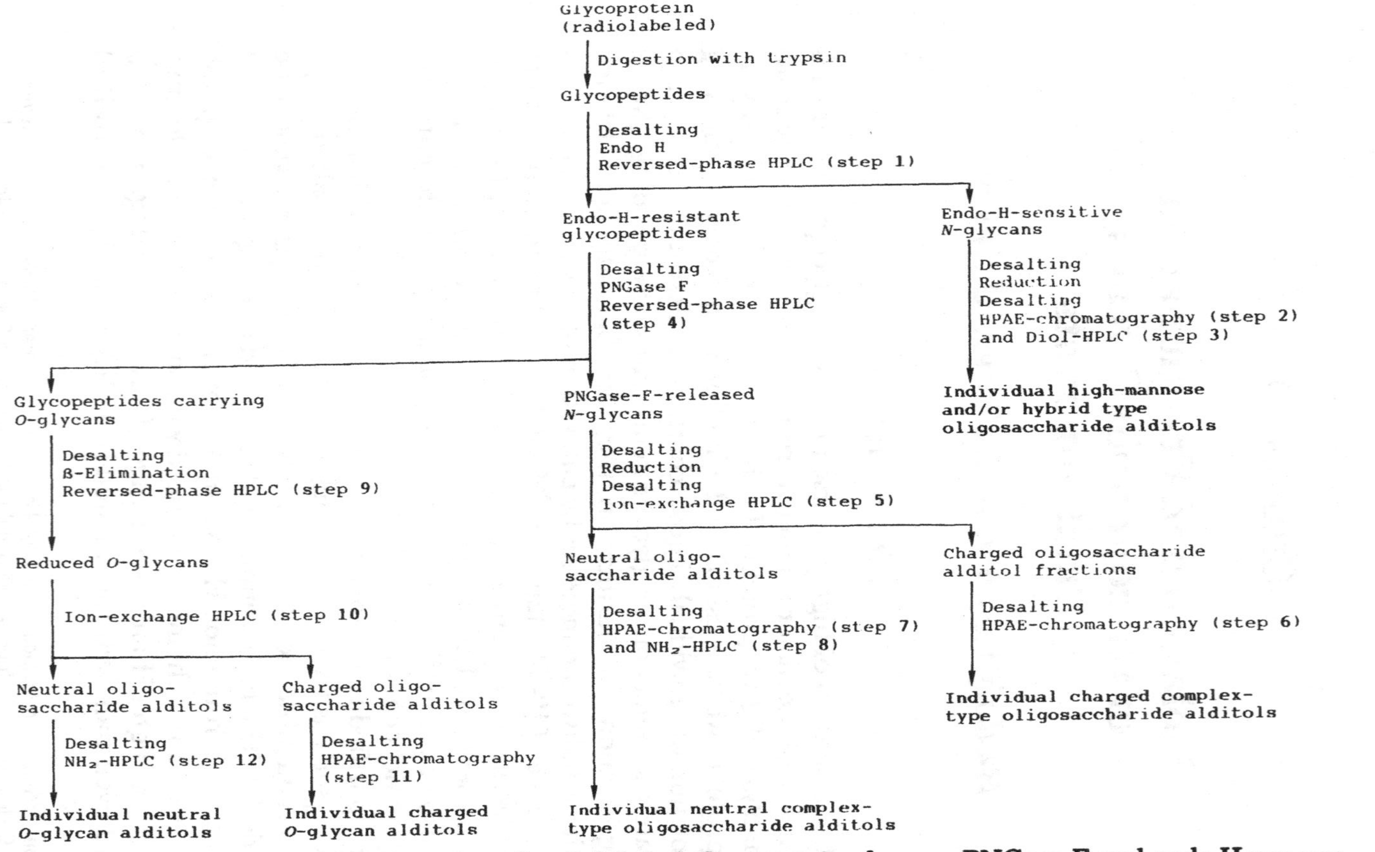

Scheme 1. Liberation and fractionation of radiolabeled glycoprotein glycans. PNGase F and endo H are peptide-N^4-(*N*-acetylglucosaminyl) asparagine amidase F and endo-β-*N*-acetylglucosaminidase H.

column (step 3). Individual high-mannose and/or hybrid-type oligosaccharide alditols representing usually single structural isomers are thus obtained.

Endo-H-resistant glycopeptides are treated with peptide-N^4-(*N*-acetyl-β-glucosaminyl)asparagine amidase F (PNGase F). Oligosaccharides liberated are separated from residual peptides and glycopeptides carrying *O*-glycans by reversed-phase HPLC (step 4). Resulting *N*-glycans are reduced, desalted, and separated according to charge by ion-exchange HPLC (step 5). Fractions containing equally charged oligosaccharides are subfractionated by HPAE chromatography (step 6). Neutral glycans obtained are separated by HPAE chromatography (step 7) and/or HPLC on a LiChrosorb-NH_2 column at pH 5.3 (step 8). In general, homogeneous complex-type oligosaccharides can thus be isolated.

O-glycans present are liberated from PNGase-F-resistant glycopeptides by β-elimination, separated from remaining peptides by reversed-phase HPLC (step 9), and fractionated by ion-exchange HPLC (step 10). Charged oligosaccharides are subfractionated by HPAE chromatography (step 11), whereas neutral species are separated by NH_2-HPLC (step 12).

In combination with authentic, radiolabeled oligosaccharide standards, the method described allows a detailed chromatographic profiling of the carbohydrate moieties present and may be generally applied to (metabolically labeled) glycoproteins available in small amounts only (*see* Note 1). Depending on the amount of starting material, oligosaccharide fractions obtained can be subsequently subjected to structural analysis.

2. Materials

1. Trypsin from bovine pancreas (sequencing grade; Boehringer Mannheim, Mannheim, FRG) dissolved in 1 m*M* HCl (5–10 μg/μL). The solution can be used for approx 1 wk, if stored at –20°C.
2. Trypsin digestion buffer: 100 m*M* Tris-HCl buffer, pH 8.5. The buffer should be filtered through a 0.45-μm filter and can be used for several months, if stored at 4°C.
3. Column (1 × 25 cm) of Bio-Gel P-2 (–400 mesh; BioRad, Munich, FRG).
4. Endo-H buffer: 0.05*M* sodium citrate-phosphate buffer, pH 5.0, 0.1*M* NaCl, 0.02% (w/v) sodium azide. The buffer is filtered and stored as described for trypsin digestion buffer.

5. Endo-β-*N*-acetylglucosaminidase H (endo H) of *Streptomyces plicatus* from *Streptomyces lividans* (Boehringer Mannheim, Mannheim, FRG) dissolved in endo-H buffer (1 mU/μL). Portions of approx 100 μL are stored at –20°C. The enzyme solution can be frozen and thawed several times without significant reduction of enzyme activity.
6. PNGase-F buffer: 20 m*M* sodium phosphate buffer, pH 7.2, 50 m*M* EDTA, 0.02% (w/v) sodium azide. The buffer is filtered and stored as previously mentioned.
7. Peptide-N^4-(*N*-acetyl-β-glucosaminyl)asparagine amidase F (PNGase F) from *Flavobacterium meningosepticum* in 20 m*M* potassium phosphate buffer, pH 7.2, 50 m*M* EDTA, 0.05% (w/v) sodium azide (approx 2 mU/μL) as supplied by the manufacturer (*N*-glycosidase F; Boehringer Mannheim, Mannheim, FRG).
8. Solution of sodium borohydride (0.8*M*) in 50 m*M* aqueous NaOH. Use solid NaOH (suprapur; Merck, Darmstadt, FRG) and bidistilled water for preparation of 50 m*M* NaOH, and store in quartz flask. Dissolve solid $NaBH_4$ directly before use.
9. Aqueous solution of sodium borohydride (10 mg/mL). Prepare fresh each time.
10. Methanol containing 1% (v/v) acetic acid.
11. Reversed-phase HPLC: 25 m*M* ammonium acetate buffer, pH 6 (eluant A), and 50% (v/v) acetonitrile in 25 m*M* ammonium acetate buffer, pH 6 (eluant B), prepared as follows: Add 3 mL of 96% (w/w) aqueous acetic acid (suprapur; Merck, Darmstadt, FRG) to approx 700 mL of bidistilled water. Adjust pH to 6.0 by dropwise addition of 25% (w/w) aqueous ammonia. Fill up to 1 L with bidistilled water. Take 500 mL of this buffer, and dilute with 500 mL of bidistilled water to prepare eluant A. Combine remaining 500 mL of buffer with 500 mL HPLC-grade acetonitrile for preparation of eluant B. Degas eluants by saturation with helium, and store the solutions in closed, pressurized vessels with helium gas sparging. Use column (250 × 4.6 mm) of silica modified with octadecylsilyl groups (ODS-Hypersil; 3 μm; Shandon, Runcorn, GB).
12. HPAE chromatography: Prepare a 50% (w/w) solution of NaOH in bidistilled water thoroughly degassed by helium sparging in order to minimize formation of sodium bicarbonate. Filter carefully through a 30-μm Pyrex filter funnel into a plastic flask. To prepare eluant A (100 m*M* aqueous NaOH), add 5.2 mL of 50% NaOH to 1 L of bidistilled degassed water. Eluant B (100 m*M* NaOH, 250 m*M* sodium acetate) is composed by adding the same amount of 50% NaOH to 1 L of 250 m*M* aqueous sodium acetate prepared with degassed water in a closed measuring flask. Both solutions are stored in closed,

pressurized vessels with helium sparging. Use CarboPac PA-1 column (250 × 4.6 mm; Dionex, Sunnyvale, CA, USA) in series with a CarboPac PA guard column.

13. Diol-HPLC: Bidistilled water and HPLC-grade acetonitrile are degassed and stored in pressurized vessels as described in Section 2., step 11. Use a diol-modified silica column (150 × 4 mm; LiChrosorb-Diol, 5 µm; Merck, Darmstadt, FRG).
14. Anion-exchange HPLC: Bidistilled water (eluant A) and aqueous 500 m*M* potassium phosphate buffer, pH 4.4 (eluant B), are filtered through a 0.45-µm filter, degassed, and stored in pressurized vessels as described in Section 2., step 11. Use a column (250 × 4.6 mm) of Micropak AX-5 (5 µm; Varian, Walnut Creek, CA, USA).
15. NH_2-HPLC: Aqueous 15 m*M* potassium phosphate buffer, pH 5.3 (eluant B), is prepared with bidistilled water and filtered through a 0.45-µm filter. Eluants A (acetonitrile), B, and C (bidistilled water) are degassed and stored in pressurized vessels as detailed earlier. Use amine-modified silica column (250 × 4.6 mm; LiChrosorb-NH_2, 5 µm; Merck, Darmstadt, FRG).
16. HPLC equipment: A gradient pump equipped with a solvent conditioner, an injector, and a fraction collector are needed for all HPLC separations. For reversed-phase HPLC, a column oven, a photometer, and an integrator are required in addition.
17. Equipment for HPAE chromatography: Dionex (Sunnyvale, CA, USA) BioLC system consisting of a gradient pump (GPM II), an eluant degas module (EDM II), a microinjection valve (LCM-2), and a pulsed amperometric detector (PAD II) containing a gold electrode to which potentials of E_1 0.1 V, E_2 0.6 V, and E_3 0.8 V are applied for duration times T_1 480 ms, T_2 120 ms, and T_3 60 ms, respectively. Chromatographic data are plotted by an integrator.
18. Liquid scintillation counting: Scintillation cocktail for aqueous solutions and a liquid scintillation counter.

All other chemicals used are purchased from the usual commercial sources in the highest grade of purity available.

3. Method

3.1. Trypsin Digestion (see *Note 2*)

1. Dissolve purified radiolabeled glycoprotein in 0.1–0.5 mL of trypsin digestion buffer in a conical Pyrex™ glass tube with ground-glass stopper. In the case of submicrogram amounts of glycoprotein, combine the samples with 500 µg of bovine serum albumin (dissolved in 100 µL of the same buffer) as carrier.

2. Add trypsin solution to a final protein:enzyme ratio of 40:1 (by weight).
3. Incubate the sample for 4 h at 37°C, boil for 3 min, cool in an ice bath, and store at –20°C until further use.

3.2. Desalting of Glycopeptides and Oligosaccharides

1. Load glycopeptide or oligosaccharide samples (100–200 µL) onto the Bio-Gel P-2 column.
2. Carry out the chromatography using bidistilled water as eluant at hydrostatic pressure.
3. Collect fractions (1 mL) at a flow rate of approx 4.5 mL/h.
4. Monitor aliquots for radioactivity by liquid scintillation counting.
5. Pool and lyophilize fractions containing radiolabeled material eluting at (or immediately after) the void volume of the column.

3.3. Liberation of N- and O-Glycans

1. Dissolve desalted glycopeptides in 50–100 µL endo H buffer in a conical Pyrex glass tube with ground-glass stopper.
2. Add 9 µL (9 mU) of endo-H solution, and incubate for 3 h at 37°C.
3. Repeat the enzyme addition, and incubate for a further 20 h at 37°C.
4. Boil for 3 min, cool in an ice bath, and apply the sample to a reversed-phase HPLC column (Section 3.5.).
5. Dissolve the endo-H-resistant glycopeptides obtained after reversed-phase HPLC, desalting (*see* Section 3.2.), and lyophilization in 50–100 µL PNGase-F buffer in a conical ground-glass stoppered Pyrex glass tube.
6. Add 10 µL of PNGase-F solution (2 mU), and incubate the sample at 37°C for 24 h.
7. Repeat the enzyme addition, and incubate for a further 24 h at 37°C.
8. Boil for 3 min, cool in an ice bath, and apply the sample to a reversed-phase HPLC column (Section 3.5.).
9. Desalt and lyophilize PNGase-F-resistant glycopeptides obtained after reversed-phase HPLC as previously described.
10. To release *O*-linked glycans, dissolve the glycopeptides in 200 µL of aqueous 0.8*M* $NaBH_4$/50 m*M* NaOH in a Pyrex glass tube with ground-glass stopper.
11. Incubate samples at 37°C for 68 h, acidify (to pH 5) by dropwise addition of 2*M* aqueous acetic acid, and dry by rotary evaporation.
12. Add 1 mL of methanol containing 1% (v/v) acetic acid, and evaporate under a stream of nitrogen. Repeat this procedure four times to remove boric acid as its volatile methyl borate.
13. Dissolve the residue in about 100 µL of bidistilled water, and subject to reversed-phase HPLC (Section 3.5.).

3.4. Reduction of Oligosaccharides

1. Desalt (*see* Section 3.2.) and lyophilize oligosaccharides obtained after reversed-phase HPLC (steps 1 and 4 in Scheme 1).
2. Dissolve the oligosaccharides in 100 µL of bidistilled water in a ground-glass stoppered Pyrex glass tube with round bottom.
3. Add aqueous sodium borohydride (10 mg/mL) in three portions of 100 µL.
4. After 12 h at room temperature, decompose residual reducing agent by dropwise addition of 2*M* acetic acid to a pH of about 5.
5. Dry the samples at room temperature by rotary evaporation. Then add 1 mL of methanol containing 1% (v/v) acetic acid, and chase the boric acid under a stream of nitrogen. Repeat this procedure four times.
6. Dissolve the residues in 100 µL of bidistilled water, and desalt as described in Section 3.2.

3.5. Reversed-Phase HPLC Separation of Oligosaccharides Obtained After Enzymic or Chemical Deglycosylation from Residual (Glyco)peptides (see *Note 3*)

1. Dissolve the samples in 50–250 µL of eluant A, and inject onto the HPLC column (*see* Note 4).
2. Elute the column at 60°C using a linear gradient from 100% eluant A to 100% eluant B in 90 min.
3. Collect fractions (1 mL) at a flow rate of 1 mL/min, and monitor for radioactivity.
4. Reequilibrate the column at starting conditions for 30 min.
5. Isolate oligosaccharides being eluted at (or close to) the void volume of the column as well as remaining glycopeptides according to the distribution of radioactivity and the profile of continuous-flow monitoring of absorbance at 220 nm.
6. Lyophilize and desalt the pooled fractions as previously described.

3.6. Fractionation of Endo-H-Sensitive Oligosaccharide Alditols by HPAE Chromatography

1. Dissolve the sample in bidistilled water (10–150 µL), and inject at room temperature (step 2 in Scheme 1).
2. Elute the column by using a gradient from 100% eluant A to 88% eluant A/12% eluant B in 36 min followed by a further increase of eluant B up to 42% within 24 min.
3. Collect fractions (250 µL) at a flow rate of 1 mL/min, and monitor for radioactivity.

4. Wash the column with 100% B for 10 min, and reequilibrate at 100% A for 20 min.
5. Pool oligosaccharide alditols according to the distribution of radioactivity and to the detector response.
6. Lyophilize and desalt as described earlier.

3.7. Subfractionation of Endo-H-Sensitive Oligosaccharide Alditols by Diol-HPLC (see *Note 5*)

1. Dissolve individual oligosaccharide alditol fractions obtained after HPAE chromatography (step 2) in 50–200 μL of 75% (v/v) aqueous acetonitrile, and inject at room temperature onto the column eluted in the same solvent.
2. Collect fractions (about 350 μL) at a flow rate of 0.5 mL/min, and monitor for radioactivity. In order to avoid chemical quenching by acetonitrile, dry fractions by speedvac concentration, and redissolve in 100 μL of water prior to addition of cocktail. (In preparative runs, oligosaccharide subfractions are pooled according to the profile of radioactivity, lyophilized, and freed from residual silica material by gel filtration.)

3.8. Fractionation According to Negative Charge by Anion Exchange HPLC (see *Note 6*)

1. Dissolve desalted PNGase-F-released oligosaccharide alditols obtained after reversed-phase HPLC (Section 3.5; step 6) and reduction or reduced *O*-glycans isolated after reversed-phase HPLC (Section 3.5., step 6) in 50–250 μL bidistilled water.
2. Inject the sample at room temperature onto a Micropak AX-5 column equilibrated with 99% eluant A and 1% eluant B.
3. Maintain the concentration of B at 1% for 4 min, and then raise this to 60% within 60 min.
4. Between each run, reequilibrate the column at starting conditions for 30 min.
5. Collect fractions (400 μL) at a flow rate of 1 mL/min, monitor for radioactivity, pool according to the profile of radioactivity, and desalt as described in Section 3.2. (*see* Note 7).

3.9. Subfractionation of Charged Oligosaccharide Alditols by HPAE Chromatography (see *Notes 8 and 9*)

Use the same experimental conditions as given in Section 3.6., except that the concentration of eluant B is linearly raised up to 50% within 100 min.

3.10. Fractionation of Neutral PNGase-F-Released Oligosaccharide Alditols by HPAE Chromatography and NH_2-HPLC

1. Follow the procedure given in Section 3.6., but use a linear gradient from 0% B to 24% B in 72 min.
2. Desalt individual oligosaccharide alditol fractions obtained as described in Section 3.2.
3. Dissolve aliquots in 50–200 µL of a mixture of 60% (v/v) eluant A/40% eluant B, and inject at room temperature onto a LiChrosorb-NH_2 column.
4. Elute the column using a linear gradient from 60% A/40% B to 40% A/60% B in 90 min.
5. Collect fractions (400 µL) at a flow rate of 1 mL/min, and monitor for radioactivity.
6. Between each run, reequilibrate the column at starting conditions for 30 min.
7. When further separation is achieved under these conditions, subject corresponding glycan samples to preparative NH_2-HPLC. Pool oligosaccharide subfractions according to the profile of radioactivity, lyophilize, and desalt.

3.11. Fractionation of Neutral O-Glycan Alditols by NH_2-HPLC (see Note 10)

1. Desalt neutral oligosaccharide alditols obtained after β-elimination, reversed-phase, and ion-exchange HPLC (Section 3.3., steps 10–13 and Section 3.8.).
2. Dissolve the samples in 50–200 µL of 75% (v/v) aqueous acetonitrile.
3. Elute the column using a gradient ranging from 75 eluant A/25% eluant C to 50% A/50% C in 90 min.
4. Collect fractions (about 350 µL) at a flow rate of 1 mL/min, and monitor for radioactivity.
5. Between each run, reequilibrate the column at starting conditions for 30 min.
6. Pool the oligosaccharide alditols according to the profile of radioactivity, lyophilize, and free from residual silica material by gel filtration using a Bio-Gel P-2 column (*see* Section 3.2.).

4. Notes

1. Metabolic labeling of the glycoproteins is a prerequisite for following this procedure. *N*-glycans can be efficiently radiolabeled with [2-^{3}H]Man or [6-^{3}H]GlcN, whereas *O*-glycans are labeled best with [^{3}H]GlcN. If the glycoprotein cannot be radiolabeled biosynthetically, oligosac-

charides released by endo H or PNGase F have to be labeled by reduction with sodium (or potassium) [^{3}H]borohydride as follows: Desalted, lyophilized samples are dissolved in 50–100 µL 10 m*M* aqueous sodium borate buffer, pH 9, in conical Pyrex glass tubes with ground-glass stoppers. After addition of 50–100 µL 10 m*M* NaOH containing 18.5–37 × 10^6 Bq of NaB^3H_4 (approx 5 × 10^{11} Bq/mmol; DuPont de Nemours, Bad Homburg, FRG), samples are incubated at room temperature for 3 h. Three portions of 100 µL of aqueous $NaBH_4$ (10 mg/mL) are added, and samples are worked up as described in Section 3.4. Acidification of samples, however, has to be carried out in a well ventilated hood in this case.

2. PNGase F liberates asparagine-linked glycans only when both the α-amino and carboxyl groups of the asparagine residue are in peptide linkages *(1)*. When tryptic cleavage sites (i.e., arginine or lysine residues) are located adjacent to asparagine residues of potential *N*-glycosylation sites, other proteolytic enzymes, such as *Staphylococcus aureus* V8 protease (endoproteinase Glu-C) or protease from *Pseudomonas fragi* mutant (endoproteinase Asp-N), have to be used to degrade the glycoprotein. This aspect has to be kept in mind when glycoproteins are studied, the amino acid sequences of which are not known. The presence of PNGase-F-resistant glycopeptides does not necessarily point to *O*-glycans, but may be owing to glycosyl asparagines with free α-amino groups. In these cases, chemical methods (hydrazinolysis, alkaline hydrolysis) have to be used to release residual *N*-glycans.
3. Oligosaccharides released by endo H may be alternatively separated from endo-H-resistant glycopeptides by gel filtration using a column (60 × 1.6 cm) of Bio-Gel P-4 (–400 mesh) at 55°C and 0.02% aqueous sodium azide as eluant at a flow rate of 0.3 mL/min. In this case, subsequent desalting of glycans and glycopeptides can be omitted. Sialylated hybrid-type oligosaccharides, however, are eluted together with glycopeptides at the void volume of the column.
4. Reversed-phase HPLC is carried out at pH 6, since previous studies have demonstrated that chromatographic separation at pH 2 (or below) results in the partial loss of fucose and sialic acid residues *(2)*.
5. Endo-H-sensitive oligosaccharide alditol fractions obtained after HPAE chromatography (step 2 in Scheme 1) are routinely analyzed by Diol-HPLC (step 3) at analytical scale. In some cases, a combination of HPAE chromatography with Diol-HPLC is required for separation of structural isomers of high mannose-type glycans *(3)*.

6. Instead of using a column of Micropak AX-5 (cf steps 5 and 10), fractionation of glycans according to their negative charge can be similarly performed by medium pressure anion-exchange chromatography on Mono Q^R *(4)* or by HPAE chromatography *(5)*.
7. Neutral oligosaccharide fractions obtained after steps 5 and 10 can be alternatively desalted by mixed-bed ion-exchange resins (e.g., Amberlite AG MB-3; Serva, Heidelberg, FRG).
8. The experimental conditions given for HPAE chromatography of charged glycans (steps 6 and 11) allow the subfractionation of monosialylated up to tetrasialylated oligosaccharides under the same conditions. In order to reduce retention times of tri- and tetrasialylated species, higher initial concentrations of sodium acetate can be employed. The steepness of the gradient, however, should be maintained at an increase of 0.5% eluant B/min. Application of nonlinear sodium acetate gradients *(5)* or NaOH gradients *(6)* may further improve the separation of negatively charged glycoprotein glycans. Alternatively, sialylated glycans may be also subfractionated by conventional HPLC using an amine-modified silica column at pH 7 *(7)*.
9. For further analysis, charged glycans obtained after ion-exchange HPLC (steps 5 and 10) or HPAE chromatography (steps 6 and 11) can be treated with sialidase. Subsequent analysis of the reaction products by ion-exchange HPLC allows conclusions to be made concerning the presence of negatively charged substituents other than sialic acid (e.g., sulfate or phosphate groups). Core structures of neutral asialo-oligosaccharide alditols obtained can be again characterized by HPAE chromatography and NH_2-HPLC (cf steps 7, 8, and 12).
10. Instead of using the NH_2-HPLC (step 12) conditions described by Lamblin et al. *(8)*, separation of neutral, reduced *O*-glycans might be also possible by HPAE-chromatography when NaOH and sodium acetate gradients are employed in the same run *(9)*.

References

1. Tarentino, A. L., Gomez, C. A., and Plummer, T. H., Jr. (1985) Deglycosylation of asparagine-linked glycans by peptide:*N*-glycosidase F. *Biochemistry* **24**,4665–4671.
2. Schlüter, M., Linder, D., and Geyer, R. (1985) Isolation of glycopeptides containing individual glycosylation sites of Friend murine leukemia virus glycoprotein: Studies on glycosylation by methylation analysis. *Carbohydr. Res.* **138**,305–314.

3. Pfeiffer, G., Geyer, H., Kalsner, I., Wendorf, P., and Geyer, R. (1990) Separation of glycoprotein-*N*-glycans by high-pH anion-exchange chromatography. *Biomed. Chromatogr.* **4**, 193–199.
4. Van Pelt, J., Damm, J. B. L., Kamerling, J. P., and Vliegenthart, J. F. G. (1987) Separation of sialyloligosaccharides by medium pressure anion-exchange chromatography on Mono Q. *Carbohydr. Res.* **169**, 43–51.
5. Townsend, R. R., Hardy, M. R., Cumming, D. A., Carver, J. P., and Bendiak, B. (1989) Separation of branched sialylated oligosaccharides using high-pH anion-exchange chromatography with pulsed amperometric detection. *Anal. Biochem.* **182**, 1–8.
6. Townsend, R. R., Hardy, M. R., Hindsgaul, O., and Lee, Y. C. (1988) High-performance anion-exchange chromatography of oligosaccharides using pellicular resins and pulsed amperometric detection. *Anal. Biochem.* **174**, 459–470.
7. Van Pelt, J., Van Kuik, A. J., Kamerling, J. P., Vliegenthart, J. F. G., Van Diggelen, O. P., and Galjaard, H. (1988) Storage of sialic acid-containing carbohydrates in the placenta of a human galactosialidosis fetus. Isolation and structural characterization of 16 sialyloligosaccharides. *Eur. J. Biochem.* **177**, 327–338.
8. Lamblin, G., Boersma, A., Lhermitte, M., Roussel, P., Mutsaers, H. G. M., Van Halbeek, H., and Vliegenthart, J. F. G. (1984) Further characterization, by a combined high-performance liquid chromatography/^{1}H-NMR approach, of the heterogeneity displayed by the neutral carbohydrate chains of human bronchial mucins. *Eur. J. Biochem.* **143**, 227–236.
9. Hardy, M. R. and Townsend, R. R. (1989) Separation of fucosylated oligosaccharides using high-pH anion-exchange chromatography with pulsed amperometric detection. *Carbohydr. Res.* **188**, 1–8.

Chapter 11

High-Mannose Chains of Mammalian Glycoproteins

Anne P. Sherblom and Rosalita M. Smagula

1. Introduction

High-mannose oligosaccharides are commonly associated with the glycoproteins of most living organisms, and a variety of techniques are available for determining the relative amounts of high-mannose and complex oligosaccharides *(1–4)*. The high-mannose chains of mammalian glycoproteins are often comprised of a series of oligosaccharides having the composition $Man_{5\text{-}8}GlcNAc_2$ (Man5-8). Although the relative amounts of each class of high-mannose oligosaccharide (i.e., Man5, Man6, and so on) have been examined during processing of some glycoproteins, relatively little work on the high-mannose "profile" of mature glycoproteins has been carried out.

The technique described in this chapter was developed in order to determine whether the profile of high-mannose chains in the human urinary glycoprotein uromodulin (Tamm-Horsfall glycoprotein) changes during pregnancy. This glycoprotein contains Man5, Man6, and Man7 when isolated from a nonpregnant source *(5,6)*. During pregnancy, the immunomodulatory activity of this glycoprotein increases *(7)*, and the activity can be recovered in a high-mannose oligosaccharide *(8)*. High-mannose chains isolated from other sources also have immunomodulatory activity, which increases with chain length *(9)*. The technique described herein yields a reproducible high-mannose profile, since duplicate analyses of a single sample differ by no more than 5% in the relative proportion of each component and by no more

From: *Methods in Molecular Biology, Vol. 14: Glycoprotein Analysis in Biomedicine*
Edited by: E. F. Hounsell

than 20% in total recovery *(8)*. The high-mannose profile of pregnant samples is significantly different from that of nonpregnant samples, showing a reduction in total high-mannose chains and a skewing toward longer chains *(8)*.

The technique involves release of high-mannose chains by endoglycosidase H (endo-H) *(10)*, following a predigestion with Pronase. Pronase is used, since the high-mannose chains of uromodulin are resistant to endo-H treatment alone *(8)*. Conditions of extraction *(11)* and labeling with NaB^3H_4 *(12)* are taken from previously described methods. Labeled high-mannose chains are isolated by Concanavalin A (Con A) affinity chromatography *(13,14)*. Following desalting by Bio-Gel P-2 chromatography, the profile of labeled high-mannose chains is determined by high-performance liquid chromatography (HPLC). This analysis yields the profile, or relative content of Man5, Man6, Man7, and Man8. Soybean agglutinin, which contains Man9 *(15)*, may be utilized as an internal standard for samples that lack Man8 or Man9, allowing quantitative comparison of high-mannose chains from different preparations.

2. Materials

1. High-performance liquid chromatograph, single pump, with detector capable of monitoring UV absorbance at 195 nm.
2. Glyco-Pak N column (Waters, Milford, MA).
3. Con A agarose (Sigma C-7911, St. Louis, MO).
4. Bio-Gel P-2 (Bio-Rad, Richmond, CA).
5. Protease buffer: 10 m*M* Tris-HCl, pH 7.5, 0.01*M* $MgCl_2$.
6. Pronase: *Streptomyces griseus* pronase (Sigma), 10 mg/mL in protease buffer; make fresh.
7. Endo H: 0.1 U/mL of endo-H (Genzyme, Boston, MA) in 50 m*M* sodium citrate, pH 6.0; store frozen.
8. Ethanol (absolute); 75% (v/v) ethanol in distilled water; both solutions should be ice-cold.
9. NaB^3H_4: 5 mCi of ^3H-$NaBH_4$ (sp. act. 600 mCi/mmol; NEN, Boston, MA) mixed with 3 mL 10 m*M* NaOH; store frozen.
10. Labeling buffer: 0.2*M* sodium borate, pH 9.8.
11. 1*M* acetic acid
12. Tris-buffered saline (TBS): 0.15*M* NaCl, 0.01*M* Tris-HCl, pH 8.0, 1 m*M* $CaCl_2$, 1 m*M* $MgCl_2$; store refrigerated.
13. Methyl α-glucopyranoside (Sigma): 10 m*M* in TBS; store refrigerated.
14. Methyl α-mannopyranoside (Sigma): 100 m*M* in TBS; store refrigerated.

15. HPLC mobile phase: 66/34 acetonitrile/H_2O, prepared by mixing 518.5 g HPLC-grade acetonitrile with 339.0 g distilled H_2O, letting solution come to room temperature, and filtering/degassing through a Millipore 0.45-μ FH filter.
16. HPLC sample solution: 1/1 (v/v) mixture of acetonitrile/H_2O, degassed.
17. HPLC standard: standard mixture of reduced high-mannose chains (MSE, Genzyme, Boston, MA); 100 μg mixed with 100 μL distilled water. Store frozen.
18. Scintillation fluid: Ecolume (ICN, Costa Mesa, CA).

3. Methods

3.1. Enzymic Release of High-Mannose Oligosaccharides and Radioactive Labeling

1. Lyophilize the sample, containing approx 10 nmol of high-mannose chains and free of buffers or salts (1 mg of uromodulin, dialyzed against water, was used).
2. Add 100 μL of protease buffer and 5 μL of protease, and incubate at 37°C for 16 h.
3. Heat-inactivate the protease by immersion in a boiling water bath for 5 min (*see* Note 1).
4. Add 100 μL of endo-H solution, and incubate at 37°C for 20 h.
5. To extract the sample, add 630 μL of ice-cold ethanol, mix, and centrifuge at 5000 rpm (Sorvall SS-34 rotor, 3000*g*) for 10 min.
6. Save the supernatant, wash pellet with 0.5 mL ice-cold 75% ethanol, pool the supernatants, and evaporate to dryness under nitrogen (*see* Note 2).
7. Radiolabel the released oligosaccharides by the addition of 0.3 mL of labeling buffer and 20 μL NaB^3H_4 solution to the dried sample, and incubation for 4 h at 30°C.
8. Add 1*M* acetic acid, until the pH is below 5; test the pH by blotting a small amount of the sample onto pH paper with a Pasteur pipet (*see* Note 3).

3.2. Isolation of High-Mannose Oligosaccharides Using Concanavalin A Affinity Chromatography and Desalting by Bio-Gel P-2 Gel Filtration

1. Prepare a 2-mL column of Con A agarose, and equilibrate with TBS at room temperature (*see* Note 4).
2. Apply the entire sample to the column; wash the sample tube with 0.5 mL TBS, and apply to the column.

3. Collect fractions of approx 2 mL. Wash the column with 20 mL TBS, followed by 20 mL of 10 m*M* methyl α-glucopyranoside solution.
4. Elute high-mannose oligosaccharides with 35 mL of 100 m*M* methyl-α-mannopyranoside solution at 60°C. This solution may be warmed in a water bath, and manually applied to the column in 5-mL aliquots.
5. Transfer 0.2 mL of each fraction to a 7-mL scintillation vial, add 5 mL Ecolume (or equivalent scintillation fluid), and determine the radioactivity in a scintillation counter.
6. Pool the fractions comprising the peak of radioactivity that elutes with methyl-α-mannopyranoside, and lyophilize the pooled fractions.
7. Desalt oligosaccharides by chromatography on a 50-mL column of Bio-Gel P-2 equilibrated with distilled water at room temperature: The lyophilized sample in 1–1.5 mL distilled water is applied to the column; fractions of 5 mL are collected (to 100 mL distilled water), counted, pooled, and lyophilized.

3.3. High-Performance Liquid Chromatography

1. Equilibrate a Waters Glyco-Pak N column with a mobile phase of acetonitrile/H_2O 66/34 at a flow rate of 1 mL/min (*see* Note 5).
2. At the beginning of each series of runs, test the system by injecting 10–20 µL of HPLC standard solution and monitoring the absorbance at 195 nm. By examining the chromatogram, you can determine whether Man5, Man6, Man7, and Man8 are sufficiently well separated. If these peaks are not sufficiently well separated, prepare fresh mobile phase of 67/33 or 68/32 acetonitrile/distilled water (v/v), and run the standard solution again (*see* Note 6).
3. Suspend the sample in 200 µL of HPLC sample solution; filter through an HV Millipore 0.45-µ filter connected to a disposable 1-mL syringe.
4. Inject 50 µL of the filtered sample; take 10 µL for scintillation counting, and freeze the remainder of the sample (if another HPLC run is required).
5. Collect fractions of 0.5 mL or 1.0 mL using timed collection. Fractions may be collected in 7-mL scintillation vials. Fraction collection may begin immediately upon injection or may be delayed until shortly before the Man5 peak emerges. Collect fractions until well after the last peak has emerged.
6. Add 5 mL Ecolume to scintillation vials, and determine radioactivity.
7. Calculate the percentage of Man5, Man6, and Man7.
   a. Plot the scintillation data on graph paper that has the same scale as the chromatogram from the UV monitor (for example, 1.0 cm/min).
   b. Align the plot with the chromatogram of standards run on the same day. To determine the calibration distance necessary to align

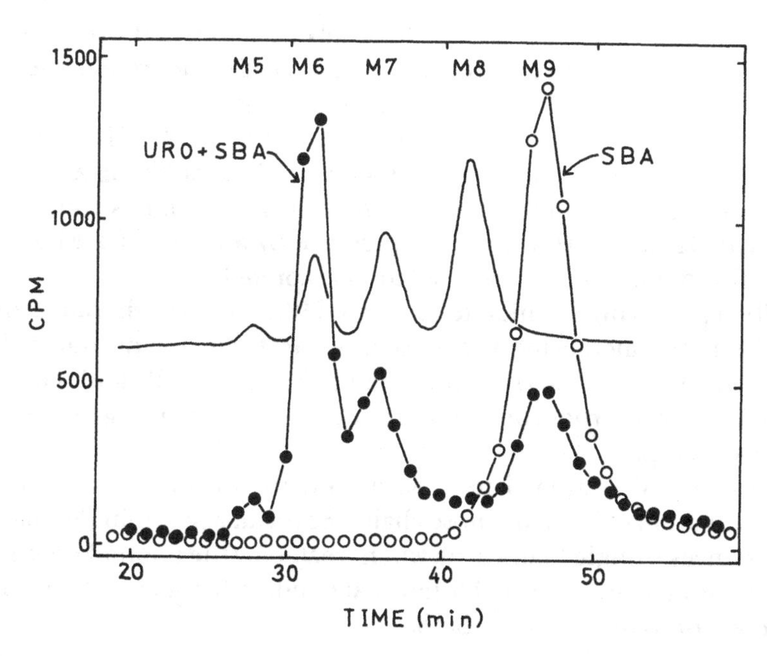

Fig. 1. HPLC of reduced high-mannose oligosaccharides. The upper solid line represents A195 (0.02 absorbance unites full-scale) for 20 µg of a standard mixture (Genzyme) containing M5, M6, M7, and M8. SBA, soybean agglutinin (0.5 mg); URO + SBA, human uromodulin (1.0 mg) plus soybean agglutinin (0.5 mg), treated as described in Section 3.

the radioactive plot with the chromatogram, inject a standard, such as L-leucine (10 µL of a 1 mg/mL solution containing 5×10^5 cpm of ^{3}H-L-leucine), and compare the resulting radioactive profile with the chromatogram.

c. Once the sample plot and the standard chromatogram have been aligned, identify each peak as Man5, Man6, Man7, and so forth (Fig. 1). Sum the radioactivity for fractions under each peak, correcting for background. Using the total for each peak summed together, you can calculate the percentage of each or determine the ratio of any two oligosaccharides (*see* Note 6).

4. Notes

1. During heat inactivation of the pronase, do not reduce the length of time or the temperature, since pronase is relatively resistant to inactivation.
2. While drying under nitrogen, the sample may be heated (35–45°C) to facilitate drying.

3. For the radioactive labeling, perform all operations in the fume hood. When testing the pH during addition of acetic acid, remember that the pH paper and Pasteur pipet are radioactive.
4. For the Con A affinity column, it is wise to "degas" the TBS buffer by subjecting it to a slight vacuum; this will reduce bubbling associated with the changes in temperature. Also, if the column is running by gravity feed, the flow rate will increase by a factor of 2 when the methyl-α-mannopyranoside solution is applied.
5. When preparing the mobile phase for HPLC, use weight rather than volume for measuring the amount of acetonitrile and water. The mobile phase will be cold after mixing the acetonitrile and water; be sure to warm to room temperature and degas before using, or bubbles will develop.
6. An internal standard may be added to the samples to aid in quantitation of high-mannose chains. Soybean agglutinin (Sigma, 0.5 mg) gives a single Man9 peak when treated by the method described in Section 3., and is a suitable internal standard for quantitating Man5, Man6, or Man7, but not Man8.

References

1. Yamashita, K., Ohkura, T., Tachibana, Y., Takasaki, S., and Kobata, A. (1984) Comparative study of the oligosaccharides released from baby hamster kidney cells and their polyoma transformant by hydrazinolysis. *J. Biol. Chem.* **259,** 10,834–10,840.
2. Swiedler, S. J., Freed, J. H., Tarentino, A. L., Plummer, T. H., Jr., and Hart, G. W. (1985) Oligosaccharide microheterogeneity of the murine major histocompatibility antigens. *J. Biol. Chem.* **260,** 4046-4054.
3. Hsieh, P., Rosner, M. R., and Robbins, P. W. (1983) Host-dependent variation of asparagine-lined oligosaccharides at individual glycosylation sites of Sindbis virus glycoproteins. *J. Biol. Chem.* **258,** 2548–2554.
4. Gyves, P. W., Gesundheit, N., Stannard, B. S., DeCherney, G. S., and Weintraub, B. (1989) Alterations in the glycosylation of secreted thyrotropin during ontogenesis. *J. Biol. Chem.* **264,** 6104–6110.
5. Serafini-Cessi, F., Dall'Olio, F, and Malagolini, N. (1984) High-mannose oligosaccharides from human Tamm-Horsfall glycoprotein. *Bioscience Rep.* **4,** 269–274.
6. Dall'Olio, F., De Kanter, F. J. J., Van den Eijnden, D. H., and Serafini-Cessi, F. (1988) Structural analysis of the preponderant high-mannose oligosaccharide of human Tamm-Horsfall glycoprotein. *Carbohyd. Res.* **178,** 327–332.
7. Hession, C., Decker, J. M., Sherblom, A. P., Kumar, S., Yue, C. C., Mattaliano, R. J., Tizard, R., Kawashima, E., Schmeissner, U., Heletky, S., Chow, E. P., Burne, C. A., Shaw, A., and Muchmore, A. V. (1987) Uromodulin (Tamm-Horsfall glycoprotein): a renal ligand for lymphokines. *Science* **237,** 1479–1484.

8. Smagula, R. M., Van Halbeek, H., Decker, J. M., Muchmore, A. V., Moody, C. E., and Sherblom, A. P. (1990) Pregnancy-associated changes in oligomannose oligosaccharides of human and bovine uromodulin (Tamm-Horsfall glycoprotein). *Glycoconjugate J.* **7,** 609–624.
9. Muchmore, A. V., Sathyamoorthy, N., Decker, J. M., and Sherblom, A. P. (1990) Evidence that specific high-mannose oligosaccharides are able to directly inhibit antigen driven T cell responses. *J. Leukocyte Biol.* **48,** 457–464.
10. Trimble, R. B. and Maley, F. (1984) Optimizing hydrolysis of N-linked high-mannose oligosaccharides by endo-beta-*N*-acetylglucosaminidase H. *Anal. Biochem.* **141,** 515–522.
11. Hirani, S., Bernasconi, R. J., and Rasmussen, J. R. (1987) Use of *N*-glycanase to release asparagine-linked oligosaccharides for structural analysis. *Anal. Biochem.* **162,** 485–492.
12. Mellis, S. J. and Baenziger, J. U. (1983) Structures of the oligosaccharides present at the three asparagine-linked glycosylation sites of human IgD. *J. Biol. Chem.* **258,** 11,546–11,556.
13. Cummings, R. D. and Kornfeld, S. (1982) Fractionation of asparagine-linked oligosaccharides by serial lectin-agarose affinity chromatography. *J. Biol. Chem.* **257,** 11,235–11,240.
14. Kornfeld, K., Reitman, M. L., and Kornfeld, R. (1981) The carbohydrate-binding specificity of pea and lentil lectins. *J. Biol. Chem.* **256,** 6633–6640.
15. Lis, H. and Sharon, N. (1978) Soybean agglutinin—a plant glycoprotein. Structue of the carbohydrate unit. *J. Biol. Chem.* **253,** 3468–3475.

[illegible]

Chapter 12

Release of *O*-Linked Glycoprotein Glycans by Endo-α-*N*-Acetylgalactosaminidase

Hitoo Iwase and Kyoko Hotta

1. Introduction

Endo-α-*N*-acetylgalactosamindase (D-galactosyl-*N*-acetamido-deoxy-α-galactoside D-galactosyl-*N*-acetamido-deoxy-D-galactohydrolase, EC 3.2.1.97, abbreviated as endo-GalNAc-ase) is a well known enzyme that catalyzes the hydrolysis of the linkage structures of most *O*-glycoside carbohydrate chains as follows:

Galβ1-3GalNAcα1-Ser/Thr (Peptide)
↓ Endo-GalNAc-ase
Galβ1-3GalNAc + Ser/Thr (Peptide)

This α-*N*-acetylgalactosaminyl serine or threonine linkage is characteristic of the so-called mucin-type carbohydrate chain. An excellent review is available on the structure and biosynthesis of this mucin-type carbohydrate chain *(1)*. Enzymes from two sources, *Diplococcus (2–4)* and *Alcaligenes (5)*, endo-GalNAc-ase-D and endo-GalNAc-ase-A, respectively, can be obtained commercially. A new enzyme from *Sreptomyces* capable of releasing oligosaccharides larger than the above disaccharide from porcine gastric mucus glycoprotein has been recently reported *(6)*. Endo-GalNAc-ase may possibly be present in mammalian tissue, but other than the secretion of mucin-type oligosaccharide in normal human urine *(7)* and mucin-type glycopeptide in mucolipidosis I *(8)* and Kanzaki disease *(9)*, there is no concrete evidence to substantiate this view.

From: *Methods in Molecular Biology, Vol. 14: Glycoprotein Analysis in Biomedicine*
Edited by: E. F. Hounsell

Endo-GalNAc-ase-D and -A, both commercially available, were used to prepare intact core protein from glycoproteins and elucidate the functions of the carbohydrate chain associated with protein. The substrates of the endoenzyme thus included membranous glycoproteins *(10,11)*, viral glycoproteins *(12,13)*, mucus glycoproteins *(14)*, enzymes *(15)*, cytokines *(16)*, and in some cases, erythrocytes *(17)* and tissue sections *(18,19)*.

The application of commercial endo-GalNAc-ase to a wide variety of glycoproteins and the new endo-GalNAc-ase from *Streptomyces* sp. OH-11242 to porcine gastric mucus glycoprotein is discussed in the following.

2. Materials

*2.1. Digestion with Endo-α-*N*-Acetylgalactosaminidase from* Diplococcus *and* Alcaligenes

1. Endo-GalNAc-ase-D (*O*-Glycanase™, Genzyme Corp., Boston, MA): The removal of glycerol added as an enzyme stabilizer from commercial enzyme by dialysis is required in one case *(20)*.
2. Endo-GalNAc-ase-A (Seikagaku Kogyo Co., Ltd, Tokyo, Japan): In contrast to endo-GalNAc-ase-D, this enzyme is not inhibited by EDTA.
3. Affinity-purified neuraminidase from *Clostridium perfringens* or *Vibrio cholera* (Sigma Chemical Co., St. Louis, MO).
4. Buffer A: 20–50 m*M* sodium phosphate, pH 6.0, containing 10 m*M* galactono-lactone as β-galactosidase inhibitor and 1–2 m*M* calcium acetate as enzyme stabilizer.
5. Buffer B: 20 m*M* Tris-maleate buffer, pH 6.0, containing 1 m*M* galactono-lactone, 0.1% SDS, and NP-40.
6. Buffer C: Sodium acetate pH 6–5 supplemented with 0.2 mg/mL bovine serum albumin and 0.1*M* NaCl.
7. RPM1 medium containing 10% human serum (cells washed three times by resuspension).
8. Buffer D: 50 m*M* phosphate buffer, pH 7.0.
9. Buffer E: 100 m*M* citrate buffer, pH 4.5.

2.2. Chromatography and Detection

2.2.1. Colorimetric Estimation (21)

1. $K_2B_4O_7$ solution: 5% potassium tetraborate adjusted to pH 9.1 by KOH.
2. DMAB solution (store in cold room): *p*-dimethylaminobenzaldehyde (8 g)/acetic acid (47.5 mL) and conc. HCl (2.5 mL).

2.2.2. Paper Chromatography (22)

1. Descending paper chromatography apparatus.
2. Solvent system: ethyl acetate:pyridine:water 12:5:4.
3. Alkaline silver nitrate reagent ($AgNO_3$). Add 0.5 mL of saturated silver nitrate to 100 mL of acetone and then H_2O dropwise until the silver nitrate is dissolved.
4. Alkaline solution: Dilute 5 mL of 10*N* NaOH with ethanol to 100 mL.
5. Sodium thiosulfate solution (5%).

2.2.3. Gel Chromatography

1. Bio-Gel P-2 (<400 mesh) column (0.9 × 64 mL).
2. Pyridine acetate (0.1*N*), pH 5.4.

2.2.4. HPLC of Pyridylamino Sugars (6,23)

1. Re-*N*-acetylated pyridylamino sugar derivatives (PA) prepared as described in Chapter 6 from 10 nmol reducing sugar.
2. Reversed-phase column (5 µm, 4.6 × 500 mm, Ultrasphere-ODS, Beckman Instruments, Inc., San Ramon, CA) equilibrated with 0.125*M* citrate buffer, pH 4.0, containing 0.75% acetonitrile.
3. TSK-GEL G2000PW column (10 µm, 7.5 × 600 mm, Toyo Soda Manufacturing Co. Ltd., Tokyo, Japan) equilibrated with 20 m*M* ammonium acetate buffer, pH 7.5.

***2.3. Endo-α-N-Acetylgalactosaminidase from* Streptomyces**

1. Crude enzyme preparation from *Streptomyces* sp. OH-11242 prepared as reported *(6)*.
2. Porcine gastric mucin (Type II, Sigma) and pepsin of porcine stomach mucosa (2500–3500 U/mg protein, Sigma).
3. Buffer F: 0.5*M* citrate phosphate buffer, pH 5.0.
4. Bio-Gel P-4, a water-jacketed thick-wall column (1.6 × 100 cm) prewarmed at 55°C, and equilibrated with deaerated distilled water. Void vol was approx 59 mL, and elution vol of maltohexaose and galactose, 128 and 175 mL, respectively.
5. Phenol-sulfuric acid assay: 5% aqueous phenol and concentrated sulfuric acid.
6. Silica gel 60 TLC plates.
7. Solvent for TLC: *n*-propanol:acetic acid:water (3:3:2).
8. Orcinol-sulfuric acid reagent: 1% orcinol in 50% sulfuric acid.
9. HCl (6*N*) and Dowex 50W × 2 (H^+ form).

3. Methods

3.1. *Endo-α-N-Acetylgalactosaminidase from* Diplococcus *or* Alcaligenes (see *Note 1)*

3.1.1. Soluble Glycoprotein

1. Predigest sample with neuraminidase at 1.0 U/mL of buffer A, since the activity of endo-GalNAc-ase is blocked by sialic acid residues on the substrate. The enzyme catalyzes the reaction:

NeuAcα2-3Galβ1-3(NeuAcα2-6)GalNAcα1-Ser/Thr
↓ Neuraminidase
2 NeuAc + Galβ1-3GalNAcα1-Ser/Thr

Sequential treatment or mixed treatment with neuraminidase can be carried out.

2. Dissolve sample at 1.5–2.5 mg/mL in buffer A containing 50–75 mU/mL of endo-GalNAc-ase A or D.
3. Incubate at 37°C for various periods of time (*see* Note 2).

3.1.2. Membrane-Bound Glycoprotein (10)

Treat samples with 80 mU/mL of endo-GalNAc-ase-D in buffer B after predigestion with neuraminidase as previously described.

3.1.3. Cells (Human Erythrocytes)

Treat erythrocytes with neuraminidase (25 mU/mL) and/or endo-GalNAc-ase-D (66 mU/mL).

3.1.4. Tissue (18,19)

1. Prepare tissue sections from porcine eye cup by fixation by 4.0% formaldehyde, embedding in acrylamide, and sectioning on a cryostat at –20°C.
2. Treat the section at 37°C for 1 h with 0.2 U/mL of endo-GalNAc-ase-D in buffer B containing 1.0 U/mL of neuraminidase.
3. Prepare tissue section from human pancreas by fixation with 10% formalin, embedding in paraffin, and cutting at 4 μm.
4. Dry deparaffinized and washed sections by blotting with filter paper.
5. Incubate at 37°C for 20–24 h in buffer D containing 0.17 U/mL endo-GalNAc-ase-D or 0.5 U/mL endo-GalNAc-ase-A in buffer E (*see* Note 3).

3.2. Analysis of Released Disaccharide

3.2.1. Colorimetric Method (21)

1. To the enzyme digest containing 10–100 nmol disaccharide/120 µL, add 100 µL of $K_2B_4O_7$ solution.
2. Stir and heat in boiling bath for 7 min.
3. Place in ice bath. Add 1 mL of acetic acid and 0.4 mL of DMAB solution.
4. Stir and incubate at 38°C for 20 min.
5. Read at 585 nm against reagent blank.

If the sample is contaminated by released *N*-acetylhexosamine and/or if the amount of disaccharide is too small to be detected, this procedure cannot be used.

3.2.2. Paper Chromatography (22)

1. To detect reducing sugar spotted onto filter paper, pull the paper through the $AgNO_3$ solution.
2. Dry the filter paper at room temperature for a few minutes.
3. Heat at 100°C for 5 min.
4. Dip the filter paper in alkaline solution.
5. Fix the paper with thiosulfate solution.
6. Wash the paper with flowing water.

3.2.3. Gel Chromatography (3)

Detect disaccharide in the 1-mL fraction from gel filtration column by the colorimetric method or radioactivity for labeled oligosaccharide.

3.3. Endo-α-N-Acetylgalactosaminidase *from* Streptomyces (6)

Because of the narrow substrate specificity of the above endo-GalNAc-ase, the removal of *O*-linked oligosaccharides other than Galβ1-3GalNAc was conducted by chemical methods, such as alkaline sodium borohydride treatment, to prepare oligosaccharides with reducing-end *N*-acetylgalactosaminitol (peptide portion is destroyed) *(2)*, and by two other methods using anhydrous hydrogen fluoride *(24)* or trifluoromethansulfonic acid *(25)* to prepare the peptide (oligosaccharide portion is destroyed) from glycoprotein. However, such methods were not always suitable for glycoprotein on cells or tissues.

The endoglycosidase in the culture medium of *Streptomyces* capable of liberating a larger oligosaccharide than the disaccharide from procine gastric mucus glycoprotein can be demonstrated as follows (*see* Note 4).

1. Make up 100 g of crude porcine gastric mucin to a final vol of 500 mL by adding 0.001% aqueous sodium azide, and adjust to pH 2.5 with HCl.
2. Add 1 g of pepsin to the suspension followed by incubation at 37°C with shaking. Add an additional 1 g of pepsin after 24 h and again at 48 h, and continue digestion to 72 h.
3. Neutralize the reaction mixture by adding NaOH followed by centrifugation at 10,000 rpm for 30 min.
4. Dissolve 5 g of cetyl pyridinium chloride in the supernatant, and allow the solution to stand overnight at room temperature.
5. Separate the clear supernatant from the precipitate by centrifugation at 10,000 rpm for 30 min, and then conduct dialysis against tap water.
6. Remove insoluble material formed at a 33% (v/v) concentration of ethanol by centrifugation, and add ethanol to a concentration of 50% followed by standing overnight at 4°C.
7. Collect the precipitate by centrifugation at 10,000 rpm for 60 min (yield 31 g, colorless powder following lyophilization).
8. Incubate 100 mg of a purified mucus glycoprotein with 5 mL of crude enzyme preparation in 1 mL of buffer F. Terminate the reaction by boiling at 100°C for 5 min.
9. Dialyze the reaction mixture against distilled water.
10. Analyze the low-mol-wt products in the dialyzate by Bio-Gel P-4 (–400 mesh) chromatography.
11. Analyze fractions by phenol-sulfuric acid reaction for hexose: Sequentially mix 0.5 mL of sample with 0.5 mL of 5% phenol and 2.5 mL of sulfuric acid, followed by standing for 20 min and monitoring at 490 nm.
12. Analyze oligosaccharide fraction further by TLC on a silica gel 60 plate developed at room temperature for 20 h (attach filter paper to top of the plate). Oligosaccharide spots were visualized with orcinol-sulfuric acid reagent.
13. Extract oligosaccharide from the spot, pyridylaminate (*see* Note 5), and hydrolyze the PA-oligosaccharide with 6*N* HCl at 100°C for 6 h in an evacuated sealed tube. Then analyze the resultant PA-monosaccharide by reversed-phase HPLC as described in Section 2.2.4. and Chapter 6 after the amino group of PA-hexosamine has been re-*N*-acetylated.
14. For carbohydrate composition determination of the oligosaccharide, pyridylaminate the passed fraction of the hydrolyzate from Dowex 50W X2 (H^+ form), and analyze the resultant PA-monosaccharide by HPLC.

4. Notes

1. Strictly speaking, commercially available endo-GalNAc-ase-D and -A are not the same enzyme with respect to specificity as reported recently *(26)*. However, minor differences between endo-GalNAc-ase-D and -A need not be considered since the enzyme is generally used in excess amount.
2. The release of the carbohydrate chain from glycoprotein was examined through detection of the reduction in its mol wt by SDS polyacrylamide gel electrophoresis.
3. In the application of the endoglycosidase to the tissue section, reduction of peanut-agglutinin-binding activity, instead of released oligosaccharide was detected.
4. Endo-GalNAc-ase from *Streptomyces* is a new endo-α-*N*-acetylgalactosaminidase presently being studied, and the substrate specificity of purified enzyme should be thoroughly clarified prior to its application.
5. Since there are many endoglycosidases capable of releasing oligosaccharide from glycoproteins, a major difficulty is detecting the release of oligosaccharides other than Galβ1-3GalNAc, and the reducing-end *N*-acetylgalactosamine of released oligosaccharide at the same time at each purification step of the enzyme. As shown in the present study, pyridylamination of the reducing terminal sugar of the released oligosaccharide is effective for eliminating this problem.

References

1. Schachter, H. (1986) Biosynthetic controls that determine the branching and microheterogeneity of protein-bound oligosaccharides. *Biochem. Cell. Biol.* **64,** 163–181.
2. Endo,Y. and Kobata, A. (1976) Partial purification and characterization of an endo-α-*N*-acetylgalactosaminidase from the culture medium of *Diplococcus pneumoniae. J. Biochem.* **80,** 1–8.
3. Bhavanandan,V. P., Umemoto, J., and Davidson, E. A. (1977) Purification and properties of an endo-α-*N*-acetylgalactosaminidase from *Diplococcus pneumoniae. Biochem. Biophys. Res. Comm.* **70,** 738–745.
4. Umemoto, J., Bhavanandan,V. P., and Davidson, E. A. (1977) Purification and properties of an endo-α-*N*-acetyl-D-galactosaminidase from *Diplococcus pneumoniae. J. Biol. Chem.* **252,** 8609–8614.
5. Fan, J.-Q., Kadowaki, S., Yamamoto, X., Kumagai, H., and Tochikura, T. (1988) Purification and characterization of endo-α-*N*-acetylgalactosaminidase from *Alcaligens* sp. *Agric. Biol. Chem.* **52,** 1715–1723.

6. Iwase, H., Ishii, I., Ishihara, X., Tanaka, Y., Ōmura, S., and Hotta, K. (1988) Release of oligosaccharides possessing reducing-end *N*-acetylgalactosamine from mucus glycoprotein in *Streptomyces* sp. OH-11242 culture medium through action of endo-type glycosidase. *Biochem. Biophys. Res. Comm.* **151,** 422–428.
7. Parkkinen, J. and Finne, J. (1987) Isolation of sialyl oligosaccharides and sialyl oligosaccharide phosphates from bovine colostrum and human urine. *Methods Enzymol.* **138,** 289–300.
8. Lecat, D., Lemonnier, M., Derappe, C., Lhermitte, M., Van Halbeek, H., Dorland, L., and Vliegenthart, J. F. G. (1984) The structure of sialyl-glycopeptides of the *O*-glycosidic type, isolated from sialidosis (mucolipidosis I) urine. *Eur. J. Biochem.* **140,** 415–420.
9. Hirabayashi, Y., Matsumoto, Y., Matsumoto, M., Toida, T., Iida, N., Matsubara, T., Kanzaki, T., Yokota, M., and Ishizuka, I. (1990) Isolation and characterization of major urinary amino acid *O*-glycosides and a dipeptide *O*-glycoside from a new lysosomal strage disorder (Kanzaki disease): Excessive excretion of serine- and threonine-linked glycan in the patient urine. *J. Biol. Chem.* **265,** 1693–1701.
10. Lublin, D. M., Krsek-Staples, J., Pangburn, M. K., and Atkinson, J. P. (1986) Biosynthesis and glycosylation of the human complement regulatory protein decay-accelerating factor. *J. Immunol.* **137,** 1629–1635.
11. Yoshimura, A., Seguchi, T., Yoshida, T., Shite, S., Waki, M., and Kuwano, M. (1988) Novel feature of metabolism of low density lipoprotein receptor in a mouse macrophage-like cell line, J774. 1. *J. Biol. Chem.* **263,** 11,935–11,942.
12. Pinter, A. and Honnen, W. J. (1988) *O*-linked glycosylation of retroviral envelope gene products. *J. Virol.* **62,** 1016–1021.
13. Lambert, D. M. (1988) Role of oligosaccharide in the structure and function of respiratory syncytial virus glycoproteins. *Virology* **164,** 458–466.
14. Eckhardt, A. E., Timpte, C. S., Abernethy, J. L., Toumaje, A., Johnson, W.C., Jr., and Hill, R. L. (1987) Structural properties of porcine submaxillary gland apomucin. *J. Biol. Chem.* **262,** 11,339–11,344.
15. Uemura, M., Winant, R. C., Sikic, B. I., and Brandt, A. E. (1988) Characterization and immunoassay of human tumor-associated galactosyltransferase isoenzyme II. *Cancer Res.* **48,** 5325–5334.
16. Gross, V., Andus, T., Castell, J., Berg, D. V., Heinrich, P. C., and Gerok, W. (1989) *O*- and *N*-glycosylation led to different molecular mass forms of human monocyte interleukin-6. *FEBS Lett.* **247,** 323–326.
17. Perkins, M. E. and Holt, E. H. (1988) Erythrocyte receptor recognition varies in plasmodium falciparum isolates. *Mol. Biochem. Parasitol.* **27,** 23–34.
18. Johnson, L. V. and Hageman, G. S. (1987) Enzymatic characterization of peanut agglutinin-binding components in the retinal interphotoreceptor matrix. *Exp. Eye Res.* **44,** 553–565.
19. Ito, N., Nishi, K., Nakajima, M., Okamura, Y., and Hirota, T. (1989) Histochemical demonstration of *O*-glycosidically linked, type 3 based ABH antigens in human pancreas using lectin staining and glycosidase digestion procedures. *Histochemistry* **92,** 307-312.

20. Bardales, R. M. and Bhavanandan, V. P. (1989) Transglycosylation and transfer reaction activities of endo-α-*N*-acetyl-D-galactosaminidase from *Diplococcus (Streptococcus) pneumoniae*. *J. Biol. Chem.* **264,** 19,893–19,897.
21. Reissig, J. L., Strominger, J. L., and Leloir, L. F. (1955) A modified colorimetric method for the estimation of *N*-acetylamino sugars. *J. Biol. Chem.* **217,** 959–966.
22. Anet, E. F. L. T. and Reynolds, T. M. (1956) Isolation of mucic acid from fruits. *Nature* **174,** 930.
23. Iwase, H., Ishii-Karakasa, I., Urata, T., Saito, T., and Hotta, K. (1990) Extraction method for preparing pyridylamino sugar derivatives and application to porcine gastric mucus glycoprotein analysis. *Anal. Biochem.* **188,** 200–202.
24. Mort, A. J. and Lamport, T. A. (1977) Anhydrous hydrogen fluoride deglycosylates glycoproteins. *Anal. Biochem.* **82,** 289–309.
25. Marianne, T., Perini, J.-M., Houvenaghel, M.-C., Tramu, G., Lamblin, G., and Roussel, P. (1986) Action of trifluoromethanesulfonic acid on highly glycosylated regions of human bronchial mucins. *Carbohydr. Res.* **151,** 7–19.
26. Fan, J.-Q., Yamamoto, K., Matsumoto, Y., Hirabayashi, Y., Kumagai, H., and Tochikura, T. (1990) Action of endo-α-*N*-acetylgalactosaminidase from Alcaligenes sp. on amino acid-*O*-glycans: Comparison with the enzyme from *Diplococcus pneumoniae*. *Biochem. Biophys. Res. Comm.* **169,** 751–757.

CHAPTER 13

Immunological Detection of Glycoproteins on Blots Based on Labeling with Digoxigenin

Anton Haselbeck and Wolfgang Hösel

1. Introduction

The analysis of macromolecules (e.g., nucleic acids, proteins, glycoconjugates) after electrophoretic separation has become increasingly important in recent years for several reasons.

1. It combines the powerful resolution capabilities of electrophoretic systems with very sensitive detection techniques. This allows the analysis of even small amounts of macromolecules present in complex mixtures without the need of previous extensive purifications.
2. Because of the availability of efficient blotting techniques the application and handling of sensitive detection techniques (e.g., based on biotin/streptavidin or immunological technologies) have become possible.
3. These analytical techniques can be used without the need of expensive equipment and special knowledge (e.g., NMR or MS techniques).

In order to apply these powerful techniques to the analysis of glycoproteins, we have developed an immunological labeling and detection system. It is based on the use of the plant steroid digoxigenin as a hapten and polyclonal antibodies with high affinity toward this hapten. (*See* Fig. 1 for the structure of digoxigenin.) Such an immunological system has some advantages compared to the biotin/(strept)avidin system: (1) reduced unspecific binding behavior compared to

From: *Methods in Molecular Biology, Vol. 14: Glycoprotein Analysis in Biomedicine*
Edited by: E. F. Hounsell

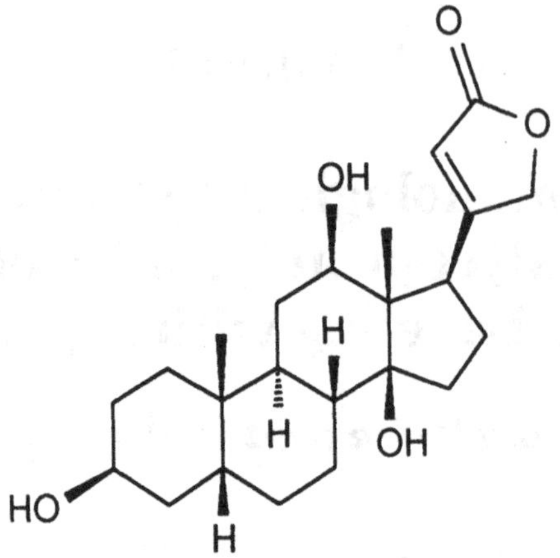

Fig. 1. Structure of digoxigenin.

(strept)avidin, and (2) a much lower number of binding sites for digoxigenin and absence of endogenous digoxigenin in most biological systems compared to biotin.

The basic feature of this particular system (in short: DIG/anti-DIG) consists of the labeling of the glycan part of glycoproteins with digoxigenin, either by chemical or enzymatical means *(1)* or by binding of digoxigenylated lectins *(2)*. The DIG hapten is then being detected by the use of an anti-DIG conjugate, especially by employing an alkaline phosphatase (AP) conjugate producing colored products on the blots.

The combination of the different labeling and detection methods described here allows us to obtain a substantial amount of information when analyzing glycoproteins on blots. The analytical potential can be increased even further when these methods are combined with the use of exo- and endoglycosidase digestions *(3,4)*.Thus, a carbohydrate detected by the general labeling method with DIG-hydrazide can be assumed to be present in a *N*-glycan linkage if there is no positive reaction after the glycoprotein has been treated with *N*-glycosidase F, which removes all types of *N*-glycan chains. Similarly, the presence of a high mannose structure detected by a positive reaction with the lectin GNA can be corroborated by showing that it is sensitive to endo H treatment. Since lectins usually react with the terminal parts of the glycan chains, it is very often possible to combine their use with that of exoglycosidases. When the positive reaction with a lectin is no longer present after treatment with exoglycosidases, e.g., neuraminidase, α-fucosidase, β-galactosidase, or α-mannosidase, the conclusion on the structural feature present is considerably substantiated. Thus, a wealth of structural information can be obtained by combining all these different methods for the analysis of blotted glycoproteins.

The use of the DIG/anti-DIG method has further been described for the very sensitive analysis of glycolipids on TLC plates *(5)* and has been applied in combination with gold labeling for the use in histochemistry *(6)*.

In the following, we describe the application of this DIG/anti-DIG system for the (structural) analysis of glycoproteins on blots. Special emphasis is being placed on the variety of the different DIG applications in order to obtain structural information concerning the glycoprotein carbohydrate chains. Prominent among them is the use of lectins with well-known specificities for carbohydrate structures.

2. Materials

1. Digoxigenin-succinyl-ε-amino caproic acid hydrochloride: 5 m*M* solution in dimethylformamide.
2. Antidigoxigenin antibody (Fab fragments), conjugated with alkaline phosphatase (750 U/mL); available from Boehringer Mannheim (Mannheim, Germany).
3. Proteins and glycoproteins: creatinase (rec. from *E. coli*), carboxypeptidase Y, fetuin, asialofetuin, transferrin, erythropoietin (recombinant from CHO cells) from Boehringer Mannheim and α_1-acid glycoprotein and lactoferrin from Sigma (Deisenhofen, Germany).
4. Mol-wt markers: Trypsin inhibitor (M_r 20,100), lactate dehydrogenase (M_r 36,500), glutamate dehydrogenase (M_r 55,400) and phosphorylase b (M_r 97,400).
5. Nitrocellulose membranes (e.g., BA 85 from Schleicher & Schüll, Dassel, Germany) or PVDF membranes (e.g., Immobilon from Millipore, Eschborn, Germany).
6. Blocking reagent: casein, special quality, Boehringer Mannheim.
7. Triton X-100, Nonidet P-40, sodium dodecylsulfate (SDS), octylglucoside, dithiothreitol, Tris, glycine (all p.a. grade).
8. Tris-buffered saline (TBS): 50 m*M* Tris-HCl, pH 7.5, 150 m*M* NaCl.
9. Phosphate-buffered saline (PBS): 50 m*M* potassium phosphate, pH 7.5 , 150 m*M* NaCl.
10. Buffer 1: 0.1*M* sodium acetate, pH 5.5.

11. Buffer 2: 0.1*M* Tris-HCl, pH 9.4, 0.05*M* $MgCl_2$.
12. Buffer 3: TBS, 1 m*M* $MgCl_2$, 1 m*M* $CaCl_2$, 1 m*M* $MnCl_2$.
13. Buffer 4: 0.1*M* potassium phosphate, pH 6.0.
14. SDS sample buffer: 0.1*M* Tris-HCl, pH 6.8, 8% SDS (w/v), 40% glycerol (v/v), 20% 2-mercaptoethanol (v/v).
15. Sodium metaperiodate solution (15 m*M*): Dissolve 3.3 mg sodium metaperiodate (e.g., Merck No. 6597, Merck, Darmstadt, Germany) in 1 mL bidistilled water.
16. Sodium disulfite solution (20 m*M*): Dissolve 3.75 mg sodium disulfite (e.g., Merck No 6528) in 1 mL bidistilled water.
17. Blocking solution: Dissolve 0.5 g blocking reagent in 100 mL TBS by heating to 50–60°C for approx 1 h; the dissolution can be accelerated by ultrasonication and incubation in a microwave oven; the solution remains turbid.
18. 5-Bromo-4-chloro-3-indolyl-phosphate solution: Dissolve 50 mg in 1 mL dimethylformamide.
19. 4-Nitroblue tetrazoliumchloride solution: Dissolve 75 mg in 70% dimethylformamide (v/v).
20. Ponceau S solution (0.2%): Dissolve 0.2 g Ponceau S in 100 mL 3% acetic acid.
21. Enzymes (Boehringer Mannheim):
    a. Galactose oxidase;
    b. Neuraminidase *(Arthrobacter ureafaciens)*;
    c. *N*-Glycosidase F;
    d. Endoglycosidase F;
    e. Endoglycosidase H.
22. Lectins, digoxigenin labeled (Boehringer Mannheim):
    a. Aleuria aurantia (AAA; specific for Fuc α1-6 GlcNAc);
    b. Concanavalin A (Con A; specific for Man/Glc);
    c. Datura stramonium (DSA; specific for Gal β 1-4 GlcNAc);
    d. Galanthus nivalis (GNA; specific for Man α1-2/3/6 Man);
    e. Maackia amurensis (MAA; specific for NeuAc α2-3 Gal);
    f. Peanut (PNA; specific for Gal β1-3 GalNAc);
    g. Phytohemagglutinin-L (PHA-L; specific for Galβ1-4GlcNacβ1-2,6Man);
    h. Ricinus communis (RCA 120; specific for β-D-Gal);
    i. Sambucus nigra (SNA; specific for NeuAc α2-6 Gal); and
    j. Wheat germ (WGA; specific for GlcNAc).
23. Standard electrophoresis apparatuses, such as the Mini systems from Bio-Rad (München, Germany) or Biometra (Göttingen, Germany) are used for SDS-PAGE; for blotting, tank blot or semi-dry blot devices

are suitable. Protein separation can also be carried out with Pharmacia Phast gels (Pharmacia LKB, Freiburg, Germany); an easy way to blot Phast gels is described by Braun and Abraham *(7)*.

When using Pharmacia Phast IEF gels for the separation, these gels have to be preequilibrated for 30 min at room temperature with the following solution: 10 mL H_2O, 0.5 mL Nonidet P-40, 9 g urea, and 2 mL Pharmalyte 3-10 (in case of a pH 3-9, IEF Phast gel; Pharmacia LKB) before use to avoid high background staining in the glycan detection reaction. Other IEF gel systems can also be used for the protein separation, such as the one described in ref. *8*. Blotting out of IEF gels is efficiently accomplished by a simple diffusion blotting procedure as described in ref. *9* (*see also* vol. 1 in this series).

3. Methods

3.1. Labeling Procedures Using Digoxigenin-Succinyl-ε-Amino Caproic Acid Hydrazide (DIG-Hydrazide)

3.1.1. General Labeling and Detection of Glycoproteins

The question of whether or not a protein contains carbohydrate or how many glycoproteins may be present in a sample can best be answered by a general, yet selective labeling procedure for carbohydrates. In this regard, periodate oxidation of vicinal diols in sugars resulting in aldehydes is the method of choice. DIG-hydrazide is then linked to these aldehyde groups. The procedure can be performed in two ways: (1) oxidation with periodate and labeling with DIG-hydrazide of the glycoproteins in solution; separation by electrophoresis and transfer onto a suitable membrane; detection of digoxigenin with a labeled anti-DIG antibody, or (2) oxidation and labeling of proteins on membranes and subsequent detection of digoxigenin. The volumes stated refer to a 50–100 cm^2 membrane; the membranes are incubated by gentle agitation at room temperature, except for color development, which should be done without shaking.

3.1.1.1. Procedure 1: Oxidation and Labeling in Solution

1. Dissolve 0.1–10 µg glycoprotein in 20 µL buffer 1 or dilute the glycoprotein solution at least 1 + 1 with buffer 1 (*see* Note 1).
2. Add 10 µL of 15 m*M* sodium metaperiodate, mix, and incubate for 20 min at room temperature in the dark (*see* Note 2).
3. Destroy excess periodate by adding 10 µL of 20 m*M* sodium disulfite.

4. After 5 min at room temperature, add 5 µL of 5 m*M* digoxigenin-succinyl-ε-amino caproic acid hydrazide, and incubate the mixture for 60 min at room temperature.
5. Add 15 µL of SDS sample buffer (yielding a final vol of 60 µL), and mix.
6. Boil the sample for 2 min, and load an aliquot or the whole sample onto an SDS gel.
7. Perform SDS-PAGE, and transfer the proteins to nitrocellulose or PVDF membranes as described in vol. 1 of this series (*see* Notes 3 and 4).
8. Stain the proteins on nitrocellulose or PVDF membranes with Ponceau S to check for transfer efficiency: Incubate the membrane for 5 min in a solution of 0.2% Ponceau S in 3% acetic acid; rinse the membrane with H_2O until the protein bands are visible; the membrane can be photographed or pencil marked at this stage for documentation of the protein pattern. The Ponceau S staining disappears again during the following incubation steps.
9. Incubate the membrane for at least 30 min with approx 20 mL of blocking solution to prevent nonspecific binding. If necessary, the glycoprotein detection process can be interrupted at this stage and the membrane kept for a longer period (e.g., overnight) at 4°C.
10. Wash three times for 5 min with approx 50 mL of TBS each.
11. Incubate the membrane with 10 µL of antidigoxigenin-AP (750 U/mL) in 10 mL of TBS for 60 min at room temperature.
12. Wash three times for 5 min with approx 50 mL of TBS.
13. Incubate the membrane with 37.5 µL of 5-bromo,4-chloro,3-indolyl-phosphate (50 mg/mL in dimethylformamide) and 50 µL of 4-nitroblue tetrazolium chloride (75 mg/mL in 70% dimethylformamide) in 10 mL of buffer 2. This staining solution has to be prepared fresh. The reaction is normally complete within a few minutes, but can be extended overnight, e.g., if very little protein is present (*see* Note 5). Glycoproteins appear as gray to black bands on an almost white background. The reaction is stopped by rinsing the membrane with H_2O. The color does not fade, and the blots can be dried and stored for documentation (*see* Note 6).

3.1.1.2. Procedure 2: Oxidation and Labeling on Membranes (*see* Note 7)

1. Separate the proteins by SDS-PAGE, and transfer to nitrocellulose or PVDF membranes (*see* Note 8).
2. Wash the membrane three times for 5 min with approx 50 mL PBS. Do not use TBS since Tris interferes with the subsequent digoxigenin labeling.

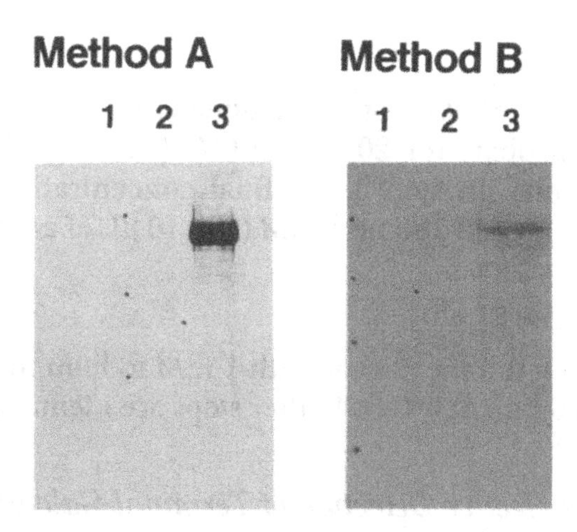

Fig. 2. The following proteins (1 µg) were loaded on an SDS gel and transferred to nitrocellulose. Lane 1, mol-wt standards (96, 55, 36, and 21 kDa); lane 2, creatinase (negative control protein); lane 3, transferrin; for procedure 1, the proteins were oxidized and digoxigenylated in solution, and for procedure 2, on the membrance as described, respectively. The little dots correspond to the position of the protein bands after Ponceau S staining.

3. Incubate the membrane with 10 m*M* sodium metaperiodate in 10 mL buffer 1 for 20 min at room temperature.
4. Wash three times for 5 min with approx 50 mL PBS.
5. Incubate the membrane with 1 µ*M* digoxigenin-succinyl-ε-amino caproic acid hydrazide (1 µL of a 5 m*M* solution in dimethylformamide) in 5 mL buffer 1, for 60 min at room temperature.
6. Wash three times for 5 min with approx 50 mL of TBS, and proceed further according to procedure 1 (from step 8).

Figure 2 illustrates these two procedures for the general detection of the glycoprotein transferrin (*see* Note 9).

3.1.2. Selective Detection of Sialic Acid Containing Glycoproteins

Since sialic acids are more susceptible to periodate oxidation than other sugars, the labeling procedure can be made specific for sialic acids by selecting the appropriate oxidation conditions. When using 1 m*M* periodate at 0°C for 20 min, only sialic acids are oxidized and subsequently labeled (*see* Note 10). This can be done either in solution or with the already transferred proteins. After the oxidation step, the protocols are exactly as described earlier.

3.1.2.1. PROCEDURE 1

Use a final concentration of 1 m*M* sodium metaperiodate in buffer in step 2, and incubate for 20 min at 0°C (add, e.g., 10 µL 3 m*M* periodate solution). In step 3, the final concentration of sodium bisulfite solution should also be 1 m*M* (add 10 µL of a 4 m*M* bisulfite solution).

3.1.2.2. PROCEDURE 2

In step 2, treat the membrane with 1 m*M* sodium metaperiodate in buffer 1 for 20 min at 0°C. All other steps are identical to the general procedure 2.

3.1.3. Selective Detection of Terminal Galactose Containing Glycoproteins

The enzyme galactose oxidase can be used to create specifically aldehyde groups at the C-6 position of terminal galactose residues of glycan chains. This allows the selective labeling of these groups with DIG-hydrazide. This particular procedure works much better on already blotted glycoproteins compared to doing it in solution. The digoxigenylation is carried out simultaneously with the oxidation reaction.

1. Incubate the membrane (dot or Western blot) for 30 min at room temperature with 20 mL of TBS, containing 1% BSA, and then wash for 5 min with 50 mL of buffer 4.
2. Incubate the membrane for 15 h at 37°C with 7.5 U galactose oxidase and 2 µL of digoxigenin-succinyl-ε-amino-caproic acid hydrazide solution (5 m*M* in dimethylformamide) dissolved in 10 mL 0.1*M* buffer 4.
3. Wash the membrane twice for 5 min with approx 50 mL TBS, and proceed with the blocking and digoxigenin detection as described earlier.

Figure 3 displays an example of using galactose oxidase for detecting terminal galactose containing glycoproteins in the range of 10–100 ng (*see* Note 11).

3.2. Structural Analysis of Glycoprotein Carbohydrate Chains Using Digoxigenin-Labeled Lectins

The suitability of employing the well-known carbohydrate binding specificities of lectins for the structural analysis of glycoproteins is well established (*see* refs. in *2*). It was logical therefore to use digoxigenylated lectins for the structural analysis of glycoproteins on membranes, thus combining the structural specificities of the lectins with a very sensitive detection system.

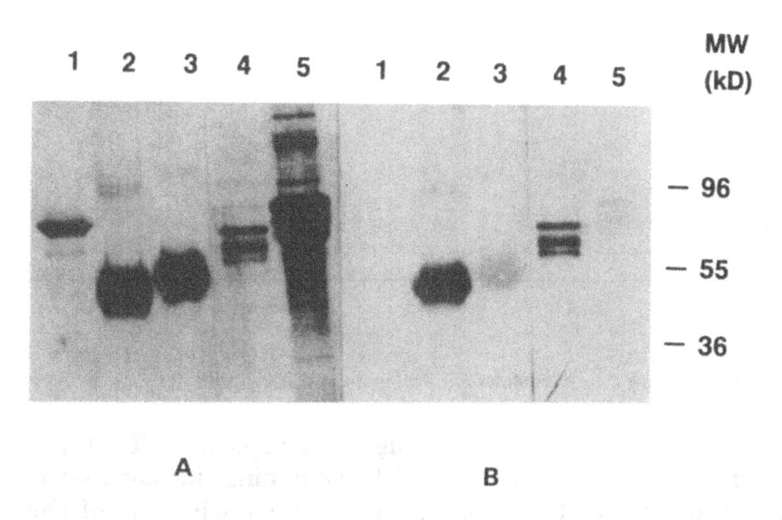

Fig. 3. The following glycoproteins (2 µg) were loaded on SDS gels and transferred to nitrocellulose. Lane 1, CpY; lane 2, α_1-acid glycoprotein, digested with neuraminidase; lane 3, α_1-acid glycoprotein; lane 4, fetuin, digested with neuraminidase; lane 5, fetuin. Blot A was oxidized chemically according to the standard procedure and blot B with galactose oxidase as described in Section 3.

Since the conclusions on the carbohydrate structures present are entirely dependent on the type of lectin used, that is, in essence on the knowledge about its binding specificity, the lectins that are used for this purpose have to be chosen carefully. Examples of suitable lectins are listed in ref. *2*, together with their respective carbohydrate binding specificities (*see* Section 2., item 22 and Note 12).

Glycoproteins to be analyzed with DIG-lectins are transferred to a membrane and incubated with the lectin of interest according to the following protocol:

1. Protein staining with Ponceau S, as described earlier, can be performed optionally prior to the lectin incubation.
2. Incubate the membrane for at least 30 min with approx 20 mL blocking solution (*see* Note 5).
3. Wash twice for 5 min with approx 50 mL TBS and once for 5 min with buffer 3.
4. Add the required amount of DIG-lectin solution to 10 mL buffer 3, and incubate the membrane for 60 min in this solution; the amounts required depend on the individual DIG-lectins. The DIG-lectins listed

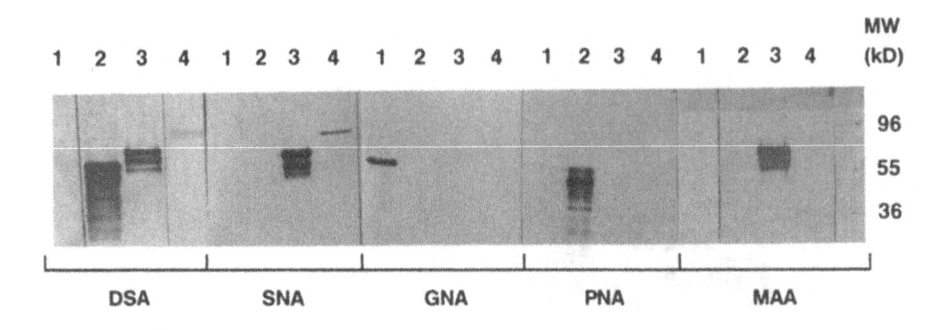

Fig. 4. The four glycoproteins (2 µg) were applied to SDS-PAGE and transferred to nitrocellulose. Five blots containing the same set of four glycoproteins were then treated individually with one of the five digoxygenylated lectins as described in Section 3.

in Section 2., item 22 are used as follows: AAA, 1 µg/mL; DSA, 1 µg/mL; GNA, 1 µg/mL; SNA, 1 µg/mL; MAA, 5 µg/mL; PNA, 10 µg/mL; ConA, 20 µg/mL; RCA-120, 10 µg/mL; PHA-L, 5 µg/mL; WGA, 10 µg/mL. For other DIG-lectins, look for the recommended concentration in the respective pack inserts.

5. Wash three times for 5 min with approx 50 mL TBS.
6. Incubate the membrane with 10 µL of antidigoxigenin-AP conjugate (750 U/mL) in 10 mL TBS for 60 min.
7. Wash three times for 5 min with approx 50 mL TBS.
8. Perform the staining reaction with BCIP and NBT as described earlier.

Figure 4 shows an example where a set of glycoproteins is incubated with five different DIG-lectins (*see* Note 13).

4. Notes

1. The presence of detergents (e.g., Triton X-100, Nonidet P-40, SDS) up to a concentration of 0.25% during oxidation and labeling does not cause interferences, with the exception of sugar containing detergents like octylglucoside. 2-Mercaptoethanol, dithiothreitol (>0.1%), glycerol (>0.01%), and amino groups containing buffers, such as Tris and glycine, have to be avoided when using procedure 1. It should also be taken into account that such substances could be present as stabilizers in enzyme preparations that might be used, e.g., for deglycosylation of the sample.

The detection limit varies for each glycoprotein. After a 60-min incubation in the staining solution, the following amounts of glycoproteins are detected (using dot blots):

a. α_1-Acid glycoprotein: 1 ng;
b. Fetuin: 1 ng;
c. Transferrin: 5 ng;
d. Carboxypeptidase Y: 10 ng.

2. It has been determined that 5 m*M* sodium metaperiodate and 5 m*M* sodium bisulfite as final concentrations in the oxidation solution result in the highest specificity of the overall labeling reaction. Oxidation with periodate has to be performed in the dark in order to prevent side reactions.
3. Glycoproteins, oxidized and labeled in solution, can also be applied directly onto membranes (dot blots); in this case, do not use detergent to denature the protein sample, since it will interfere with protein binding to the membrane. The use of nitrocellulose membranes for this purpose is also recommended, since PVDF membranes have to be prewetted before use, which will cause diffusion of the protein spots.
4. Because of the higher protein binding capacity, PVDF-type membranes are very well suited for this type of analysis, especially for detecting low amounts of glycoproteins. Nylon membranes are not recommended with this procedure, since they generate too much background.
5. To reduce background upon prolonged incubation with the staining solution when analyzing small amounts of glycoproteins, the membrane should be kept in the blocking solution (step 1) for longer than 30 min (e.g., overnight).
6. The alkaline phosphatase detection system with BCIP and NBT as substrates worked in our hands very satisfactorily and is one of the most sensitive visible detection systems available so far for the analysis of membranes. However, other anti-DIG conjugates (e.g., anti-DIG-peroxidase (POD), anti-DIG labeled with fluorescence markers) can also be used and are commercially available, if these systems are preferred for some reason. When using anti-DIG-POD, for instance, the same protocols as described above for anti-DIG-AP can be applied, and various substrates for the peroxidase reaction producing all sorts of different colors are available. There are also chemoluminescent substrates available, both for AP and POD, which can be used and are even more sensitive for certain applications than the colored products.
7. The advantages of procedure 2 are (a) Components in the glycoprotein solution that could prevent labeling in solution or lead to unspecific labeling of nonglycoproteins are removed by performing the

SDS-PAGE prior to the oxidation and labeling steps, and (b) the mobility of the proteins in SDS-PAGE is not altered, thus retaining the high resolution power of the SDS-PAGE. (The digoxigenylation of glycoproteins prior to electrophoresis causes a broadening of the protein bands.)

Nonglycosylated proteins can appear as white bands or spots on a slight pink to gray background with this procedure. This is caused by a certain degree of digoxigenylation of the membrane, which does not take place at protein covered areas. When using method 2, sialic acids containing glycoproteins (e.g., fetuin) are generally more easily detectable than those without sialic acids.

8. Procedure 2 can be used with transfer and dot blots; when using dot blots, the first washing step can be omitted. After blotting with Tris-glycine buffers, the membrane has to be washed carefully to remove Tris and glycine, which interfere with the digoxigenylation step.

   In general, the sensitivity of procedure 2 is lower as compared to procedure 1. This holds true especially for glycoproteins not containing sialic acids. Approximate detection limits for the following glycoproteins on dot blots are:
   a. α_1-Acid glycoprotein: 5 ng;
   b. Fetuin: 5 ng;
   c. Transferrin: 25 ng;
   d. Carboxypeptidase Y: 50 ng.
9. Positive and negative controls should always be included in all of these experiments. Purified *E. coli* proteins (e.g., recombinant creatinase) are suitable as negative control, as are glycoproteins, such as carboxypeptidase Y or transferrin, after they have been completely deglycosylated with *N*-glycosidase F. Be aware of the presence of nonenzymatically linked glucose or fructose on many serum proteins (e.g., BSA), which will react positive especially when using Procedure 1.
10. As positive and negative controls for the sialic acid specific detection procedure, fetuin and asialofetuin can be used. Only fetuin will react under these conditions.
11. Sialic acid containing glycoproteins, such as fetuin and α_1-acid glycoprotein, can be rendered sensitive to galactose oxidase by incubation with neuraminidase (e.g., from *Arthrobacter ureafaciens*); for this purpose, neuraminidase (e.g., 0.1 U/mL) should be included in step 2 of Section 3.1.3.
12. Positive and negative control proteins should always be included in the analysis to verify the carbohydrate specifities of the individual lectins. The lectins used for this type of carbohydrate analysis are

tested primarily with glycoproteins from yeast and animal sources; when analyzing glycoproteins from plants or bacteria, different carbohydrate structures may be recognized also by the various lectins.

13. The sensitivity depends greatly on the respective glycoprotein and also varies somewhat between the individual lectins. Generally, the detection limit ranges between 1 and 10 ng glycoprotein for dot blot samples.

References

1. Haselbeck, A. and Hösel, W. (1990) Description and application of an immunological detection system for analyzing glycoproteins on blots. *Glycoconjugate J.* **7,** 63–74.
2. Haselbeck, A., Schickaneder, E., von der Eltz, H., and Hösel, W. (1990) Structural characterization of glycoprotein carbohydrate chains by using digoxigenin-labeled lectins on blots. *Anal. Biochem.* **191,** 25–30.
3. Maley, F., Trimble, R. B., Tarentino, A. L., and Plummer, T. H., Jr. (1989) Characterization of glycoproteins and their associated oligosaccharides through the use of endoglycosidases. *Anal. Biochem.* **180,** 195–204.
4. Tarentino, A. L., Trimble, R. B., and Plummer, T. H., Jr. (1989) Enzymatic approaches for studying the structure, synthesis and processing of glycoproteins. *Methods Cell Biol.* **32,** 111–139.
5. Kniep, B. and Mühlradt, P. F. (1990) Immunochemical detection of glycosphingolipids on thin layer chromatograms. *Anal. Biochem.* **188,** 5–8.
6. Sata, T., Zuber, J., and Roth, J. (1990) Lectin-digoxigenin conjugates: A new hapten system for glycoconjugate cytochemistry. *Histochemistry* **94,** 1–11.
7. Braun, W. and Abraham, R. (1989) Modified diffusion blotting for rapid and efficient protein transfer with Phast System. *Electrophoresis* **10,** 249–253.
8. Schipper, H. J., Kruse, H., and Reiber, H. (1984) Silver staining of oligoclonal immunoglobulin G subfractions in cerebrospinal fluid after IEF in thin layer polyacrylamide gels. *Sci. Tools* **31,** 5,6.
9. Bowen, B., Steinberg, J., Laemmli, U. K., and Weintraub, H. (1980) The detection of DNA-binding proteins by protein blotting. *Nucleic Acids Res.* **8,** 1–3.

CHAPTER 14

Glycosyltransferases as Tools in Cell Biological Studies

Robert S. Haltiwanger and Gerald W. Hart

1. Introduction

Complex carbohydrates consist of an amazingly diverse array of highly branched structures *(1,2)*. This branching structure precludes the use of a linear template for assembly, as in the case of protein and nucleic acid biosynthesis. In contrast, the biosynthesis of these molecules is dependent on a series of highly specific enzymes, glycosyltransferases, which elongate growing saccharide chains sequentially *(1,2)*. The product of each reaction becomes the substrate for the next. Thus, these enzymes must specifically recognize the structure of the acceptor carbohydrate and add a monosaccharide in a particular linkage at a precise location. In most cases, one enzyme exists for each glycosidic bond that is formed *(3)*. The high level of specificity displayed by glycosyltransferases allows them to synthesize complex structures with a high degree of fidelity. Regulating this fidelity appears to be crucial for the biological functions of complex carbohydrates in vivo *(4)*.

The specificity of these enzymes also makes them attractive as probes of saccharide structure *(1,5)*. They offer many advantages in addition to their exquisite substrate specificity over other probes, such as lectins or chemical methods. For instance, transferases catalyze the formation of a covalent bond by transferring a monosaccharide from a high-energy sugar donor to an acceptor oligosaccharide, lipid, or protein. The generalized reaction is:

From: *Methods in Molecular Biology, Vol. 14: Glycoprotein Analysis in Biomedicine*
Edited by: E. F. Hounsell

$$\text{Sugar nucleotide + acceptor} \xrightarrow{\text{Transferase}} \text{Acceptor sugar + nucleotide} \quad (1)$$

When radiolabeled sugar nucleotides are employed in the reaction, the acceptors are labeled and can then be analyzed by numerous means. Depending on the specific activity of the radiolabel employed, fmoles of acceptor substrates can be detected with transferases. Using the proper controls, the amounts of various acceptors can also be quantified based on the specific activity of the radiolabel. Another advantage of glycosyltransferases as structural probes is that both the enzymes and the sugar nucleotides are impermeant to membranes. Thus, cell-surface labeling of intact cells or latency studies on isolated, sealed membrane fractions may be performed. When used in conjunction with glycosidases, transferases can be used to analyze penultimate saccharide structures. For instance, when cells are treated with *V. cholera* sialidase, which removes both α2,3- and α2,6-linked sialic acids, the sialyltransferases that add sialic acids in these linkages can be used to determine the relative amounts of the structures to which these sialic acids were linked *(6)*. Finally, because labeling with transferases is a biological method and no denaturation or digestion is required, the structures recognized are often those that are most interesting biologically.

There are numerous examples in the literature of how transferases have been used as structural probes. They have been used to detect and analyze specific *O*-linked and *N*-linked saccharide structures on cell surfaces *(5–13)*, viruses *(14)*, intracellular organelles *(15)*, or on purified proteins *(16,17)*. Transferases have also been used to label specific cell-surface oligosaccharides in order to elucidate intracellular membrane trafficking of glycoconjugates *(13,18,19)*. In addition, glycosyltransferases have proven useful in evaluating the roles of glycoconjugates in specific recognition events, such as adhesion of viruses to cells *(9,20,21)* or the binding of antibodies *(8)*. Labeling with galactosyltransferase resulted in the identification of a novel form of glycosylation found in the nucleus and cytoplasm, *O*-linked *N*-acetylglucosamine, which was undetected by conventional carbohydrate labeling methods *(10,15)*. Several reviews on the use of glycosyltransferases as probes of saccharide structure have been published elsewhere *(1,5,19,21–23)*.

These techniques were not possible several years ago because of the difficulty in obtaining sufficient quantities of these enzymes. Recent development of nucleotide-affinity adsorbents has made possible the large-scale purification of glycosyltransferases, allowing the development of these procedures *(24,25)*. Numerous transferases have been purified *(24)* and several have been cloned *(26)*, some of which are being overexpressed *(27,28)*. Currently, three transferases are available commercially, but undoubtedly as increasing numbers are stably overexpressed, that number will grow.

2. Materials

1. Glycosyltransferases (*see* Table 1 for details and Note 1).
2. Glycosidases: (a) *Vibrio cholera* sialidase—Calbiochem, Boehringer Mannheim. (b) Peptide *N*-glycosidase F, abbreviated PNGase (also called *N*-glycanase, Genzyme, Cambridge, MA; Glycopeptidase F, Boehringer Mannheim, Indianapolis, IN; and Calbiochem, San Diego, CA).
3. Nucleotides: These are available from New England Nuclear (Boston, MA), American Radiolabeled Chemicals, Inc. (St. Louis, MO), and a few from Amersham (Amersham, UK). Most are shipped and stored in 50–70% ethanol at –20°C. The appropriate amount of label should be evaporated to dryness in a SpeedVac (Savant) *immediately* before use and resuspended in 10 µL of the appropriate labeling buffer (*see* Section 2.7.).
   a. Cytidine 5'-monophospho[9-^{3}H]sialic acid (CMP-NANA: 10–35 Ci/mmol from NEN or ARC.
   b. Uridine diphospho-D-[6-^{3}H]galactose (UDP-Gal): 5–20 Ci/mmol irom ARC and Amersham; UDP[1-^{3}H]Gal (10–25 Ci/mmol) and UDP[4,5^{3}H]-Gal (30–50 Ci/mmol) are available irom NEN.
   c. Uridine diphospho-D-[6-^{3}H]*N*-acetyl-D-glucosamine (UDP-GlcNAc): 2–25 Ci/mmol from ARC or NEN.
4. Phosphate-buffered saline (PBS): 10 m*M* sodium phosphate, pH 7.4, 0.15*M* NaCl.
5. PBS/sucrose: 30% sucrose in PBS.
6. Cell preparation buffer (CPB): 10 m*M* HEPES-NaOH, pH 7.3, 24 m*M* $NaHCO_3$, 137 m*M* NaCl, and 5 m*M* sodium pyruvate.
7. Labeling buffers: These are unique to each of the transferases. *(See* Note 2.)
   a. Sialyltransferase buffer (STB): 25 m*M* HEPES-NaOH, pH 6.5, 75 m*M* NaCl, 100 m*M* glucose, and 10 mg/mL BSA (Fraction V, fatty acid free from Sigma, St. Louis, MO).
   b. Galactosyltransferase buffer (GTB): CPB containing 7% aprotinin (v/v), 10 m*M* galactose, and 5 m*M* $MnCl_2$. 5'-Adenosine monophosphate (2.5 m*M* final) is added to this fresh from a 25-m*M* stock solution as a phosphatase inhibitor.

Table 1
Glycosyltransferases Used as Probes

Enzyme[a]	Sugar donor,[a] K_m	Acceptor sequence	Commercial source	Purification	Storage conditions
Rat liver β-Gal α2, 6 ST	CMP-NeuAc ($8.5 \times 10^{-5} M$)	[b]Galβ1,4GlcNAc-R	Boehringer-Mannheim, Genzyme, Calbiochem, Sigma	*(29)*	36 m*M* Na cacodylate 0.4*M* NaCl, pH 6.5, 50% glycerol (v/v), 0.1% Triton CF-54 (w/v), –20°C
Porcine submaxillary β-Gal α2, 3 ST	CMP-NeuAc ($3.0 \times 10^{-6} M$)	[b]Galβ1,3GalNAc-Ser/Thr	Boehringer-Mannheim, Genzyme	*(30)*	7 m*M* Na cacodylate, 71 m*M* NaCl, pH 6.5, 0.7% Triton X-100 (w/v), 50% glycerol (v/v), –20°C
Bovine milk GlcNAc β1,4 GT	UDP-Gal ($6.0 \times 10^{-5} M$)	[b]GlcNAc-R	Sigma, Boehringer-Mannheim	*(31)*	25 m*M* HEPES, pH 7.3, 5 m*M* $MnCl_2$, 50% glycerol (v/v), –20°C
Porcine liver GlcNAc TF I	UDP-GlcNAc ($3.84 \times 10^{-5} M$)	Manα1,6 and Manα1,3 → Manα1,6 → Man-R; [b]Manα1,3 → Man-R	None	*(32,33)*	50 m*M* Na cacodylate, 10 m*M* $MnCl_2$, pH 6.5, 0.5% Triton X-100 (w/v), 20% glycerol (v/v), –20°C

[a]Abbreviation: β-Gal α2,6 ST: β-galactoside α2,6-sialyltransferase; β-Gal α2,3 ST: β-galactoside α2,3-sialyltransferase; GlcNAc β1,4 GT: *N*-acetylglucosaminide β1,4-galactosyltransferse; GlcNAc TF I: α1,3-mannoside β1,2-*N*-acetylglucosaminyltransferase; CMP-NeuAc: cytidine 5'-monophospho-*N*-acetylneurminic acid; UDP-Gal: uridine 5'-diphospho-galactose; UDP-GlcNAc: uridine 5'-diphospho-*N*-acetylglucosamine.

[b]Position to which sugar is added.

   c. GlcNAc transferase I buffer (GNTB): 25 m*M* HEPES-NaOH, pH 6.5, 75 m*M* NaCl, 100 m*M* glucose, 10 mg/mL BSA, and 1 m*M* $MnCl_2$.
8. Cell lysis buffer: 0.1*M* Tris-HCl, pH 7.2, 0.15*M* NaCl, 1.5 m*M* $MgCl_2$, 1% aprotinin (v/v), 0.5% Nonidet P-40 (v/v), and 1 m*M* phenylmethylsulfonyl fluoride (added just before use).
9. Desalting columns: Sephadex G-50 (coarse, 1 × 30 cm) equilibrated in 50 m*M* ammonium formate, 0.1% SDS, and 0.02% sodium azide.
10. Reagents for removal of detergent from membrane-bound transferases:
    a. Column: Siliconized 0.3 × 30 cm glass column containing 25 cm of Sephadex G-50 (fine) overlaid with 1.3 cm of the appropriate nucleotide affinity resin, equilibrated in labeling buffer containing 0.25% octyl-β-glucopyranoside (w/v). For sialyltransferases, the nucleotide resin is CDP-hexanolamine Sepharose 4B (10–13 μmol/mL), and the buffer is STB. For GlcNAc transferase I, the resin is UDP-hexanolamine Sepharose 4B (10–14 μmol/mL), and the buffer GNTB. CDP-Hexanolamine Sepharose is available from Genzyme (Cat. No. CDPH-1) or can be synthesized as described *(30)*. UDP-Hexanolamine (Sigma, Cat. No. U-2627) can be coupled to Sepharose 4B with cyanogen bromide as described *(30)*.
    b. Elution buffer: Appropriate column buffer (STB or GNTB) containing 0.25% octyl-β-glucopyranoside (w/v), and 1.5*M* NaCl (sialyltransferases) or 3*M* NaCl (GlcNAc transferase I).
    c. Marker buffer: Elution buffer containing 5 mg/mL Blue Dextran.
11. Glycosidase buffers:
    a. Sialidase buffer: 25 m*M* HEPES-KOH, pH 6.75, 150 m*M* NaCl, 0.1 m*M* $MgCl_2$, and 1 m*M* $CaCl_2$.
    b. PNGase buffer A: 1% SDS, 1% β-mercaptoethanol.
    c. PNGase buffer B: 150 m*M* Na phosphate, pH 8.6, 15 m*M* EDTA, 5% NP-40 (v/v).

3. Methods

3.1. Detergent Removal (see *Note 3)*

1. Dilute a sufficient amount of the appropriate glycosyltransferase (up to 0.5 mL) 1:4 in column buffer, and slowly apply to the detergent-removing column. (This should be done at 2–4°C to minimize loss of enzyme activity.)
2. Wash the column with 4 mL of appropriate buffer (STB or GNTB). This washes away the detergent.
3. Layer marker buffer (60 μL) on top of the column.
4. Elute the enzyme with 5 mL elution buffer, collecting 250-μL fractions.

5. Pool fractions containing Blue Dextran. Enzyme activity is assayed as described (sialytransferases: *[6]*; galactosyltransferase: *[31]*; GlcNAc transferase I: *[32]*), and should be used within 24 h.

3.2. Glycosidase Treatment of Cells (see *Note 4)*

The *V. cholera* sialidase used here is equally effective at 4 or 37°C. This is usually done in the cold (0–4°C) if cell surface proteins are being examined in order to prevent membrane recycling.

1. Prepare a single cell suspension of cells, and wash several times with sialidase buffer.
2. Incubate the cells (up to 10^8 cells/mL) with 200 mU/mL of *V. cholera* sialidase for 45 min at 4°C.
3. Pellet the cells at 350*g* for 10 min. The supernatant can be used to quantitate the amount of sialic acid released using an HPLC modification of the thiobarbituric acid assay *(34)*.
4. Wash the cells four times in PBS to remove sialidase. Cells are now ready to label with sialyltransferase.

3.3. Cell Labeling (see *Note 5)*

Again, labelings of intact cells are usually performed at 0–4°C to prevent membrane recycling.

1. Prepare a single cell suspension of up to 10^8 cells (*see* Note 6), and wash several times with either STB (for sialyltransferases), CPB (for galactosyltransferase), or PBS (GlcNAc transferase I). For labeling with galactosyltransferase and GlcNAc transferase I, the final wash should be with appropriate labeling buffer (GTB or GNTB, respectively).
2. Resuspend the final pellet in 50 µL of the appropriate labeling buffer (STB, GTB, or GNTB).
3. Add the transferase (typically 0–20 mU [U = µmol/min] of sialyltransferases or GlcNAc transferase I, and up to 200 mU of galactosyltransferase/10^7 cells, but this must be determined empirically for each cell type) diluted in the appropriate labeling buffer. Final vol should be 100 µL. Controls should always be performed where only cells are added (because of endogenous glycosyltransferase activity on the cell surfaces) and where only enzyme is added (because enzymes often label themselves to a small extent).
4. Labeling is initiated by addition of 1–4 µCi of the appropriate sugar nucleotide (*see* Note 7). If product analysis is to be performed, then the sugar nucleotide should not be isotopically diluted, so that suffi-

cient radiolabel is incorporated to allow analysis. If quantitation of the acceptor substrates is to be performed, then the radiolabel should be diluted to about five times K_m for the nucleotide sugar (*see* Table 1 for K_m values of nucleotide sugars).

5. After an appropriate length of time, the reaction is stopped by the addition of 1 mL of ice-cold labeling buffer.
6. Pellet the cells, 350*g* for 10 min. The supernatant should be saved, so that the breakdown of the sugar nucleotide can be monitored (*see* Note 7).
7. Resuspend the cells in 100 µL of labeling buffer, and layer over 1 mL of PBS/sucrose. Pellet the cells through the sucrose at 1500*g* for 10 min. Discard supernatant.
8. Resuspend the cell pellet in 1 mL of lysis buffer, vortex, and incubate on ice for 45 min.
9. Pellet insoluble material at 1500*g* for 10 min. Discard the pellet.
10. Add 100 µL of 20% SDS to the supernatant (2% final), and boil for 5 min.
11. Apply sample to desalting column. Collect 1-mL fractions. Locate and pool macromolecular material by counting aliquots of each fraction. Amounts of acceptor molecules are calculated based on the radioactivity incorporated into macromolecular material.
12. Lyophilize macromolecular material. Resuspend the lyophilized material in 1 mL water, and precipitate the protein by adding 8 mL of cold acetone (–20°C). Allow to precipitate at –20 °C for 5 h and collect the precipitate by centrifugation at 1500*g* for 10 min.

3.4. *Product Analysis* (see *Note 8*)

3.4.1. *SDS-PAGE*

Dissolve the acetone-precipitated sample in SDS sample buffer, and separate by SDS-PAGE. The radiolabeled proteins are then visualized by autofluorography after En^3Hance (NEN) treatment.

3.4.2. *Determination of* N-*Linked Oligosaccharides*

1. Resuspend the acetone-precipitated sample in 100 µL of PNGase buffer A, and boil for 5 min (*see* Note 9).
2. After cooling, add 200 µL of PNGase buffer B, and vortex the sample.
3. Add PNGase (100 mU as defined by Boehringer Mannheim) and allow the digestion to proceed for 24 h. A second addition of PNGase is allowed to proceed for another 24 h.
4. Stop the reaction by the addition of 500 µL of 20% SDS.

5. Separate the released oligosaccharides from the resistant material by gel-filtration chromatography on the G-50 desalting columns. Determine the resistant material (void), and released oligosaccharides (included) as before. Lyophilize each separately.
6. Analyze the resistant material (proteins) by SDS-PAGE as before. The released oligosaccharides can be analyzed by a number of techniques: gel filtration chromatography *(35)*, anion-exchange chromatography (Mono Q FPLC *[36]* or Dionex HPLC *[37]*), or paper chromatography *(38)*.

3.4.3. Determination of O-Linkage by β-Elimination

1. Resuspend the acetone-precipitated sample in 0.5 mL of 0.1*M* NaOH, and 1*M* $NaBH_4$. Incubate at 37°C for 24–48 h. Stop by neutralizing with 2*M* acetic acid on ice.
2. Separate the released oligosaccharides from resistant material by gel filtration on desalting columns as before. Discard the macromolecular material. Proteins are partially hydrolyzed by this procedure.
3. Analyze the released oligosaccharides by conventional techniques as described earlier.

4. Notes

1. Commercial galactosyltransferase (Sigma) contains proteins with terminal GlcNAc residues that serve as acceptors for the enzyme. To reduce this background incorporation, the enzyme should be autogalactosylated with unlabeled UDP-galactose before it is used as a reagent *(10,15)*. Briefly, this is done by incubating 25 U in 1 mL of 50 m*M* Tris-HCl, pH 7.3, 5 m*M* $MgCl_2$, 1 m*M* β-mercaptoethanol, 1% aprotinin (v/v), and 0.4 m*M* UDP-galactose for 30 min at 37°C. The enzyme is then concentrated by precipitation with 85% saturated ammonium sulfate, washed one time with 85% saturated ammonium sulfate, and resuspended in 1 mL of 25 m*M* Hepes-NaOH, pH 7.3, 5 m*M* $MnCl_2$, 50% glycerol. The activity is stable for at least 1 yr when stored in this manner at –20°C.
2. Most of the transferases are stable in nonionic detergents, such as Triton X-100, Triton CF-54, or Nonidet P-40, up to concentrations of at least 1%. The manganese requiring transferases (galactosyltransferase and GlcNAc transferase I) should not be used with buffers containing EDTA, EGTA, or buffers that complex divalent cations (e.g., phosphate or citrate). Galactosyltransferase can tolerate all concentrations up to 0.5*M* NaCl, whereas the other transferases are more strongly inhibited by salt. All buffers containing $MnCl_2$ should be

adjusted to 0.1 pH unit above the final before the $MnCl_2$ may be added. The addition of the $MnCl_2$ will lower the pH slightly. HCl may be used to bring the pH down further if necessary. Manganese hydroxide will precipitate if NaOH is added directly to buffers containing $MnCl_2$.

3. Detergent removal is performed only if a membrane-bound transferase is being used (sialyltransferase and GlcNAc transferase I), and if intact cells or membranes are being labeled. The method given is a modification of the procedure developed by Sadler et al. *(8)*. The same basic procedure is used for all of the sialyltransferases and GlcNAc transferase I. Detergent can be removed from transferases by a number of alternative techniques, in addition to the method given in Section 3. Paulson and Rogers *(21)* have used a virgin column of Sephadex G-50 to remove Triton detergents. Bio-Beads SM-2 have also been used successfully in selected cases *(21)*. We have found that complete removal of detergents can easily be monitored by lysis of red blood cells under isotonic conditions *(5)*. When soluble substrates, such as glycoproteins, are to be labeled with transferases, removal of detergent is not necessary. In this case, the labeling buffers described earlier can be used, except that for the sialyltransferases and GlcNAc transferase I, 0.1% (v/v) Triton X-100 should be included. For examples, *see* refs. *14–17,39.*
4. Glycosidase treatment of cells is performed only if penultimate structures are to be examined. Most often, this is done in conjunction with the sialyltransferases, so the procedure given is for sialidase treatment of cells.
5. The procedure given shows how to label cells in the general case. In order to quantitate the amount of acceptor in a sample, several controls must be performed. The dependence of the labeling on time, enzyme concentration, and sugar nucleotide concentration must be performed to show that the acceptor sites are being saturated with label under the conditions of the experiment. In addition, a dependence on acceptor concentration (e.g., cell number) should be performed to show that the acceptor is within the linear portion of the assay. The number of cells, amount of enzyme, and amount of nucleotide sugar need to be optimized for each cell type examined. Some amounts are included in the method as guidelines from experiments using murine T-lymphocytes. For further examples, *see* refs. *5,6,* and *12.*
6. Do not use trypsin/EDTA or other protease methods for making single-cell suspensions, since they will remove many of the glycoproteins from the cell surface.

7. The stability of the radiolabeled sugar nucleotide should be examined for each cell type or substrate examined by established techniques *(38)*. This is particularly important when labeling crude subcellular fractions that may be contaminated with pyrophosphatases. Several inhibitors of phosphatases may be used *(40,41)*. We have found sodium fluoride (25 m*M* final concentration) to be particularly effective.
8. Product analysis typically involves examination of the labeled products by SDS-PAGE, followed by autofluorography. It may also include a determination of how the labeled oligosaccharides are linked to protein. PNGase is used to remove *N*-linked structures *(42)*, and mild base induced β-elimination is used to release *O*-linked structures *(43)*. After treatment with PNGase, both the protein and the released oligosaccharides may be analyzed, whereas the protein is largely destroyed by the β-elimination procedure permitting analysis of the oligosaccharide only.
9. Native glycoproteins are very poor substrates for PNGase, and must be denatured or proteolyzed to make the sugars accessible to the enzyme. Boiling in SDS with mercaptoethanol works well for most proteins. The SDS-denatured protein is then mixed with an excess of a nonionic detergent, such as Triton X-100 (*see* PNGase buffer B), which effectively removes the SDS from solution. The sample is then ready for the addition of the enzyme.

Acknowledgment

This work was supported by National Institutes of Health Grant CA 42486.

References

1. Beyer, T. A., Sadler, J. E., Rearick, J. I., Paulson, J. C., and Hill, R. L. (1981) Glycosyltransferases and their use in assessing oligosaccharide structure and structure-function relationships. *Adv. Enzymol.* **52,** 23–175.
2. Kornfeld, R. and Kornfeld, S. (1985) Assembly of asparagine-linked oligosaccharides. *Annu. Rev. Biochem.* **54,** 631–664.
3. Roseman, S. (1970) The synthesis of complex carbohydrates by multiglycosyltransferase systems and their potential function in intercellular adhesion. *Chem. Phys. Lipids.* **5,** 270–297.
4. Rademacher, T. W., Parekh, R. B., and Dwek, R. A. (1988) Glycobiology. *Annu. Rev. Biochem.* **57,** 785–838.
5. Whiteheart, S. W. and Hart, G. W. (1987) Sialyltransferases as specific cell surface probes of terminal and penultimate saccharide structures on living cells. *Anal. Biochem.* **163,** 123–135.

6. Passaniti, A. and Hart, G. W. (1988) Cell surface sialylation and tumor metastasis. Metastatic potential of B16 melanoma variants correlates with their relative numbers of specific penultimate oligosaccharide structures. *J. Biol. Chem.* **263,** 7591–7603.
7. Powell, L. D., Whiteheart, S. W., and Hart, G. W. (1987) Cell surface sialic acid influences tumor cell recognition in the mixed lymphocyte reaction. *J. Immunol.* **139,** 262–270.
8. Sadler, J. E., Paulson, J. C., and Hill, R. L. (1979) The role of sialic acid in the expression of human MN blood group antigens. *J. Biol. Chem.* **254,** 2112–2119.
9. Rogers, G. N., Herrler, G., Paulson, J. C., and Klenk, H. D. (1986) Influenza C virus uses 9-*O*-acetyl-*N*-acetylneuraminic acid as a high affinity receptor determinant for attachment to cells. *J. Biol. Chem.* **261,** 5947–5951.
10. Torres, C-R. and Hart, G. W. (1984) Topography and polypeptide distribution of terminal *N*-acetylglucosamine residues on the surfaces of intact lymphocytes. *J. Biol. Chem.* **259,5,** 3308–3317.
11. Viitala, J. and Finne, J. (1984) Specific cell-surface labeling of polyglycosyl chains in human erythrocytes and HL-60 cells using endo-beta-galactosidase and galactosyltransferase. *Eur. J. Biochem.* **138,** 393–397.
12. Passaniti, A. and Hart, G. W. (1990) Metastasis-associated murine melanoma cell surface galactosyltransferase: Characterization of enzyme activity and identification of the major surface substrates. *Cancer Res.* **50,** 7261–7271.
13. Reichner, J. S., Whiteheart, S. W., and Hart, G. W. (1988) Intracellular trafficking of cell surface sialoglycoconjugates. *J. Biol. Chem.* **263**, 16,316–16,326.
14. Benko, D. M., Haltiwanger, R. S., Hart, G. W., and Gibson, W. (1988) Virion basic phosphoprotein from human cytomegalovirus contains *O*-linked *N*-acetylglucosamine. *Proc. Natl. Acad. Sci. USA* **85,** 2573–2577.
15. Holt, G. D. and Hart, G. W. (1986) The subcellular distribution of terminal *N*-acetylglucosamine moietics. Localization of a novel protein-saccharide linkage, *O*-linked GlcNAc. *J. Biol. Chem.* **261,** 8049–8057.
16. Holt, G. D., Haltiwanger, R. S., Torres, C. R., and Hart, G. W. (1987) Erythrocytes contain cytoplasmic glycoproteins. *O*-linked GlcNAc on Band 4.1. *J. Biol. Chem.* **262,** 14,847–14,850.
17. Machamer, C. E. and Cresswell, P. (1984) Monensin prevents terminal glycosylation of the *N*- and *O*-linked oligosaccharides of the HLA-DR-associated invariant chain and inhibits its dissociation from the α- βchain complex. *Proc. Natl. Acad. Sci. USA* **81,** 1287–1291.
18. Duncan, J. R. and Kornfeld, S. (1988) Intracellular movement of two mannose 6-phosphate receptors: return to the Golgi apparatus. *J. Cell Biol.* **106,** 617–628.
19. Thilo, L. (1983) Labeling of plasma membrane glycoconjugates by terminal glycosylation (galactosyltransferase and glycosidase). *Methods Enzymol.* **98,** 415–420.
20. Paulson, J. C., Sadler, J. E., and Hill, R. L. (1979) Restoration of specific myxovirus receptors to asialoerythrocytes by incorporation of sialic acid with pure sialyltransferases. *J. Biol. Chem.* **254,** 2120–2124.

21. Paulson, J. C. and Rogers, G. N. (1987) Resialylated erythrocytes for assessment of the specificity of sialyloligosaccharide binding proteins. *Methods Enzymol.* **138,** 162–168.
22. Hill, R. L., Beyer, T. A., Paulson, J. C., Prieels, J. P., Rearick, J. I., and Sadler, J. E. (1980) Glycosyl transferases in oligosaccharide biosynthesis and their use in structure-function analysis of glycoproteins, in *Frontiers of Bioorganic Chemistry and Molecular Biology* (Ananchenko, S. N., ed.), Pergamon, Oxford and New York, pp. 63–71.
23. Whiteheart, S. W., Passaniti, A., Reichner, J. S., Holt, G. D., Haltiwanger, R. S., and Hart, G. W. (1989) Glycosyltransferase probes. *Methods Enzymol.* **179,** 82–95.
24. Sadler, J. E., Beyer, T. A., Oppenheimer, C. L., Paulson, J. C., Prieels, J. P., Rearick, J. I., and Hill, R. L. (1982) Purification of mammalian glycosyltransferases. *Methods Enzymol.* **83,** 458–514.
25. Sadler, J. E., Beyer, T. A., and Hill, R. L. (1981) Affinity chromatography of glycosyltransferases. *J. Chromatogr.* **215,** 181–194.
26. Paulson, J. C. and Colley, K. J. (1989) Glycosyltransferases. Structure, localization, and control of cell type-specific glycosylation. *J. Biol. Chem.* **264,** 17,615–17,618.
27. Colley, K. J., Lee, E. U., Adler, B., Browne, J. K., and Paulson, J. C. (1989) Conversion of a Golgi apparatus sialyltransferase to a secretory protein by replacement of the NH_2-terminal signal anchor with a signal peptide. *J. Biol. Chem.* **264,** 17,619–17,622.
28. Larsen, R. D., Rajan, V. P., Ruff, M. M., Kukowska-Latallo, J., Cummings, R. D., and Lowe, J. B. (1989) Isolation of a cDNA encoding a murine UDPgalactose: β-D-galactosyl- 1,4-*N*-acetyl-D-glucosaminide α-1,3-galactosyltransferase: Expression cloning by gene transfer. *Proc. Natl. Acad. Sci. USA* **86,** 8227–8231.
29. Weinstein, J., de Souza-e-Silva, U., and Paulson, J. C. (1982) Purification of a Gal β1,4GlcNAc α2,6 sialyltransferase and a Gal β1,3(4)GlcNAc α2,3 sialyltransferase to homogeneity from rat liver. *J. Biol. Chem.* **257,** 13,835–13,844.
30. Sadler, J. E., Rearick, J. I., Paulson, J. C., and Hill, R. L. (1979) Purification to homogeneity of a β-galactoside α2,3 sialytransferase and partial purification of an α-*N*-acetygalactosaminide α2,6 sialytransferase from porcine submaxillary glands. *J. Biol. Chem.* **254,** 4434–4443.
31. Trayer, I. P. and Hill, R. L. (1971) The purification and properties of the A protein of lactose synthetase. *J. Biol. Chem.* **246,** 6666–6675.
32. Oppenheimer, C. L. and Hill, R. L. (1981) Purification and characterization of a rabbit liver α1,3 mannoside β1,2 *N*-acetylglucosaminyltransferase. *J. Biol. Chem.* **256,** 799–804.
33. Nishikawa, Y., Pegg, W., Paulsen, H., and Schachter, H. (1988) Control of glycoprotein synthesis: Purification and characterization of rabbit liver UDP-*N*-acetylglucosamine: α-3-D-mannoside β-1,2-*N*-acetylglucosaminyltransferase I. *J. Biol. Chem.* **263,** 8270–8281.

34. Powell, L. D. and Hart, G. W. (1986) Quantitation of picomole levels of *N*-acetyl- and *N*-glycolylneuraminic acids by a HPLC-adaptation of the thiobarbituric acid assay. *Anal. Biochem.* **157,** 179–185.
35. Yamashita, K., Mizuochi, T., and Kobata, A. (1982) Analysis of oligosaccharides by gel filtration. *Methods Enzymol.* **83,** 105–126.
36. Van Pelt, J., Damm, J. B., Kamerling, J. P., and Vliegenthart, J. F. (1987) Separation of sialyl-oligosaccharides by medium pressure anion-exchange chromatography on Mono Q. *Carbohydr. Res.* **169,** 43–51.
37. Hardy, M. R. and Townsend, R. R. (1988) Separation of positional isomers of oligosaccharides and glycopeptides by high-performance anion-exchange chromatography with pulsed amperometric detection. *Proc. Natl. Acad. Sci. USA* **85,** 3289–3293.
38. Porzig, E. F. (1978) Galactosyltransferase activity of intact neural retinal cells from the embryonic chicken. *Dev. Biol.* **67,** 114–126.
39. Kearse, K. P. and Hart, G. W. (1991) Lymphocyte activation induces rapid changes in nuclear and cytoplasmic glycoproteins. *Proc. Natl. Acad. Sci. USA* **88,** 1701–1705.
40. Lau, J. T. Y. and Carlson, D. M. (1981) Galactosyltransferase activities in rat intestinal mucosa: Inhibition of nucleotide pyrophosphatase. *J. Biol. Chem.* **256,** 7142–7145.
41. Faltynek, C. R., Silbert, J. E., and Hof, L. (1981) Inhibition of the action of pyrophosphatase and phosphatase on sugar nucleotides. *J. Biol. Chem.* **256,** 7139–7141.
42. Tarentino, A. L., Gomez, C. M., and Plummer, T. H., Jr. (1985) Deglycosylation of asparagine-linked glycans by peptide: N-glycosidase F. *Biochemistry* **24,** 4665–4671.
43. Spiro, R. G. (1972) Study of the carbohydrates of glycoproteins. *Methods Enzymol.* **28,** 3–43.

CHAPTER 15

Purification of the EGF Receptor for Oligosaccharide Studies

Mary Gregoriou

1. Introduction

The epidermal growth factor receptor is a membrane glycoprotein expressed in a large variety of higher eukaryotic cells. Its role in growth and development as well as its high levels of expression in certain types of cancer have stimulated wide-ranging studies at the tissue, cell, and molecular levels. The receptor is a 175-kDa single polypeptide protein consisting of an extracellular EGF binding domain, a transmembrane peptide, and an intracellular protein tyrosine kinase domain *(1,2)*. The extracellular domain contains 12 potential glycosylation sites and is heavily glycosylated *(3–5)*. The structures of some of these carbohydrate antigens have been identified by either determining the specificity of monoclonal antibodies raised against the receptor or, conversely, by investigating the reactivity of unrelated monoclonals of known antigenic specificity with purified receptor *(6,7)*.

This chapter describes the classical approach for purification of biologically active EGF receptor from the overexpressing A431 epidermoid carcinoma cells, using EGF-affinity chromatography essentially as first reported by Cohen et al. (1980) *(8; see also 9,10)*. The reader is also referred to alternative purification methods, most of which employ immunoaffinity chromatography (e.g., *11,12)* and would, therefore, require availability of the corresponding purified antibody in milligram quantities. Methods for the preparation of mouse EGF

From: *Methods in Molecular Biology, Vol. 14: Glycoprotein Analysis in Biomedicine*
Edited by: E. F. Hounsell

(13) and the EGF-affinity column are described, along with protocols for biosynthetic labeling of the EGF receptor of A431 cells, and its extraction, purification, and analysis.

2. Materials

1. Submaxillary glands from adult male mice (30 g minimum body wt). The glands are immediately stored on dry ice (or at –70°C for long-term storage).
2. Stock solutions: NaCl, 0.5*M*. PBS, pH 7.4: 0.14*M* NaCl, 0.0015*M* KH_2PO_4, 0.0081*M* $Na_2PO_4 \cdot 12H_2O$, and 0.0027*M* KCl. 0.2*M* NaCl/PBS. PBS/0.005*M* EDTA. Acetic acid, pH 3.0, 0.05*M*. Hydrochloric acid, 1.0*M*. Glycine/HCl, pH 2.5, 0.2*M*. Sodium hydroxide, 0.5*M*. Sodium bicarbonate, pH 8.2, 1.0*M*. Ammonium acetate, pH 5.7, 0.02*M*. Potassium/sodium phosphate buffer, pH 8.0, 0.3*M*. Tris/HCl buffer, pH 7.5, 1*M*. TBS (Tris-buffered saline): 0.05*M* Tris/HCl, 0.150*M* NaCl, pH 7.5. Ethanolamine, pH 8.2, 1*M*; pH 9.7, 0.005*M*.
3. Extraction buffer: PBS, 1% (v/v) Triton X-100, 10 % (v/v) glycerol, 20 m*M* benzamidine, and 1 m*M* EDTA (*see* Note 1).
4. Reagents: Epidermal growth factor (100 µg, receptor grade). Iodogen. Amplify™ (Amersham, Intl. plc, Amersham, UK). Sodium iodide, [^{125}-I]-NaI, 100 mCi/mL. D-[6-^{3}H]glucosamine/HCl (35 Ci/mmol) or L-[6-^{3}H]fucose (25 Ci/mmol). Isopropyl alcohol.
5. Protein purification materials: Biogel P-10 (200–400 mesh, Bio-Rad, Herts, UK), in 0.15*M* NaCl/0.05*M* HCl, packed in a 5 × 90 cm column at 4°C. Affi-Gel 10 (Bio-Rad), DE52 anion exchanger in NaCl/PBS packed in a 0.7 × 50 cm column. SpectraPor 6 dialysis tubing (2000 mol wt cutoff, 38 mm). Ultrafiltration concentrator (50 mL), fitted with an XM 50 membrane (e.g., amicon). Fraction collector.
6. Cell culture: A431 cells, DMEM medium supplemented with 5% fetal calf serum, penicillin (30 mg/L), and streptomycin (75 mg/L). Humidified 5% CO_2/95% air incubator at 37°C.

3. Methods

The EGF receptor is conveniently isolated from Triton X-100 extracts of A431 cells in a single affinity-purification step. To prepare the affinity column, EGF (10 mg) is first purified from male mouse submaxillary glands *(13)*, assayed for receptor binding activity, and coupled to Affigel-10 via amide linkages of primary amino groups on EGF *(8)*. A431 cells are cultured in 175-cm^2 flasks and labeled biosynthetically with tritiated monosaccharides. Alternatively, they can be grown in roller bottles on a larger scale. The cells are harvested. Clarified detergent extracts are prepared and chromatographed on the Affi-Gel-EGF column. Bound EGF receptor is eluted at high pH, neutralized, and concentrated for analysis.

3.1. Preparation of EGF

3.1.1. Purification

1. Thaw glands (30 g) at 4°C and homogenize with 0.05*M* acetic acid (120 mL) in a blender at high speed for 3 min at 4°C.
2. Shell freeze the homogenate in a round flask in an isopropanol/dry-ice mixture, followed by thawing in a 37°C water bath. (The temperature of the homogenate should not exceed 4°C.)
3. Obtain a clear soluble fraction by centrifugation of the crude extract at 100,000*g* for 30 min and filtration of the supernatant through glasswool. Lyophilize the clear fraction overnight.
4. Resuspend the preparation in 10 mL 1*M* HCl, and dissolve by dilution with 32 mL 0.05*M* HCl.
5. Load all the extract onto the P-10 column, and develop at 1 mL/min in equilibration buffer collecting 16-mL fractions. EGF is identified by a radioligand displacement receptor binding assay: Incubate 2-µL samples from every third fraction mixed with [^{125}I]-EGF (20,000 cpm) in 100 µL binding buffer with A431 cell monolayers in a 96-multiwell plate, and assay as described later. EGF elutes in a small A_{280} peak after approx 2 column vol (Fig. 1a; *see* Note 2)
6. Pool active fractions, and adjust pH to 5.7 with ammonia. Dialyze in SpectraPor 6 dialysis tubing against 2 L of 1 m*M* ammonium acetate, pH 5.7, with two buffer changes overnight.
7. Lyophilize the preparation, dissolve in 10–20 mL of water, and dialyze against 1 L of 20 m*M* ammonium acetate, pH 5.7, with two to three changes overnight.
8. Apply the sample to the DE52 column, and develop with equilibration buffer, collecting 10-mL fractions, until the A_{280} of the fractions is down to baseline (Fig. 1b). Apply a linear gradient 0.02–0.2*M* ammonium acetate buffer, pH 5.7, in 400 mL total vol, and assay for EGF as for the P-10 column. Pool fractions around the major peak of activity, lyophilize, and store dry at 4°C. It is normally sufficient to quantitate a small fraction by assuming that at 1 mg/mL E_{280} = 3. Quantitation of biologically active EGF using a receptor binding assay is appropriate, since the preparation of EGF is to be used as an affinity ligand for receptor purification and is described later. EGF (10–20 mg) is usually obtained from 30 g of submaxillary glands.

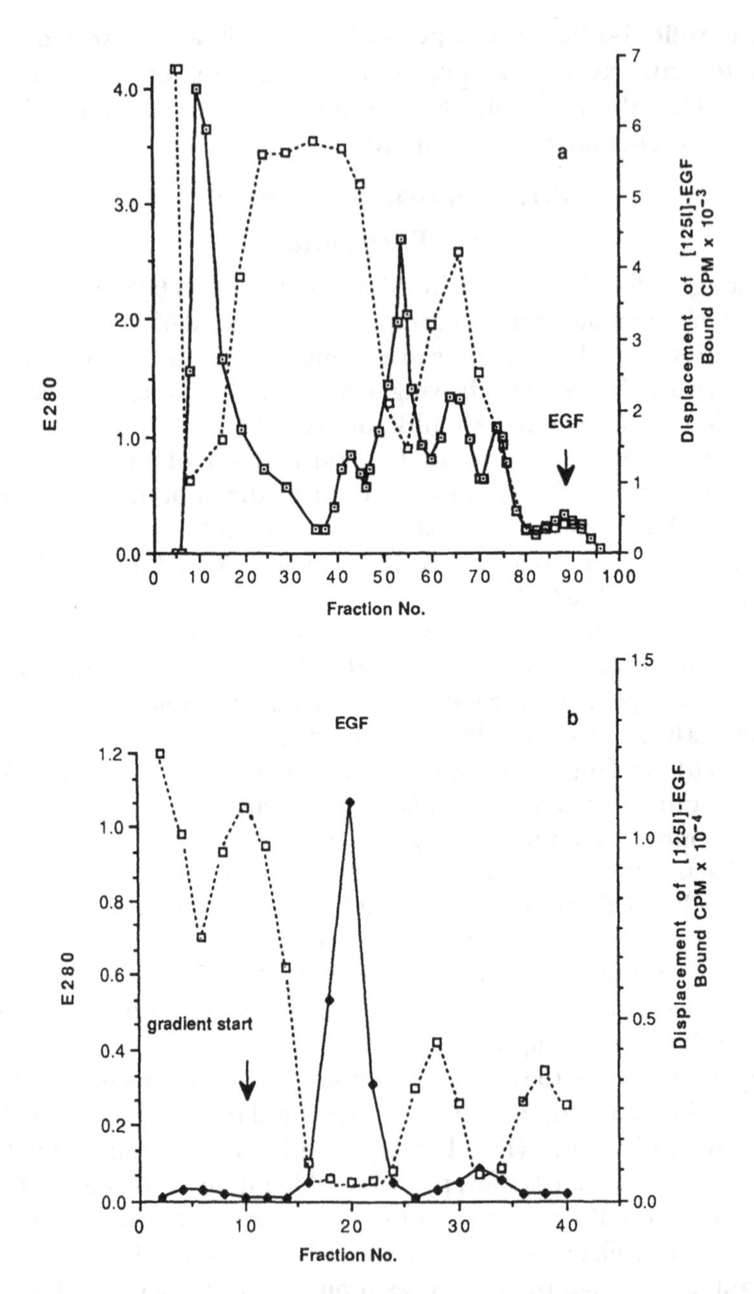

Fig. 1. Purification of mouse EGF. (a) P-10 protein (▣) and activity (-□-) profiles, obtained as detailed in text. (b) Chromatography of the EGF pool on DE52: Protein (◆) and activity (-□-) profiles.

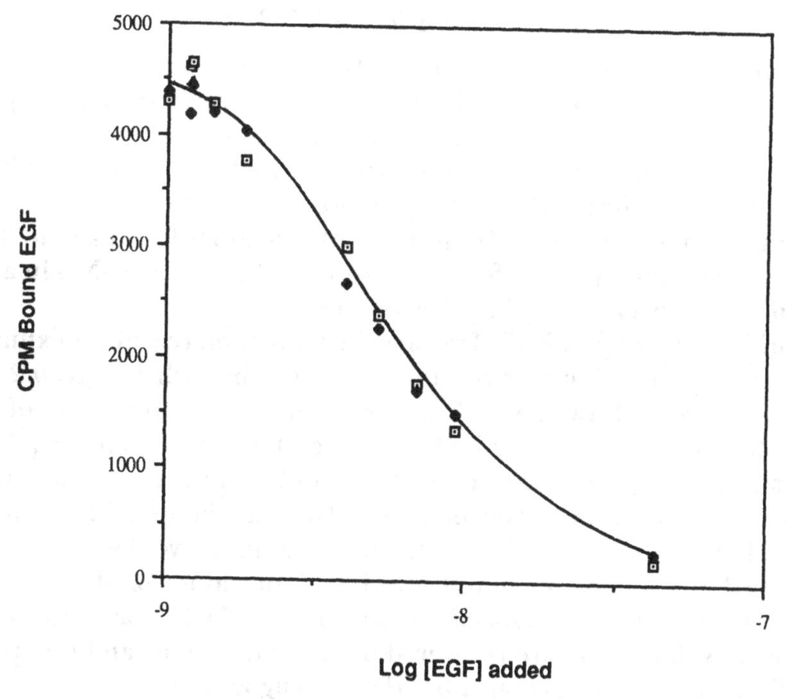

Fig. 2. Quantitation of purified EGF. Increasing concentrations of purified EGF (◘) in the presence of a constant amount of [^{125}I]-EGF are incubated with A431 cell monolayers and bound radioactivity determined as described. A standard curve of bound CPM against log[EGF] is constructed using commercially available EGF (◆). The two curves coincide, indicating that the preparation has high specific activity and its concentration corresponds to that of standard EGF.

3.1.2. Quantitation of Biologically Active EGF

The amount of active EGF in the preparation may be quantitated in a radioligand displacement assay using authentic (commercially available) EGF. A standard binding curve is constructed in which A431 cells are incubated with increasing concentrations of authentic EGF in the presence of a constant amount of [^{125}I]-EGF. The amount of bound [^{125}I]-EGF is determined by scintillation counting and is related to the concentration of added EGF in a standard curve. The EGF preparation is quantitated by comparison of a similarly constructed plot in which the authentic radiolabeled EGF is displaced from A431 cells by increasing concentrations of the purified preparation (Fig. 2).

3.1.3. Iodination of EGF

1. Preparation of iodogen-coated tubes: Dissolve iodogen in chloroform at 0.5 mg/mL and distribute 100-µL aliquots to clean glass tubes (e.g., 1 × 7.5 cm). Evaporate the solvent slowly under a stream of nitrogen; store dried tubes at room temperature. Rinse tubes once with PBS and twice with distilled water. Dry before use.
2. Iodination: Dissolve EGF (10 µg) in 5 µL 0.05*M* HCl, and add to 30 µL K/Na buffer, pH 8.0, followed by 10 µL (1 mCi) [^{125}I]-NaI in an iodogen tube on ice, and agitate for 15 min.
3. Purification of [^{125}I]-EGF: Transfer the iodination reaction mixture to an Eppendorf tube, rinse the reaction tube with 100 µL 0.2*M* NaCl/PBS, and transfer with the reaction mixture to the top of a Sephadex G-10 column (0.7 × 50 cm). Develop the column in equilibration buffer, collect 200-µL fractions, and count 2-µL samples in a γ scintillation counter. The iodinated EGF elutes in the void volume; fractions representing 95% of the total cpm in the void volume are pooled, the volume measured, and 100-µL portions stored at –20°C in lead containers. The specific activity of [^{125}I]-EGF is calculated on the basis that 9.5 µg are recovered in the void volume, and the specific activity is not less than 1.6×10^5 cpm/ng EGF.

3.1.4. EGF Binding Assay

1. A431 cell monolayers are prepared in 96-multiwell plates. Cells are subcultured at 10^4 cells/0.1 mL medium/well, incubated for 1–2 d, and used at subconfluence.
2. Prepare two dilution series of EGF in binding buffer using (a) commercially obtained EGF and (b) the EGF preparation to be quantitated (assuming $A_{280} = 3$ at 1 mg /mL) starting with stock solutions at 1 µg/mL ($1.67 \times 10^{-7}M$). An appropriate dilution range would be, e.g., 0, $3.34 \times 10^{-10}M$, $4.17 \times 10^{-10}M$, $8.35 \times 10^{-10}M$, $1.67 \times 10^{-9}M$, $5.84 \times 10^{-9}M$, $8.35 \times 10^{-9}M$, $1.17 \times 10^{-8}M$, $1.67 \times 10^{-8}M$, $8.35 \times 10^{-8}M$, and $1.67 \times 10^{-7}M$. Using these stock solutions, prepare the series of working solutions by mixing 100 µL with 100 µL [^{125}I]-EGF ($2 \times 10^{-9}M$) also diluted in binding buffer.
3. Aspirate the culture medium, wash cells twice with 100 µL binding buffer, and incubate for 1 h at 37°C in binding buffer. Transfer to ice, replace buffer with [^{125}I]-EGF working solutions containing increasing amounts of unlabeled EGF, and place at 4°C for 4–16 h. Remove unbound EGF, and wash monolayers three times with 100 µL ice-cold buffer. Add 100 µL 0.5*M* NaOH, and leave at 37°C for 1 h. Transfer cell lysates to tubes, and count bound radioactivity in a γ scintillation counter.

4. Plot the amount of bound [^{125}I]-EGF (in CPM) after subtracting nonspecific binding (obtained at the highest cold EGF concentration) against total concentration of EGF added. The curves obtained with authentic and prepared EGF should be superimposable, indicating that the preparation is fully active with respect to receptor binding and that its specific activity is identical to that of authentic EGF.

3.2. Preparation of the EGF Affinity Column

1. Dialyze a solution of EGF (10 mg dissolved in 4 mL 0.01*M* HCl and adjusted to pH 8 with 1*M* $NaHCO_3$) against 3 × 1 L 0.1*M* $NaHCO_3$ buffer, pH 8.2, and quantitate the amount of protein from A_{280} measurements.
2. Remove approx 5 g packed Affi-Gel 10, and wash quickly on a sintered funnel with isopropanol, ice-cold water, and 0.1*M* $NaHCO_3$ (3 bed vol of each) without letting the gel go dry. The gel is weighed, transferred into a polypropylene screw-cap tube on ice, and mixed with the dialyzed EGF solution. Coupling is carried out with gentle agitation using an end-over-end mixer, at 4°C overnight.
3. Collect unreacted EGF by filtration through a small sintered funnel, and wash gel with 2 × 10 mL 0.2*M* glycine/HCl, pH 2.5. Combine the washes with the filtrate, adjust pH to 2.5, and estimate the amount of unbound EGF from the A_{280}. The efficiency of coupling is typically 80–90%. In the meantime, transfer the gel to the original coupling tube with 1*M* ethanolamine, pH 8.2, and mix for 2 h at room temperature to block reactive groups.
4. Transfer the gel back to the funnel, and wash successively with 10 × 10 mL each of 0.2*M* glycine/HCl, pH 2.5, 0.005*M* ethanolamine, pH 9.7, and 0.1*M* $NaHCO_3$ buffer, pH 8.2. Wash with 15 × 10 mL TBS, and store at 4°C with 0.05% sodium azide (*see* Note 3).

3.3. Cell Culture

A-431 cells are grown in DMEM medium supplemented with 5% (v/v) newborn calf serum, penicillin, and streptomycin, under an atmosphere of 5% CO_2/95% air, at 37°C. Stock cultures are maintained at 0.6×10^6–3×10^6 cells/5 mL medium/25 cm^2 flask, and are expanded initially to two 75-cm^2 flasks (seeded at 3×10^6 cells/15 mL medium/flask) and subsequently four 175-cm^2 flasks (seeded at 5×10^6 cells/25 mL medium/flask).

1. Biosynthetic labeling: Set up four A431 cultures in 175-cm^2 flasks, and grow to ~50% confluence. Replace medium with fresh DMEM/serum containing either D-[6-^{3}H]glucosamine/HCl or L-[6-^{3}H]fucose at 5 µCi/mL for 48 h.

2. Cell harvesting: Remove medium, wash the flasks briefly with 10 mL ice-cold PBS/EDTA, and leave in 10 mL fresh PBS/EDTA for 5–10 min. Collect detached cells by centrifugation at 2000*g* for 5 min, and either store at –70°C until required or proceed with extraction (*see* Note 4).

3.4. Purification of the EGF Receptor

1. Resuspend the cells in 30 mL ice-cold extraction buffer, and stir for 20 min. Spin at 3000*g* for 5 min at 4°C, and clarify extract at 100,000*g* for 60 min at 4°C.
2. Transfer the clear supernatant to a 50 mL polypropylene tube containing 5 mL packed Affi-Gel-EGF, and mix gently for 1 h at room temperature. Transfer the gel suspension into a 10-mL column, and when the gel is packed, wash with 50 mL of ice-cold extraction buffer without benzamidine. Elute the receptor with two 50-mL washes of 0.005*M* ethanolamine/HCl, pH 9.7, adjust the pH 7.5–8 with 1*M* Tris-HCl, pH 7.5, and concentrate the two fractions 20-fold by ultrafiltration in an amicon concentrator fitted with an XM-50 membrane.
3. Analyze the ethanolamine fractions for EGF receptor by SDS-PAGE combined with silver staining and autoradiography (Fig. 3). The gel intended for autoradiography should be equilibrated in 5 vol of Amplify™ for 30 min prior to drying, in order to enhance the ^{3}H signal.
4. The purity of the receptor preparation is >95%, as judged by silver staining of SDS gels. The receptor may be quantitated as described by Gullick et al. *(14)* using [^{125}I]-EGF binding in a radioimmunoassay in which the receptor-EGF complex is separated from free [^{125}I]-EGF by immunoprecipitation with a monoclonal antireceptor antibody and protein A Sepharose. This method is preferable to conventional protein quantitation methods because it is least wasteful. From 2×10^{8} cells (harvested from four 175-cm^{2} flasks), 10 µg of purified receptor is obtained after concentration of the preparation. The receptor retains full biological activity for several months if stored with 50% glycerol at –70°C.

4. Notes

1. Detergents other than Triton X-100 may be used. Sodium deoxycholate, *n*-octylglucoside, and NP-40 have been used without adverse effects on the activity or purification of the EGF receptor preparation.
2. Negligible amounts of [^{125}I]-EGF displacing activity are detected in the void and after 1 column vol during P-10 chromatography. This activity is believed to be associated with storage or carrier proteins,

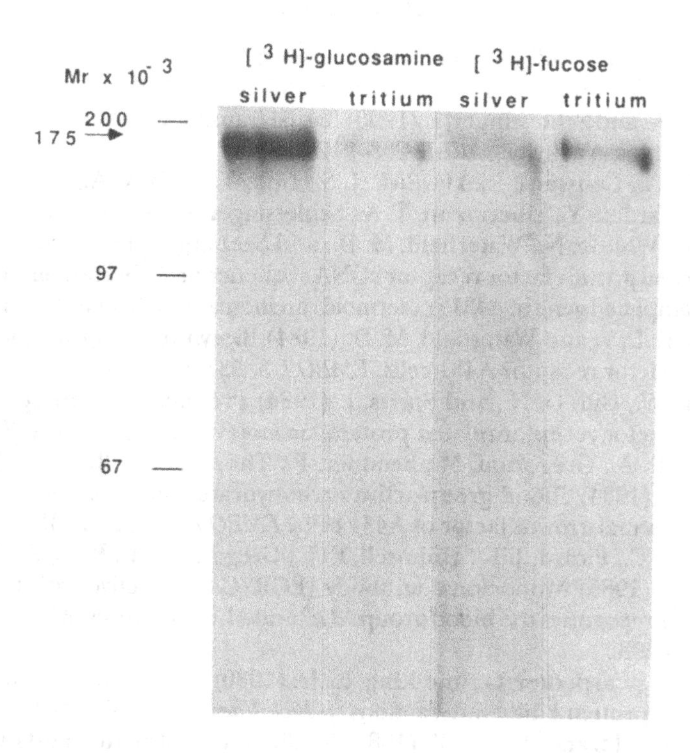

Fig. 3. SDS PAGE analysis of purified EGF receptor biosynthetically labeled with [^{3}H]-glucosamine and [^{3}H]-fucose. The receptor was detected by silver staining of protein and by autoradiography after enhancement of the tritium signal by treatment of the gel with Amplify. The receptor corresponds to the 175-kDa band indicated.

and is discarded. Although it is suggested for convenience that 2-µL EGF fractions be assayed to obtain chromatographic activity profiles, this amount is excessive, resulting in underestimation of the amount of activity in the major EGF peaks. Activity profiles thus obtained are qualitative rather than quantitative.

3. Affi-Gel-EGF columns are stable at least 1 yr (usually longer) at 4°C. After use, wash with 20 vol of TBS/0.05% sodium azide, close the column at both ends, and store at 4°C. Extensive use of the column may eventually result in nonspecific binding, leading to impurities in the receptor preparation.
4. It is preferable to harvest cells in PBS/EDTA, rather than by scraping, which is known to cause calcium-activated receptor proteolysis.

References

1. Carpenter, G. (1987) Receptors for epidermal growth factor and other polypeptide mitogens. *Ann. Rev. Biochem.* **56,** 881–914.
2. Ullrich, A. and Schlessinger, J. (1990) Signal transduction by receptors with tyrosine kinase activity. *Cell* **61,** 203–212.
3. Ullrich, A., Coussens, L., Haytlick. J. S., Dull, T. J., Gray, A., Tam, A. W., Lee, J., Yarden, Y., Libermann, T. A., Schlessinger, J., Downward, J., Mayes, E. L. V., Whittle, N., Waterfield, M. D., and Seeburg, P. H. (1984) Human epidermal growth factor receptor cDNA sequence and aberrant expression of the amplified gene in A431 epidermoid carcinoma cells. *Nature* **309,** 418–425.
4. Mayes, E .L. V. and Waterfield, M. D. (1984) Biosynthesis of the epidermal growth factor receptor A431 cells. *EMBO J.* **3,** 531–537.
5. Weber, W., Gill, G. N., and Spiess, J. (1984) Production of an epidermal growth factor receptor-related protein. *Science (Wash.)* **224,** 294–297.
6. Childs, R. A., Gregoriou, M., Scudder, P., Thorpe, S. J., Rees, A. R., and Feizi, T. (1984) Blood group-active carbohydrate chains on the receptor for epidermal growth factor of A431 cells. *EMBO J.* **3,** 2227–2233.
7. Gooi, H. C., Picard, J. K., Hounsell, E. F., Gregoriou, M., Rees, A. R., and Feizi, T. (1985) Monoclonal antibody (EGR/G49) reactive with the EGF receptor recognises the blood group ALe^b and ALe^y structures. *Mol. Immunol.* **22,** 689–693.
8. Cohen, S., Carpenter, G., and King, L., Jr. (1980) Epidermal growth factor-receptor-protein kinase interactions. *J. Biol. Chem.* **255,** 4834–4842.
9. Gregoriou, M. and Rees, A. R. (1984) Studies on the structure and function of the EGF receptor. *Biochem. Soc. Transactions* **12,** 160–165.
10. Panayotou, G. and Gregoriou, M. (1990) Purification of the epidermal growth factor receptor from A431 cells, in *Receptor Biochemistry, The Practical Approach Series* (Hulme, E. C., ed.), IRL, Oxford, UK, pp. 203–211.
11. Parker, P. J., Young, S., Gullick, W. J., Mayes, E. L. V., Bennett, P., and Waterfield, M. D. (1984) Monoclonal antibodies against the human epidermal growth factor receptor from A431 cells: isolation, characterisation and use in the purification of active epidermal growth factor receptor. *J. Biol. Chem.* **259,** 9906–9912.
12. Weber, W., Bertics, P. J., and Gill, G. N. (1984) Immunoaffinity purification of the epidermal growth factor receptor. *J. Biol. Chem.* **259,** 14,631–14,636.
13. Savage, C. R., Jr. and Cohen, S. (1972) Epidermal growth factor and a new derivative. *J. Biol. Chem.* **247,** 7609–7611.
14. Gullick, W .J., Downward, D. J. H., Marsden, J. J., and Waterfield, M. D. (1984) A radioimmunoassay for human epidermal growth factor receptor. *Anal. Biochem.* **141,** 253–261.

Chapter 16

Analysis of Sulfated Polysaccharide Conjugates

Christopher C. Rider

1. Introduction

1.1. Metabolic Labeling with [^{35}S]Sulfate

Glycosaminoglycan and proteoglycan biosynthesis appears to be a ubiquitous function in animal cells. Some biological sources, notably connective tissue, produce large quantities of proteoglycans and glycosaminoglycans, which can be readily detected by colorimetric assays and therefore investigated by well established techniques, which are fully described elsewhere *(1,2)*. However, most tissues and culture systems will contain submilligram amounts of these macromolecules. This means that their investigation requires radiolabeling. [^{35}S]sulfate is relatively inexpensive, and is efficiently detected by both liquid scintillation counting and fluorography. A major disadvantage is the relatively short half-life of ^{35}S, 88 d, which limits the time available for postincorporation analysis. However, an advantage of radiolabeling is that only those macromolecules synthesized during the labeling period will be studied. Previously synthesized macromolecules that may be partially degraded will not be detectable.

After metabolic labeling, it is necessary to remove unincorporated [^{35}S]sulfate. This can readily be achieved by gel-filtration chromatography or by exhaustive dialysis. Gel filtration is more convenient from the point of view of disposal of radioisotope, an important consideration, since by far the majority of radiolabels employed will remain unincorporated.

From: *Methods in Molecular Biology, Vol. 14: Glycoprotein Analysis in Biomedicine*
Edited by: E. F. Hounsell

1.2. Anion-Exchange Purification of Glycosaminoglycans and Proteoglycans

The desalted preparation will include macromolecules with nonglycosaminoglycan sulfate radiolabel, in particular proteins with sulfated tyrosines or sulfated oligosaccharides. Glycosaminoglycans and proteoglycans will have much higher anionic charge densities, and therefore, their separation from other macromolecules is effectively achieved by anion-exchange chromatography. Anion-exchange chromatography of proteoglycans, together with their extraction from labeled cell cultures, has been fully covered elsewhere *(3)*. However, recently several manufacturers have produced ion-exchange membrane cartridges, which offer the advantages of fast flow rate, high capacity in a low bed volume, and rigid bed to facilitate column washing. Since anion-exchange separation is on the basis of charge density, for cells producing multiple glycosaminoglycan chain types, the resulting chromatographic profiles may show resolution of the bound fraction into multiple peaks.

1.3. Selective Enzymic and Chemical Degradation of Glycosaminoglycans

It is often of interest to liberate intact glycosaminoglycan chains from their polypeptide cores in order to determine their size and other characteristics. This may be achieved by alkaline β-elimination, a procedure resulting in the release of all glycans that are *O*-linked to serine or threonine. If required, the free glycosaminoglycans can be separated from liberated *O*-linked oligosaccharides by gel filtration. The procedure described here uses a relatively low concentration of alkali, which should minimize release of *N*-linked oligosaccharides. Nonetheless, some cleavage of polypeptide chains may occur.

The class of glycosaminoglycan present may be established through the use of several well established selective enzymic and chemical methods of degradation. Susceptibility or resistance to these methods differentiates between the glycosaminoglycan types. Moreover, the glycosaminoglycan fragments so obtained may be subject to further analysis so as to yield structural information on the chains from which they have been derived. Chondroitin sulfate and dermatan sulfate may be degraded with chondroitin ABCase, which degrades both, or chondroitin ACase, which is specific for chondroitin sulfate and does not digest the iduronic acid-rich chains of dermatan sulfate (formerly termed chondroitin B). Use of chondroitin ABCase is described later in the chapter.

The selective partial degradation of heparin and heparan sulfate may be achieved using one or more of the commercially available heparinase or heparitinase preparations. The description, use, and substrate specificity of these enzymes is fully covered elsewhere *(4)*. Heparin and heparan sulfate differ from other glycosaminoglycans by the possession of unacetylated glucosamine residues. This hexose, irrespective of whether it is *N*-sulfated or not, is subject to deaminative cleavage in the presence of nitrous acid. Such treatment of a glycosaminoglycan will therefore result in chain cleavage at the site of each unacetylated glucosamine, and the extent of depolymerization will depend on the frequency and distribution of this particular saccharide.

1.4. Determination of the Size of Proteoglycans and Glycosaminoglycans

The size distribution of either intact proteoglycans or eliminated glycosaminoglycans has been conventionally determined by gel filtration under dissociating conditions (4*M* guanidine hydrochloride) on an appropriate grade of Sepharose CL *(1,2)*. This is a reliable technique, but also laborious, especially when multiple samples are to be compared. Large proteoglycans, such as the cartilage chondroitin sulfate proteoglycan (M_r 1–4 × 10^6), cannot be resolved by electrophoresis on polyacrylamide gels because of their size. However, a polyacrylamide-agarose gel method has been devised for this purpose by McDevitt and Muir *(5)*, and a full description of this procedure is available elsewhere *(1)*. Proteoglycans synthesized by many cell types have smaller hydrodynamic sizes, approaching those of large polypeptides. Therefore, conventional sodium dodecyl sulfate polyacrylamide slab gel electrophoresis methods may be used. In this laboratory, we routinely employ the protocol of Laemmli, a complete account of which is to be found elsewhere (*6* and *see* vol. 1 of this series). However, most laboratories will have their own variant procedure, which is likely to prove satisfactory (*see* Note 1).

With proteoglycans, determination of molecular weight is not straightforward, because even proteoglycan molecules with the same core polypeptide will display considerable size heterogeneity. This arises from disperse lengths of the glycosaminoglycan chain and even variation in the number of chains carried. Dispersity is particularly evident on gel electrophoresis with its high resolution of molecular weight. Proteoglycans will separate as a smeared band extending over a considerable apparent M_r range.

The size of liberated glycosaminoglycan chains may be determined by gel filtration. Compared to proteins, free glycosaminoglycan chains migrate anomalously rapidly through discontinuous SDS polyacrylamide gels. However, size separation of glycosaminoglycans can be performed by the continuous polyacrylamide gel electrophoresis method of Hampson and Gallagher *(7)*, which is described here adapted to the minigel format.

2. Materials

2.1. Metabolic Labeling of Glycosaminoglycans and Proteoglycans Synthesized by Lymphocytes and Lymphoma Cells

1. Sterile, tissue-culture grade 24-well plates.
2. Sterile plastic Pasteur pipets.
3. Sterile 50-mL conical centrifuge tubes.
4. Powdered, sulfate-free RPMI 1640 Dutch Modification medium, containing HEPES, but without magnesium sulfate or sodium bicarbonate (Gibco, Paisley, Scotland, UK). The powdered medium for 1 L is dissolved in 950 mL of high-purity water containing $MgCl_2 \cdot 6H_2O$, 82.5 mg, and $NaHCO_3$, 2 g. (The pH is routinely 7.3, but should be adjusted with HCl or NaOH if necessary.) The volume is made up to 1 L, and the medium is sterile-filtered into presterilized media bottles. Samples from the beginning and end of the filtration may be collected into small sterile Petri dishes and incubated at 37°C for 48 h to check for sterility. This reconstituted medium has a shelf-life of 3 mo at 5°C.
5. Heat-inactivated fetal calf serum, L-glutamine, penicillin-G, and streptomycin sulfate.
6. [^{35}S]sulfuric acid in water (ICN Biomedicals, High Wycombe, UK).
7. Desalting column of Bio-Gel P-6DG or Sephadex G-50 (medium), of bed volume at least 10 times the sample volume. Prepacked disposable columns may be used if desired. The column is equilibrated before use with elution buffer, which may be selected according to the subsequent analyses to be applied to the sample (*see* Notes 2 and 3).

2.2. Anion-Exchange Purification of Glycosaminoglycans and Proteoglycans

1. DEAE MemSep 1000 ion-exchange cartridge (Millipore Corp., Bedford, MA).
2. Filter device (0.22 µm).
3. HPLC grade water.

4. Sodium acetate buffer (50 m*M*), pH 5.8, containing 180 m*M* NaCl, 6*M* urea, and 0.1% (w/v) Zwittergent 3-08 (Calbiochem-Novabiochem Corp., Nottingham, UK) (*see* Notes 2 and 3).
5. In-line 280-nm absorbance monitor and peristalic pump.

2.3. Selective Degradation of Glycosaminoglycans and Proteoglycans

1. HCl (0.1*M*).
2. Sodium borohydride (*see* Note 4).
3. Glacial acetic acid.
4. NaOH (0.1*M*).
5. 1-Butyl nitrite (butyl nitrite, Aldrich, Gillingham, UK, stored at 5°C) freshly diluted in 4 vol ethanol on day of use (potential carcinogen, *see* Note 5).
6. Cetylpyridium chloride (10% [w/v]) in distilled water.
7. Chondroitin ABCase (chondroitin ABC lyase) from *Proteus vulgaris* (ICN Biochemicals, Seikagaku Kogyco Co., Tokyo, Japan, or Sigma Chemical Co., Poole, UK). The enzyme is to be reconstituted in 0.01*M* Tris-HCl, pH 8.0, containing 50% glycerol and will remain active in storage at –20°C for at least 1 yr.
8. Standard heparin and chondroitin sulfate preparations made up at 2 mg/mL in distilled water.

2.4. Size Determination of Glycosaminoglycan Chains by Gel Electrophoresis

1. Vertical slab minigel apparatus (Mini-Protean II, Bio-Rad Laboratories, Hemel Hempstead, UK, or equivalent) and constant voltage DC power supply.
2. Glycine buffer (0.2*M*), containing 2.5 m*M* EDTA and 5 m*M* NaN_3 adjusted to pH 8.9 by addition of solid Tris base. This is 2X electrophoresis buffer. Store at 5°C.
3. Acrylamide (40 g) and 0.13 g bisacrylamide dissolved in 100 mL distilled water. Store at 5°C.
4. Stain; 0.08% (w/v) aqueous azure A.
5. Ammonium persulfate.
6. TEMED (*N,N,N′,N′*,-tetramethylethylendiamine).
7. Size standard heparin and chondroitin sulfate preparations. (Most commercial supplies will provide size data on request, but these are not quoted here because of likely batch-to-batch variation.)
8. Sample buffer comprising 2.5 mL Tris-glycine buffer, 2.0 mL distilled water, 0.5 mL analar glycerol, and sufficient Bromophenol Blue and Phenol Red to provide an intense color.

3. Methods

3.1. Metabolic Labeling of Glycosaminoglycans and Proteoglycans Synthesized by Lymphocytes and Lymphoma Cells

1. In the week of use, prepare the complete labeling medium by supplementing the reconstituted, sulfate-free medium with L-glutamine, 2 m*M*, 2-mercaptoethanol, 14 µL/L, and penicillin-G, 60 mg/L. At this stage, two sources of sulfate are also added, streptomycin sulfate, 1.55 mg/L, and 1% fetal calf serum (*see* Notes 6 and 7).
2. Harvest the cells by spinning at 500*g* for 10 min at room temperature. After resuspension in labeling medium, count the cells and assess viability by dye exclusion. Populations with viabilities <90% are unsuitable for study.
3. Wash the cells in labeling medium, and finally resuspend at 10^7/mL. (Verify this density with a further count.)
4. Place 150 µCi [^{35}S]sulfate in each labeling well, add 1 mL of cell suspension/well, and immediately transfer the plate to an incubator at 37°C with a humidified atmosphere with 5% CO_2. (At least one well is incubated without label, so that cell viability may be reassessed during the labeling incubation.)
5. After 4 h, terminate the incubation by transferring the well contents to ice-cold centrifuge tubes after thorough resuspension of the cells with a Pasteur pipet. Spin these cell suspensions immediately at 500*g* for 10 min at 4°C, wash the cell pellets twice with ice-cold phosphate-buffered saline, and combine these washings with the first supernatant.
6. Desalt labeled cell extracts and labeling supernatants by gel filtration to separate macromolecular incorporated sulfate from unincorporated radiolabel. The resulting elution profile will possess a minor void volume peak of macromolecular incorporated radiolabel followed by a major peak of free sulfate at the bed volume. Baseline separation is not usually obtained, because some incorporation will be into molecules of intermediate size. Notable among these are the sulfated glycolipids. Selected peak fractions should therefore avoid such material.

3.2. Anion-Exchange Purification of Glycosaminoglycans and Proteoglycans

1. Filter and degas all solvents and buffers on the day of use.
2. Flush the cartridge with 10 mL of filtered HPLC grade water at a flow rate of 10 mL/h, and then equilibrate with 10 mL of chromatography buffer.
3. Filter the desalted proteoglycan sample dissolved in, or dialyzed against, chromatography buffer, through a 0.22-µm membrane, apply to

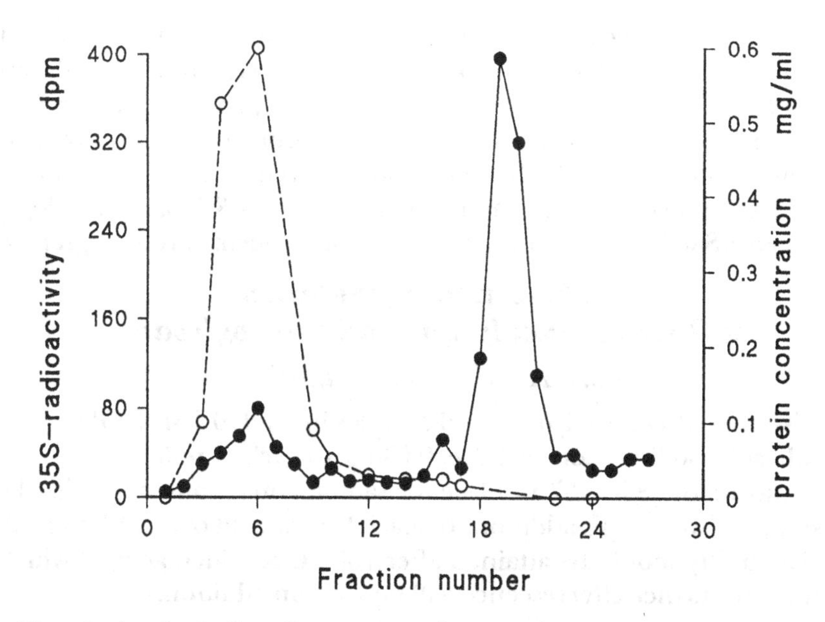

Fig. 1. Analytical-scale anion-exchange chromatography on DEAE MemSep 1000 cartridge of crude proteoglycan preparation obtained from the culture medium conditioned by a murine lymphoma cell line. The separation was performed as described in the text. A linear salt gradient from 0.15–1.8*M* NaCl was started at fraction 13. The fraction volume was 1.8 mL. Protein concentration,○; [^{35}S]radioactivity, ●.

the cartridge, and elute with 20 mL of buffer. The conditions of chromatography are such that only highly anionic molecules bind to the column.

4. Once the eluted radioactivity and 280-nm absorbance have fallen back to low plateaus near zero, elute the bound proteoglycan with a 40 mL linear salt gradient rising to 1.8*M* NaCl in chromatography buffer.
5. Wash the cartridge in sequence with the following filtered solutions (*see* Note 8).
   a. HCl (0.5*M*) 10 mL.
   b. HPLC-grade water until pH is >4.0.
   c. NaOH (0.5*M*) 10 mL.
   d. HPLC-grade water until pH is <8.0.
   e. Sodium azide (0.02% [w/v]).

   The unit may then be stored at 5°C.
6. An example of the separation of proteoglycan by ion-exchange chromatography on a DEAE Memsep cartridge is shown in Fig. 1. The bound peak, fractions 18–21, is proteoglycan, whereas the minor breakthrough peak, fractions 3–8, is sulfated protein. These separations readily

resolve proteoglycan from protein as shown in Fig. 1, and manyfold purification is achieved in one step. However, the proteoglycan fraction is still likely to contain contaminating proteins. Further ion-exchange separations under similar conditions will be required to remove these. It will be beneficial to alter the nature of the anion-exchange medium, and Mono Q (Pharmacia LKB Biotechnology, Uppsala, Sweden) is an attractive choice for subsequent chromatography.

3.3. Selective Degradation of Glycosaminoglycans and Proteoglycans

3.3.1. Alkaline β-Elimination

1. Dissolve buffer-free, lyophilized samples in 1 mL 0.1*M* NaOH.
2. Add solid sodium borohydride, 0.038 g, to each sample.
3. Incubate for 18 h at 37°C. Place the samples on ice and neutralize by slow, stepwise 20-µL additions of glacial acetic acid over a 1 h period. (Neutrality should be attained after 160 µL of added acid, at which time no further effervescence will be seen on addition.)

3.3.2. Nitrous Acid Cleavage of Heparin/Heparan Sulfate

1. Dissolve salt-free sample to be treated in 1 mL distilled water (*see* Note 9). Set up controls, comprising two samples of heparin and two samples of chondroitin sulfate, containing 2 mg glycosaminoglycan in 1 mL distilled water in clear glass tubes. (One of each pair will serve as treated control, with the other as untreated control.)
2. Add 0.5 mL 1*M* HCl to each sample and control.
3. Add 0.5 mL of ethanolic butyl nitrite to each sample and treated control. Add 0.5 mL ethanol to untreated controls.
4. Incubate for 2 h at room temperature with brief vortex mixing every 15–20 min, and stop the reactions by neutralization with 0.5 mL 1*M* NaOH (*see* Note 10).
5. Check that selective cleavage of heparin/heparan sulfate glycosaminoglycans has occurred by adding 100 µL 10% (w/v) cetylpyridium chloride solution to each of treated and untreated controls. Flocculant precipitation of intact glycosaminoglycan will occur within 15 min.

3.3.3. Chondroitin ABCase Digestion of Glycosaminoglycans

1. Prepare the glycosaminoglycan samples and controls in distilled water, as for nitrous acid cleavage above (*see* Note 11).
2. Set up 1 mL of each sample and control in individual screw-capped plastic centrifuge tube. Add 50 µL of 0.2*M* Tris-HCl buffer, pH 8.0.

3. Reconstitute the lyophilized enzyme at a concentration of 2 U/mL in 0.01*M* Tris-HCl, pH 8.0, containing 50% (v/v) glycerol, where 1 U is the enzyme activity capable of liberating 1 µmol of disaccharide/min at 37°C. Add 100 µL of this enzyme solution/digestion, and 100 µL of enzyme buffer to undigested controls.
4. Add 1 drop of toluene/tube, seal the cap tight, and incubate for 24 h at room temperature.
5. Check controls for digestion as described for nitrous acid degradation using cetylpyridinium chloride precipitation (*see* Section 3.3.2., step 5).
6. The digested sample should be boiled for 20 min to destroy enzyme activity.

3.3.4. Size Determination of Glycosaminoglycan Chains by Gel Electrophoresis

1. Cast the gel by thoroughly mixing 6.25 mL Tris-glycine buffer, 6.25 mL acrylamide/bisacrylamide solution, 0.55 mL freshly prepared 10% (w/v) aqueous ammonium persulfate, and 11.75 µL TEMED. Leave to polymerize; the cast gel may be stored for several days at 5°C provided it is sealed to prevent it from drying out.
2. Fill the assembled gel apparatus with Tris-glycine buffer diluted with an equal volume of distilled water.
3. Pre-electrophorese the gel for 30 min at 150 V.
4. Fill sample wells with 20 µg glycosaminoglycan in 10 µL sample buffer.
5. Run at 200 V. After approx 1 h, when the phenol red is approaching the bottom of the gel, switch off the power supply, remove the gel from tank, and measure the migration distances of the two dyes. (The migration distance of the bromophenol blue should be 3/4 that of the phenol red.)
6. Immediately transfer the gel to the stain solution, and leave on a rotatory shaker for 15 min. Destain in several changes of distilled water, and dry the gel as soon as background is sufficiently clear (*see* Note 12).

4. Notes

1. Gel-filtration columns and electrophoresis gels are calibrated with globular protein standards, either in native conformation or denatured according to the system. The highly anionic linear chains of glycosaminoglycan adopt extended conformations in solution. Therefore, proteoglycans, even in the presence of denaturants, will run anomalously when compared to standards.
2. The presence of either 4*M* guanidine hydrochloride or 6*M* urea in the elution buffer will minimize binding of contaminating proteins to the glycosaminoglycan and proteoglycan. Guanidine hydrochloride is likely to be more effective in this regard, but will interfere with

any subsequent anion-exchange purification and is more likely to denature the core polypeptide, thus destroying its biological activity. It is also a skin irritant. The buffer should also contain a detergent to minimize nonspecific binding. The zwitterionic detergent, Zwittergent 3-08 (Calbiochem), at a concentration of 0.1% (wt/vol), is suitable.

3. To avoid proteolysis, we routinely add on the day of use 10 m*M* 6-aminohexanoic acid, 10 m*M* *N*-ethylmaleimide, 1 m*M* benzamidine HCl, and 0.2 m*M* phenylmethylsulfonyl flouride. The latter must be introduced by dissolving in 50 µL dimethylsulfoxide, which is then added to 100 mL of eluant buffer with vigorous stirring. Where proteolysis may be a particular problem, a supplementary cocktail of antipain, leupeptin, aprotinin, and chymostatin is employed. These are made up as a mixture, each 0.5 mg/mL in buffer, stored frozen, and added 1% (v/v) on day of use.
4. Sodium borohydride may undergo explosive hydration on exposure to a humid atmosphere and is also highly toxic. It must be stored under dry nitrogen and opened only in a fume hood with the sash down as far as possible. Safety glasses should be worn, as should laboratory overalls with the cuffs tucked inside rubber gloves. The smallest bottle supplied, 10 g (Sigma Chemical Company), is adequate for treatment of many samples. Sodium borohydride is not essential for the elimination reaction, but serves to reduce the liberated glycan in order to prevent its degradation by "peeling reactions."
5. Butyl nitrite is a potential carcinogen; thus, it and solutions containing it should only be handled in a fume hood. Gloves should be worn.
6. The labeling medium employed should be a sulfate-free variant of the most appropriate medium for the particular cells or tissue preparation being studied. Usually, magnesium chloride is substituted for magnesium sulfate in the formulation. Undialyzed fetal calf serum is likely to have a sulfate concentration of around 1 m*M*. Therefore, in the case of cells that require concentrations of 5% or above for survival during labeling, it may be worthwhile trying dialyzed serum.
7. The aforementioned protocol uses low-sulfate medium to give high incorporations of radiolabel. A major potential drawback is that cells exposed to sulfate starvation may become deficient in the synthesis of sulfated polysaccharides. It is therefore important to check that the rate of sulfate incorporation is linear throughout, and ideally beyond, the labeling period. However, this will not rule out the possibility of qualitative changes in sulfation induced by sulfate deprivation. Silbert's laboratory has shown that medium sulfate concentrations of <0.1 m*M* result in undersulfation of glycosaminoglycan chains,

which remain of normal length. Such undersulfated chains may have further structural deficiencies. In particular, the undersulfated dermatan/chondroitin sulfate glycosaminoglycans synthesized by sulfate-deprived fibroblasts were found to have a markedly reduced degree of epimerization of glucuronic acid to iduronic acid *(8)*. Therefore if detailed structural studies of sulfated products are envisaged, it may be preferable to raise the sulfate content of the labeling incubation and sacrifice high incorporations of radiolabel.

8. Refer to manufacturer's instructions. It is particularly important that the cartridge is not allowed to dry out. The maximum operating pressure is 1.3 bar (18 psi) which should allow a maximal flow rate of 20 mL/min. The aforementioned flow rate is therefore highly conservative, and flow rates of 2–5 mL/min are recommended by the manufacturer. However, since the nominal bed vol is 1.4 mL, 10 mL/h allows several bed volumes of eluate change per hour.
9. This procedure may, where necessary, be scaled down to volumes and amounts one-fifth those given.
10. For subsequent gel-filtration chromatography, the degraded sample may be mixed with an equal volume of 2X chromatography buffer. Gel electrophoresis of treated samples will be subject to interference from the relatively high salt concentrations. Electrophoretic samples may be desalted using small ultrafiltration units, e.g., Centricon microconcentrators (Amicon, Stonehouse, UK).
11. The zwitterionic detergent, Zwittergent 3-08, is suitable for solubilizing proteoglycans, and, at concentrations up to 1% (w/v), does not inhibit chondroitin ABCase digestion.
12. This method is capable of resolving quite small oligosaccharides. Those migrating with the phenol red marker dye are likely to be hexasaccharides. Excessive destaining will tend to wash small oligosaccharides out of the gel. Higher resolution separations may be achieved on large gradient gels, and counterstaining with ammoniacal silver increases the sensitivity of detection *(9,10)*.

References

1. Carney, S. L. (1986) Proteoglycans, in *Carbohydrate Analysis—a Practical Approach* (Chaplin, M. F. and Kennedy, J. F., eds.), IRL Press, Oxford, pp. 97–142.
2. Beeley, J. G. (1987) *Glycoprotein and Proteoglycan Techniques*. Elsevier, Amsterdam.
3. Yanagishita, M., Midura, R. J., and Hascall, V. C. (1987) Proteoglycans: isolation and purification from tissue culture, in *Methods in Enzymology*, vol. 138 (Ginsburg, V., ed.), Academic, Orlando, pp. 279–289.

4. Lindhart, R. J., Turnbull, J. E., Wang, H. M., Longanathan, D., and Gallagher, J. T. (1990) Examination of the substrate specificity of heparin and heparan sulfate lyases. *Biochemistry* **29**, 2611–2617.
5. McDevitt, C. A. and Muir, H. (1971) Gel electrophoresis of proteoglycans and glycosaminoglycans on large-pore composite polyacrylamide-agarose gels. *Anal. Biochem.* **44**, 612–622.
6. Hames, B. D. (1981) An introduction to polyacrylamide gel electrophoresis, in *Gel Electrophoresis—a Practical Approach* (Hames, B. D. and Rickwood, D., eds.), IRL Press, Oxford, pp. 1–92.
7. Hampson, I. N. and Gallagher, J. T. (1984) Separation of radiolabelled glycosaminoglycan oligosaccharides by polyacrylamide-gel electrophoresis. *Biochem. J.* **221**, 697–705.
8. Silbert, J. E., Palmer, M. E., Humphries, D. E., and Silbert, C. K. (1986) Formation of dermatan sulfate by cultured human skin fibroblasts. *J. Biol. Chem.* **261**, 13,397–13,400.
9. Turnbull, J. E. and Gallagher, J. T. (1988) Oligosaccharide mapping of heparan sulphate by polyacrylamide-gradient-gel electrophoresis and electrotransfer to nylon membrane. *Biochem. J.* **251**, 597–608.
10. Lyon, M. and Gallagher, J. T. (1990) A general method for the detection and mapping of submicrogram quantities of glycosaminoglycan oligosaccharides on polyacrylamide gels by sequential staining with azure A and ammoniacal silver. *Anal. Biochem.* **185**, 63–70.

CHAPTER 17

Secreted Mucus Glycoproteins in Cell and Organ Culture

Anthony P. Corfield and Christos Paraskeva

1. Introduction

Mucus glycoproteins (mucins) are major differentiated products of colonic mucosal cells, and are thus important markers of normal and disease development in this tissue. The use of human colonic cell lines representing different stages in the progression of malignant disease is of particular interest since they allow changes in the expression of mucins to be studied during the development and progression of disease *(1,2)*. In the same way, tissue obtained from patients as biopsies or at surgery can be placed in short-term primary or organ culture to study similar changes in disease *(3,4)*.

Glycoproteins are expressed by all cells as secretory products, intracellular, and membrane components. The mucins are characteristically located in the vesicles of Goblet cells in vivo *(5–8)*. Histochemical analysis of the mucin content of Goblet cells has indicated that these molecules are changed during many mucosal diseases *(5,6)*. Biochemical analysis of the mucin changes has been limited because of the paucity of material available from mucosae in general and the difficulty in obtaining normal material for comparison *(7)*. Improvements in the study of glycoproteins, especially mucins, have been achieved through (1) the use of defined human mucosal cells that can be grown in long-term culture *(1,2,9)* and (2) application of metabolic labeling techniques *(3,4,10)*. The radioactive methods allow relatively small numbers of cells and tissue fragments (biopsies) to be analyzed, and cell culture also gives access to larger amounts of the mucins produced by

From: *Methods in Molecular Biology, Vol. 14: Glycoprotein Analysis in Biomedicine*
Edited by: E. F. Hounsell

the individual cell lines *(1,2,9,11)*. The mucins are very high mol-wt glycoproteins that aggregate and form gels on secretion *(6–8,12)*. They contain typically 70–80% of their dry weight as carbohydrate in *O*-glycosidic linkage to serine and threonine residues on the polypeptide backbone. Most of the polypeptide is resistant to proteolytic attack because of the high substitution with oligosaccharide units. However, short "naked" peptide regions are susceptible to proteases, and these contain the bulk of any disulfide bridges present. Many mucins are made up of native macromolecules composed of subunits linked to each other through the disulfide bridges. Reduction and alkylation of these disulfide bridges allow separation of the individual subunits, and together with proteolytic digestion, results in loss of the aggregate structure and its associated viscoelastic properties *(7,8,12)*.

There is a choice of radioactive monosaccharide and amino acid precursors that label the mucin oligosaccharide chains or the polypeptide backbone *(3,4,12)*. Metabolic labeling experiments yield many labeled products throughout the cell, and the mucins require separation before analysis. Antimucin antibodies have proved particularly valuable in conjunction with metabolic labeling methods in providing a rapid and specific method for precipitation of radiolabeled mucins *(13,14)*.

Separation methods for mucins have relied on the properties of these molecules, typically their very high mol wt on gel filtration and SDS-PAGE, and their buoyant density in density gradients *(7,12)*. These separation methods have been applied to microscale radiolabeled mucins *(3,4,10)*, and to larger amounts of mucins from cell culture or resected tissue *(7,11,12)*.

The methods described here cover the detection, metabolic labeling, and isolation of mucins from cultured cells (primary cultures and cell lines), organ culture, and macroscopic tissue samples. The data refer to human colorectal cells and tissue, but similar systems and principles apply to a wide range of other tissues where mucins are produced *(7,12)*.

2. Materials

1. Biopsies or tissue from surgery are placed on lens tissue soaked in culture medium at room temperature before incubation. The tissue is cut with a scalpel to give a size of approx 2–4 mm^2. These are placed on a steel grid in culture dishes.
2. Calf serum—batch testing is essential.
3. Cell culture media.

a. Standard growth medium: Dulbecco's modified Eagles medium (DMEM) containing 2 m*M* glutamine, 1 µg/mL hydrocortisone sodium succinate, 0.2 U/mL insulin, 100 U/mL penicillin, 100 µg/mL streptomycin, and 20% fetal bovine serum (FBS).
b. Washing medium: The same as standard growth medium, but with 5% FBS, double the concentration of penicillin and streptomycin, and 50 µg/mL gentamycin.
c. Digestion solution: The same as the washing medium, but containing 5% FBS together with 1.5 mg/mL collagenase (Worthington type 4) and 0.25 mg/mL hyaluronidase Sigma, Poole, UK; type 1).
d. 3T3 Conditioned medium: DMEM is supplemented with 10% FBS, 2 m*M* glutamine, 100 U/mL penicillin, and 100 µg/mL streptomycin, and put onto 24-h postconfluent 3T3 cell layers for 24 h. After conditioning, the medium is filtered through a 0.2-µm filter (Nalgene, Milton Keynes, UK) and further supplemented to give 20% FBS, 1 µg/mL hydrocortisone sodium succinate, and 0.2 U/mL insulin.

4. Collagen (human placental, type 4 Sigma) at 1 mg/mL is prepared in 1 part glacial acetic acid to 1000 parts sterile tissue culture grade distilled water, and stored at 4°C.
5. Dispase solution: Dispase (Boehringer, Lewes, UK; grade 1) is prepared in DMEM containing 10% FBS, sterile-filtered, and stored at –20°C.
6. Dithiothreitol (Sigma).
7. Fixed *Staphylococcus aureus* cells (commercial preparation). Pellets for preclearing and specific antibody binding are prepared by washing the cells from a 10% suspension at 100 µL for each 10 µL of rabbit serum used to preclear (maximum). The suspension is pelleted, washed with immunoprecipitation buffer, and recentrifuged. Use the pellet.
8. FBS—batch testing is essential.
9. Guanidine hydrochloride, approx 7*M* stock solution in PBS, treated with charcoal, passed through filter paper, and finally through a 0.45-µm Millipore (Watford, UK) filter. The stock solution can be stored at room temperature, and dilutions prepared using PBS.
10. Immunoprecipitation buffer—50 m*M* Tris, 5 m*M* EDTA, 0.5% SDS, 1% Triton X-100, 0.5% sodium desoxycholate, 1% bovine serum albumin, 1 m*M* phenylmethylsulfonylfluoride (dissolved in a small volume of propan-2-ol), 0.1 mg/mL soybean trypsin inhibitor, and 5 m*M* *N*-ethylmaleimide adjusted to pH 7.4. Make up fresh before use.

11. Mol-wt markers used are the high-mol-wt range of Rainbow markers (Amersham International plc., Amersham, UK), Maximum 200 kDa (myosin), and also murine laminin (400 kDa) and IgM (970 kDa), both from Sigma.
12. Organ culture medium—Minimal Eagles Medium, containing 10% FBS, 10 m*M* sodium bicarbonate, 2 m*M* glutamine, 50 U/mL streptomycin, 50 U/mL penicillin, 50 µg/mL gentamycin, and 20 m*M* HEPES, pH 7.2.
13. PBS/inhibitor cocktail in PBS. 1 m*M* phenylmethylsulfonylfluoride, 5 m*M* EDTA, 0.1 mg/mL soybean trypsin inhibitor, 5 m*M* *N*-ethylmaleimide, 10 m*M* benzamidine, and 0.02% sodium azide. Prepare inhibitor cocktail fresh as required.
14. Periodic acid-Schiff (PAS) reagent, commercial solution (Sigma).
15. Sepharose CL 2B and CL 4B (Pharmacia, Milton Keynes, UK). Use: 1 × 30 cm and 2.5 × 80 cm in all glass columns equilibrated in PBS/4*M* guanidine hydrochloride.
16. SDS-polyacrylamide gels for electrophoresis: Gels are prepared as usual, except that piperazine diacrylamide (PDA, Bio-Rad, Hemel Hempstead, UK) is used in place of *N,N'*-methylenebisacrylamide *(bis)* at the same concentration. Stacking gels of 2.5% and running gels of 3% are prepared at 1.5 mm thickness.
17. SDS-PAGE sample buffer: 0.065*M* Tris-HCl, 10% glycerol, 2% SDS, 0.05 bromophenol blue, pH 6.8, without β-mercaptoethanol or dithiothreitol.
18. Sodium iodoacetamide (Sigma).
19. Swiss 3T3 cells are obtained from the American Tissue Type Culture Collection (ATCC No CCL92).
20. Trypsin 0.1% by weight in 0.1% EDTA.

3. Methods

3.1. Cell and Organ Culture (see *Notes 1–3*)

1. To prepare collagen-coated flasks, coat tissue culture flasks (T25 25 cm^2) with a thin layer of collagen solution (Section 2., item 4; 0.2 mg/flask), and allow to dry at room temperature in a laminar flow hood for 2–4 h *(see* Note 4).
2. Grow Swiss 3T3 cells (Section 2., item 19) on collagen in 3T3-conditioned medium until they are 24 h postconfluent.
3. Lethally irradiate the cells with 60 KGray (6 mrads) of γ radiation, or treat with 10 µg/mL mitomycin C (Sigma) for 2 h.
4. Wash the cells, and produce a single-cell suspension by pipeting. The cells can either be used immediately as feeders or stored at 4°C as a single-cell suspension for up to 4 d *(see* Note 5).

3.1.1. Primary Culture—Enzyme Digestion (see Notes 3, 6, and 7)

1. Wash the tumor specimens (adenoma and carcinoma) four times in washing medium (Section 2., item 3b) and cut with surgical blades to approx 1 mm in a small vol of the same medium.
2. Wash the tissue four times in washing medium, and place in digestion solution (Section 2, item 3c). Roughly 1 cm^3 is put in 20–40 mL of solution.
3. Rotate at 37°C overnight (12–16 h).
4. Mix the suspension by pipeting to improve the separation of the epithelial elements from the stroma resulting from enzymic digestion.
5. Filter the suspension through 50-mm mesh nylon gauze, or repeatedly allow to settle out under gravity and collect the pellets. The large clumps of cells and epithelial tubules (organoids that contain the majority of the epithelial cells) are separated from single cells (mostly from the blood and stroma) and cell debris.
6. Wash the cell pellets three times, and place in culture on collagen-coated T25 flasks in the presence of Swiss mouse 3T3 feeders (approx 1×10^4 cells/cm^2) at 37°C in a 5% CO_2 in air incubator *(15)*. In some situations, 3T3-conditioned medium can be used instead of adding mouse 3T3 cells directly to cultures (*see* Note 5).

3.1.2. Long-Term Culture of Adenoma Cell Lines (see *Note 8)*

1. Prepare culture conditions for adenoma cell lines as previously described for primary cultures.
2. Passage of adenoma cells is carried out as clumps of cells using sufficient dispase solution just to cover the cells, and incubate for 30 min at 37°C. The cells are removed as a sheet, and are pipeted to remove them from the flask and to break up the sheets into small clumps of cells *(15)*.
3. Wash the clumps of cells, and replate under standard culture conditions. Reattachment of cells may take several days, and during medium changing, any floating clumps of cells must be centrifuged and replated with the fresh medium.

3.1.3. Long-Term Culture of Carcinoma Cell Lines (see *Note 8)*

1. Grow the carcinoma cell lines in tissue culture plastics without collagen coating and without 3T3 feeders in DMEM supplemented with 10% FBS and 1 m*M* glutamine.
2. Passage as single cells using 0.1% trypsin in 0.1% EDTA.

3.1.4. *Organ Culture* (see *Note 2*)

Place the biopsies or tissue pieces of 2–4 mm^2 singly or up to six on lens tissue over a steel grid in culture dishes with a central well containing 2 mL of medium (Section 2., item 12). The orientation of the tissue is with the luminal surface uppermost.

3.2. Collection of Secreted and Cellular Material

3.2.1. *Collection of Radioactive Fractions After Metabolic Labeling in Cell Culture* (see *Note 9*)

1. Add the radioactive precursor (e.g., [^{14}C]-glucosamine) to the standard growth medium, and incubate for times between 4 and 96 h, depending on the type of radioactive precursor and the nature of the cells (*see* Notes 10–12).
2. Collect the medium, and wash the cells with a further 5 mL of fresh nonradioactive medium.
3. Irrigate the flasks with 5 mL of PBS/inhibitor cocktail (*see* Note 13) containing 10 m*M* dithiothreitol, and scrape the cells off with a "cell scraper."
4. Wash the cells twice with PBS/inhibitor cocktail containing 10 m*M* dithiothreitol, and pool the total washings.
5. Adjust the dithiothreitol washings to a 2.5*M* excess with sodium iodoacetamide, and incubate for 15 h at room temperature in the dark.
6. Dialyze the secreted medium and dithiothreitol wash material extensively against three changes of distilled water, each of 5 L at 4°C over 48 h.
7. Homogenize the washed cell pellet in 1 mL of PBS/inhibitor cocktail with an ultraturrax for 10 s at maximum setting on ice.
8. Centrifuge the homogenate at 100,000*g* for 60 min, and decant the supernatant. Resuspend the membrane fraction in 1 mL of PBS/inhibitor cocktail.

3.2.2. *Collection of Radioactive Fractions After Metabolic Labeling in Organ Culture* (see *Note 9*)

1. Add the radioactive precursor to the organ culture medium in the plastic culture dishes and place in an incubator or Mackintosh jar at 37°C in an atmosphere of 95% air or oxygen/5% carbon dioxide. Continue the incubation for periods of up to 24 h for colonic tissue (*see* Notes 10–12).

2. After incubation, remove the medium from the central well and wash the tissue and dish with 1 mL of PBS. Pool the medium and washings, and dialyze against three changes of 5 L of distilled water at 4°C for 48 h.
3. Homogenize the tissue in 1 mL of PBS/inhibitor cocktail in an all-glass Potter homogenizer ensuring complete disruption of the mucosal cells (about 20 strokes). Remove connective tissue if present.
4. Centrifuge the homogenate at 12,000*g* for 10 min, and separate the supernatant soluble fraction from the membrane pellet.
5. Resuspend the membrane pellet in 1 mL of PBS/inhibitor cocktail.

3.2.3. Collection of Fractions from Nonradioactive Cell Cultures

1. Remove the medium from the flasks and scrape the cells and gel layer into PBS/inhibitor cocktail containing 6*M* guanidine hydrochloride (Section 2., step 9).
2. Sediment the cells and cell debris by centrifugation at 100,000*g* for 30 min, and aspirate the soluble and gel layers from the pellet.
3. Solubilize the gel layer by continued stirring in 6*M* guanidine hydrochloride at 4°C, **or** by incubation with 10 m*M* dithiothreitol and 5 m*M* EDTA for 15 h at 37°C followed by the addition of a 2.5*M* excess of iodoacetamide over dithiothreitol, and incubate at room temperature for 15 h in the dark.
4. Centrifuge soluble fractions for 60 min at 100,000*g* and discard the pellets.

3.2.4. Collection of Fractions from Nonradioactive Surgical Tissue Specimens

1. Pin out the surgical specimen on a dissection board, and irrigate with PBS/inhibitor cocktail.
2. Scrape the mucosal surface using a glass slide, and wash the combined scrapings into PBS/inhibitor cocktail containing 6*M* guanidine hydrochloride.
3. Homogenize the scrapings in an all-glass Potter homogenizer to achieve complete disruption of the mucosal cells, and centrifuge the suspension at 100,000*g* for 60 min to sediment all membranes.
4. Separate the soluble and gel (if present) fractions, and solubilize the gel samples using dithiothreitol as described in Section 3.2.3., step 3.

3.3. Separation of Mucus Glycoproteins from the Fractions Obtained After Culture (see *Note 14)*

3.3.1. Density Gradient Centrifugation

1. Make up the samples from the fractions prepared in Sections 3.2.1.–3.2.4. in 4*M* guanidine hydrochloride/PBS to a concentration of approx 1–5 mg/mL or containing a suitable amount of radioactivity (e.g., >10,000 cpm) for subsequent analytical techniques. Add CsCl to give a density of about 1.4 g/L in a final vol of approx 8 mL.
2. Load the samples into centrifuge tubes and centrifuge at, e.g., 100,000*g* for 48 h at 10°C to obtain a CsCl density gradient.
3. Aspirate samples of 0.5 mL from the top of the tube by pipet, or drain from the bottom after piercing the tubes. Weigh the samples to obtain the density of each fraction.
4. Slot blot aliquots (50–200 µL) of each fraction onto nitrocellulose, and visualize with the PAS stain *(16)* (*see* Note 15). Quantify the results using a densitometer, if necessary. For the radioactive samples, each slot blot is cut out and placed in scintillation cocktail for quantitation (*see* Note 16).
5. Pool the carbohydrate containing fractions located at densities between 1.35 and 1.5 g/mL.

3.3.2. Gel Filtration (see *Notes 14 and 17)*

1. Make up samples in PBS/4*M* guanidine hydrochloride buffer to give concentrations of 1–5 mg/mL or by radioactivity (e.g., >5000 cpm), and load onto columns of Sepharose Cl 2B or CL 4B (Section 2., step 15). Elute the column with the same buffer, and collect fractions (1–5 mL) in plastic or glass tubes.
2. Slot blot aliquots of the fractions onto nitrocellulose, and detect the carbohydrate with the PAS stain *(16)* (*see* Notes 15 and 16). Alternatively, the radioactivity may be detected on slot blots or by direct measurement of fraction aliquots.
3. Pool and concentrate the mucin fractions identified in excluded or included volumes.

3.3.3. SDS-Polyacrylamide Gel Electrophoresis

1. Mix samples containing 10–500 µg mucin, or >10,000 cpm with 50 µL sample buffer (Section 2., item 17) and heat for 5 min at 95°C.
2. Load the samples onto the SDS-gels (Section 2., item 16), and electrophorese at 140 V with water cooling for approx 4 h (16 cm gels). Run mol-wt markers for calibration.

3. Fix the gels in 10% acetic acid, 40% methanol, and stain using commercial silver stain or carbohydrate staining methods (*see* Notes 18 and 19). Air-dry gels for autoradiography onto Whatman 3MM paper without staining. Handle the gels on polyester mesh supports during fixing, staining, and air-drying. The resulting dried paper/gel is taken for autoradiography or for filing.
4. Remove the gels after electrophoresis (*see* Note 19), and take for Western blotting in a standard (e.g., Bio-Rad Transblot) "wet" blot apparatus using an overnight (16 h) transfer at a constant voltage of 30 V followed by 4–5 h at 70 V. The efficiency of the blot is seen directly if Rainbow markers are present.
5. Probe the blots with stains or antibodies using standard methods for visualization.

3.3.4. Immunoprecipitation with Antimucin Antibodies (see *Note 20)*

1. Incubate the soluble fractions (containing sufficient radioactivity to allow detection in the final analytical procedure, e.g., SDS-PAGE) in immunoprecipitation buffer (Section 2., item 10) with normal rabbit serum at 10 µL/200 µL of sample, for 60 min at 4°C to preclear the fraction.
2. Mix the incubation solution containing rabbit serum with a pellet of fixed *S. aureus* cells (Section 2., item 7 and *see* Note 20), and incubate for 30 min at 4°C.
3. Centrifuge the suspension, remove the supernatant and use it to suspend a second pellet of fixed *S. aureus* cells. Incubate the suspension for 30 min at 4°C and centrifuge for 15 min at 12,000*g*. Collect the supernatant.
4. Add the specific antibody as serum, ascites fluid, or hybridoma culture supernatant (1–100 µL). Incubate the solution for 1–2 h at 4°C.
5. Add 50 µL of a 10% solution of fixed *S. aureus* cells in immunoprecipitation buffer and incubate for 30 min at 4°C. Centrifuge at 12,000*g* for 1 min and remove the supernatant.
6. Resuspend the pellet carefully in 0.5 mL of immunoprecipitation buffer, and recentrifuge. Repeat the washing twice, and ensure efficient removal of the final wash supernatant.
7. Use the pellet directly for analysis by SDS-PAGE and autoradiography.

3.3.5. Concentration of Mucin Samples After Purification

1. Dialyze the samples against four to five changes of 5 L of distilled water at 4°C for 48 h to remove all salt, and lyophilize.
2. Concentrate radioactive samples by immunoprecipitation with antimucin antibodies, or by dialysis and lyophilization.

4. Notes

1. The production and secretion of mucus glycoproteins by cultured colonic cells should be examined using cells at different stages of confluency, because this may alter the differentiation properties of the cells and, therefore, the amount and type of mucin produced.
2. When using organ and primary cultures, it is important to consider the heterogenous nature of the cell types in the tissue (i.e., stromal elements and lymphoid cells in addition to the colonic epithelial cells). It may not be clear which cell type is producing the glycoproteins. However, the complex cell:cell interaction may be necessary for the cells to differentiate.
3. The primary culture techniques described can be used for normal adult colon. However, these are not as reproducible as those with the adenomas and carcinomas, and there are more problems from the contaminating stromal elements. There are at present no normal adult colonic epithelial cell lines, only adenoma and carcinoma cell lines *(9,15,17)*.
4. Collagen-coated flasks are necessary to obtain efficient attachment of primary cultures and some adenoma and carcinoma cell lines to the flasks, and to retain the optimum differentiated characteristics of the cells *(15)*.
5. The use of 3T3 cell feeders requires controls to determine which cell type is producing the glycoproteins of interest. This can be achieved using 3T3-conditioned medium where mucin production is being assessed or removing the 3T3 feeders from the flask once the epithelium has grown.
6. Colorectal adenomas invariably need digestion with enzymes because of their organization into well differentiated glandular structures. With carcinomas, it is possible to adopt a nonenzymic approach with surgical blades to release small clumps of tumor cells that can be cultured *(18)*.
7. When using colonic cell lines, it is important to check the true colonic epithelial nature using a battery of markers, including antikeratin antibodies, ultrastructural analysis showing the presence of desmosomes, and other colonic differentiation markers *(9)*.
8. Although many tumor cell lines, especially colon carcinomas, can be grown in simple medium without 3T3 feeders and without collagen coating, the colonic cells retain better differentiated phenotypes when using the more complex culture conditions described for primary cultures.
9. In radiolabeling experiments, the total amount of mucin is often small, and significant losses owing to nonspecific adsorption on plastic and

glass vessels, silicon rubber, and dialysis tubing may occur. Treatment of all vessels and tubing with 1 mg/mL ovalbumin in PBS before use improves yields.

10. The choice of radioactive precursor is important. Radioactive glucosamine is most commonly used, since it is incorporated into *N*-acetylglucosamine, *N*-acetylgalactosamine, and sialic acids, major monosaccharide components of mucins. In tissues other than liver, such labels as mannose and fucose may be randomized to other monosaccharides before they are transferred to glycoproteins.
11. The metabolic rate of the cells should be considered for optimum labeling with the radioactive precursor. The type of isotope will govern the amount added, typically [^{14}C] and [^{35}S] precursors in the range of 185–1850 kBq/experiment and [^{3}H] precursors in the range of 370–1850 kBq/experiment. Short-term labeling of 2–4 h may not result in labeling of secreted material and may require higher doses of radioactive precursor (1.85–3.7 MBq/experiment). Longer labeling periods may reflect synthesis, catabolism, and recycling of glycoproteins. Dual labeling experiments with, e.g., [^{14}C] and [^{3}H], precursors need to be planned such that the relative incorporation of each label is readily detectable in the isolated product; thus, consideration of Notes 10 and 12 is necessary. Organ culture experiments should be controlled by histochemical criteria to ensure the integrity of the tissue during incubation. Diseased tissue may shows signs of degradation during acceptable times for normal samples.
12. Increased incorporation of radioactive precursors may be achieved by reduction of the concentration of the same nonlabeled compound in the medium (monosaccharides or amino-acids) for the labeling period. However, this should be balanced against any changes in the growth of the cells or tissue under these "depleted" conditions.
13. A protease inhibitor cocktail is needed to avoid the degradation of mucins owing to bacterial enzymes and in cell homogenates. Soybean trypsin inhibitor and PMSF appear to be most important for colonic tissue *(8)*.
14. The sequence of purification steps in mucin isolation is important. If density gradient separation is followed by gel filtration, the lower mol-wt subunits or degradation products will be identified.
15. The PAS stain for carbohydrate is well suited to blotting on nitrocellulose because of the high monosaccharide content of mucins and the removal of salt, which otherwise interferes with this assay. A rapid and sensitive method has been described *(16)*.

16. Owing to the high salt concentrations in density gradient experiments, many colorimetric assays and some radioactive scintillation cocktails are inefficient. Extensive dialysis of each fraction may allow colorimetric assays to be performed, but with small amounts of metabolically labeled material, this results in significant losses (*see* Note 9). The blotting technique mentioned in Note 15 is more reliable for colorimetric and radioactive detection.
17. Gel filtration is the most rapid and convenient method to obtain a mucin-enriched fraction from crude culture fractions for comparative studies *(10)*. It is also a starting point for preparation of native mucins for further purification and analysis (*see* Note 14).
18. Protein stains for mucins are usually very poor. Coomassie blue and silver stain frequently give negative results. Stains for carbohydrate, lectin conjugates, or specific antibodies are most useful.
19. Low-percentage gels crosslinked with PDA stick to paper, nitrocellulose, and lab gloves. The effective handling of PDA gels requires wetting of gloves with buffer or fixer. The use of mesh mats for transferring gels during staining and washing avoids problems. Because of the adhesive nature of gels containing PDA a sheet of Whatman No. 1 filter paper is used to separate the gel and the nitrocellulose in the "sandwich."
20. Suspensions of fixed *S. aureus* cells are used to bind antibodies owing to the presence of protein A in the cellular membranes of these bacteria. Immunoprecipitation can be carried out without preclearing. Individual antibodies should be tested by titration for optimum conditions with low nonspecific background. The procedure described can be followed from Section 3.3.4., step 4 onward.

References

1. Paraskeva, C., Corfield, A. P., Harper, S., Hague, A., Audcent, K., and Williams, A. C. (1990) Colorectal carcinogenesis: Sequential steps in the in vitro immortalization and transformation of human colonic epithelial cells (review). *Anticancer Res.* **10,** 1189–1200.
2. Williams, A. C., Harper, S., and Paraskeva, C. (1990) Neoplastic transformation of a human colonic epithelial cell line: In vitro evidence for the adenoma to carcinoma sequence. *Cancer Res.* **50,** 4724–4730
3. Smith, A. C. and Podolsky, D. K. (1987) Biosynthesis and secretion of human colonic glycoproteins. *J. Clin. Invest.* **80,** 300–307.
4. Corfield, A. P., Casey, A. D., Wagner, S. A., Cox, M., do Amaral Corfield, C., and Clamp, J. R. (1989) Selection of radioactive precursors for metabolic labelling of mucus glycoproteins. *Biochem. Soc. Transact.* **17,** 1037,1038.

5. Filipe, M. I. (1989) The histochemistry of intestinal mucins. changes in disease, in *Gastrointestinal and Oesophageal Pathology* (Whitehead, R., ed.), Churchill Livingstone, Edinburgh, pp. 65–89.
6. Smith, A. C. and Podolsky, D. K. (1986) Colonic mucin glycoproteins in health and disease. *Clin. Gastroenterol.* **5**, 815–837.
7. Neutra, M. R. and Forstner, J. F. (1987) Gastrointestinal mucus: Synthesis, secretion, function, in *Physiology of the Gastrointestinal Tract*, vol. 2, 4th ed. (Johnson, L. R., ed.) Raven, New York, pp. 975–1009.
8. Allen, A., Hutton, D. A., Pearson, J. P., and Sellars, L. A. (1990) The colonic mucus barrier: Structure, gel formation and degradation, in *The Cell Biology of Inflammation in the Gastrointestinal Tract* (Peters, T. J., ed.), Corners, Hull, UK, pp. 113–125.
9. Laboisse, C. L. (1989) Differentiation of colon cells in culture, in *The Cell and Molecular Biology of Colon Cancer* (Augenlicht, L. H., ed.), CRC Press, Boca Raton, FL, pp. 27–43.
10. Corfield, A. P., Warren, B. F., and Bartolo, D. C. C. (1990) Colonic metaplasia following restorative proctocolectomy monitored using a new metabolic labelling technique for mucin. *J. Pathol.* **160**, 170A.
11. Corfield, A. P., Clamp, J. R., Casey, A. D., and Paraskeva, C. (1990) Characterization of a sialic acid-rich mucus glycoprotein secreted by a premalignant human colorectal adenoma cell line. *Int. J. Cancer* **46**, 1059–1065.
12. Carlstedt, I., Sheehan, J. K., Corfield, A. P., and Gallagher, J. T. (1985) Mucous glycoproteins: A gel of a problem. *Essays Biochem* **20**, 40–76.
13. Dekker, J., Van Beurden-Lamers, W. M. O., and Strous, G. J. (1989) Biosynthesis of gastric mucus glycoprotein of the rat. *J. Biol. Chem.* **264**, 10,431–10,437.
14. Hilkens, J. and Buijs, F. (1988) Biosynthesis of MAM-6, an epithelial sialomucin. *J. Biol. Chem.* **263**, 4215–4222.
15. Paraskeva, C., Buckle, B. G., Sheer, D., and Wigley, C. B. (1984) The isolation and characterization of colorectal epithelial cell lines at different stages in malignant transformation from familial polyposis coli patients. *Int. J. Cancer* **34**, 49–56.
16. Thornton, D. J., Holmes, D. F., Sheehan, J. K., and Carlstedt, I. (1989) Quantitation of mucus glycoproteins blotted onto nitrocellulose membranes. *Anal. Biochem.* **182**, 160–164.
17. Willson, J. K. V., Bittner, G. N., Oberley, T. D., Meissner, L. F., and Weese, J. L. (1987) Cell culture of human colon adenomas and carcinomas. *Cancer Res.* **47**, 2704–2713.
18. Leibovitz, A., Stinson, J. C., McComb, W. B., McCoy, C. E., Mazur, K. C., and Mabry, N. D. (1976) Classification of human colorectal adenocarcinoma cell lines. *Cancer Res.* **36**, 3562–3569.

CHAPTER 18

Preparation of Polyclonal Antibodies to Native and Modified Mucin Antigens

Ismat Khatri, Gordon Forstner, and Janet Forstner

1. Introduction

Antibodies directed against mucin epitopes have a wide number of uses, including the construction of immunoassays for measuring secretion *(1)*, preparing immunoprecipitates for measuring radioisotope incorporation *(2)*, for use in tissue and cellular immunocytochemistry *(1,3)*, for characterizing mucin subdomains *(4)*, as markers of malignancy *(5)*, and for the screening of cDNA expression libraries *(6)*.

Intestinal mucins separate into two major components when reduced and alkylated, a denser, highly glycosylated, high-mol-wt glycopeptide (GP) enriched in serine, threonine, and proline, and a smaller, less well glycosylated 118-kDa glycopeptide enriched in aspartic and glutamic acid, glycine, alanine, and cysteine. The former contains the tightly packed *O*-glycosidic-linked oligosaccharide branches typical of mucins, whereas the latter contains *N*- as well as *O*-linked oligosaccharides *(7)*. Because of its release by reduction and depolymerization, the 118-kDa glycopeptide has been dubbed a "link" glycopeptide, although vigorous proof for this assertion is not yet available. We describe methods for making these components and their deglycosylated products, as well as methods for production of polyclonal antibodies to each.

From: *Methods in Molecular Biology, Vol. 14: Glycoprotein Analysis in Biomedicine*
Edited by: E. F. Hounsell

2. Materials

1. Alkylating reagent: iodoacetamide stored in the dark at 4°C.
2. BCA (Bicinchoninic Acid) protein assay reagents obtained from Pierce Chemical Company, IL.
3. Homogenization buffer: 5 m*M* EDTA (ethylene-diamine tetraacetic acid), pH 7.4, 1 m*M* PMSF (phenylmethyl sulfonyl fluoride), 5 m*M* NEM (*N*-ethyl-maleimide), and 0.05% sodium azide (*see* Note 1).
4. Phosphate-buffered saline (PBS): 10 m*M* phosphate buffer, pH 7.4, 0.15*M* NaCl, and 0.05% sodium azide.
5. Quenching solution: Solution A = 200 m*M* sodium iodide in 100 m*M* sodium tricarbonate. Solution B = 800 m*M* sodium metabisulfite ($Na_2S_2O_3$). Mix 1:1 of Solution A and Solution B.
6. Tris-buffer: 10 m*M* Tris-HCl buffer, pH 7.5, 1 m*M* PMSF, and 0.05% sodium azide.
7. CsCl density gradient ultracentrifugation: To each 100 mL homogenate add 60 g cesium chloride, while stirring continuously. Measure the density, and add more CsCl if needed to achieve a final density of 1.40 g/mL.
8. SDS-PAGE: All solutions for SDS-PAGE gels were prepared by the method of Laemmli *(8)*, and all reagents were of electrophoresis grade, purchased from Bio-Rad, Richmond, CA. Minigels were cast and run using the MiniGel apparatus of Bio-Rad according to the manufacturer's instructions.
9. Tris-buffered saline (TBS): 10 m*M* Tris-HCl and 150 m*M* NaCl buffer, pH 8.0.
10. Tris-buffered saline/Tween (TBST): 10 m*M* Tris-HCl, 150 m*M* NaCl, 0.05% Tween 20, pH 8.0.
11. Freund's adjuvant (complete).
12. Freund's adjuvant (incomplete).
13. Bovine serum albumin (Cat #A7030).
14. Anisole, 99%, Trifluoromethanesulfonic acid (TFMS), Tween 20, Pyridine acetate, diethylether, sodium iodide, sodium metaperiodate, and sodium metabisulfite, Guanidinum HCl, dithiothreitol (DTT), and mercaptoethanol.
15. *O*-glycanase (Endo-α-*N*-acetylgalactosaminidase of *Diplococcus pneumoniae*, Genzyme Corporation, Boston, MA).
16. Sepharose CL-2B.
17. Goat antirabbit alkaline phosphatase.
18. NBT and BCIP reagents (Promega Corporation, Madison, WI).
19. Hollow-fiber cartridge concentrator (Cat #HIP 10043) purchased from Amicon, Beverly, MA.

20. Beckman L8-55 ultracentrifuge, Beckman J2-21 centrifuge, 50.2 Ti rotor, JA-14 rotor, Polyallomer Quick-seal centrifuge tubes (Beckman Cat. #342414), and Tube Sealer were purchased from Beckman Instruments Inc., Palo Alto, CA.
21. Pyrex Brand Borosilicate glass tubes with Teflon™-lined screw-caps.
22. Spectrapor dialysis tubing 12K, 5K, and lK.
23. A minidialysis system capable of dialyzing 200 µL to 1 mL vol over a short period of time (3–4 h) and that maintains a continuous flow of dialysis buffer (water, low salt solution, and so on) during the dialysis (e.g., Mini Dialysis System, Gibco BRL).

3. Methods

3.1. Preparation of Antigen

3.1.1. Native Mucin

1. Fast 10 male Wistar rats each weighing ~200 g for 16 h. Sacrifice one rat at a time by cervical dislocation, and quickly remove the small intestine from its mesentery by gentle dissection with scissors. Open each intestinal segment (~6 cm/segment), and place the serosal side down on a cold glass plate over crushed ice.
2. Remove the mucosa by gentle scraping, with a glass slide, into an ice-cold weighing boat, weigh quickly, and add the contents to a small volume of cold homogenizing buffer in a beaker in crushed ice. When the total mucosa is collected, homogenize the scrapings in cold homogenization buffer (75 mL buffer/g mucosa) for 25 s at two-thirds its full speed in a Waring blender.
3. Centrifuge the homogenate at 30,000*g* for 30 min at 4°C in a Beckman J2-21 centrifuge or similar instrument.
4. Concentrate the supernatant fluid to one-third its original volume by passing through a Hollow Fibre Cartridge concentrator (mol wt cut-off 100 kDa) at 4°C. Only proteins >100 kDa are concentrated during circulation through the hollow fiber, whereas the rest of the solute is removed (*see* Note 2).
5. To 100 mL of concentrated sample add ~60 g of CsCl to achieve a density of 1.4 g/mL. Adjust with CsCl powder if required.
6. Centrifuge the sample at 150,000*g* for 48 h at 4–8°C in Polyallomer Quick-Seal Centrifuge tubes (~40-mL capacity).
7. Remove eight equal fractions of 4.8 mL using a long needle and tubing connected to a peristaltic pump. Introduce the needle carefully to the bottom of the tube numbering fractions 1–8 starting from the bottom. Fractionate the tubes as quickly as possible. Pool fractions 2,

3, and 4. Measure buoyant density, and adjust to 1.4 g/mL by adding CsCl powder or water as required.
8. Centrifuge the pooled fraction, repeating step 6.
9. Divide into eight equal aliquots as in step 7, and measure the density.
10. Remove 1 mL from each fraction, and dialyze for 3 h against water at room temperature using the Mini Dialysis System and a 12 kDa cutoff dialysis membrane (obtained from BRL).
11. Using 50–200 µL of the dialyzed sample, measure protein, nucleic acid, and carbohydrate (*see* Note 3). Pool fractions with density 1.44–1.50 g/mL (usually fractions 2, 3, and 4). These fractions should be rich in carbohydrate and free of nucleic acid. (*See* Note 4.)
12. Dialyze the pooled fraction at 4°C using a standard 12 kDa cutoff dialyses membrane against two changes of distilled water followed by three changes with 10 m*M* Tris-HCl buffer, pH 7.5, and a final change with distilled water. Transfer the sample to a 50-mL Falcon polypropylene tube, and lyophilize to ~200–250 µg/mL Adjust to 1 m*M* with PMSF and 0.05% with sodium azide, analyze for nucleic acid, protein, and carbohydrate (*see* Note 3), and store at –20°C.

For optional purification, following 2X CsCl density gradient ultracentrifugation, the mucin sample can be added to a Sepharose CL-2B column as follows:

1. Dialyze fractions with density 1.44–1.5 g/mL against two changes of distilled water followed by three changes with PBS and a final change with distilled water over 48 h at 4°C.
2. Concentrate to ~250 µg/mL protein, and apply about 5–6 mg to a 2.6 × 100 cm column, equilibrated with PBS at a flow rate of 0.5 mL/min. Collect 5-mL fractions, and assay for protein and carbohydrate (*see* Note 3). Mucin appears in the void volume as a sharp peak that can be collected, pooled, concentrated, and stored at –20°C in the presence of 1 m*M* PMSF and 0.05% sodium azide (*see* Note 5).

3.1.2. High-Mol-Wt and the "Link" Glycopeptide

3.1.2.1. Reduction

1. To purified native mucin (5–10 mg protein) add 50 mL of 0.1*M* Tris-HCl buffer, pH 8.5, in a 150-mL flask.
2. Weigh and add guanidinum HCl to achieve a final concentration of 6*M*.
3. Weigh and add DTT (final concentration 10 m*M*) or β-mercaptoethanol (final concentration 0.2*M*).
4. Place the flask covered with a loose rubber stopper in a boiling water bath for 5 min.

3.1.2.2. Alkylation

1. Cool the reduced mucin sample to room temperature.
2. Add Iodoacetamide to make a final concentration of 20 m*M* (if reduced with DTT) and concentration of 40 m*M* (if reduced with β-mercaptoethanol).
3. Keep at 4°C for 4 h in the dark.

3.1.2.3. Separation of High-Mol-Wt and "Link" Glycopeptides

1. Dialyze against 5 m*M* Tris-HCl, 4*M* guanidinum HCl, pH 7.5, at 4°C (six changes in 48 h) using a standard 12 kDa dialysis tubing.
2. Transfer the sample to a 250-mL beaker, and add dry CsCl to a density of 1.4 g/mL.
3. Place in Polyallomer Quick Seal Tubes, and centrifuge at 150,000*g* for 48 h at 4–8°C.
4. Carefully introduce a long needle into the tube without disturbing the gradient. Using a peristaltic pump connected to the needle, collect 20 equal fractions of ~2 mL each starting from the bottom of the tube. Measure the density of all fractions.
5. Aliquot a 0.5-mL sample from each fraction, and dialyze for 3 h against water using the Mini Dialysis System and a 12 kDa cutoff dialysis membrane.
6. Use 100 μL of each dialyzed fraction for carbohydrate analysis (Note 3), and subject 100–200 μL to 7.5% SDS-PAGE using minigels.
7. To obtain the high-mol-wt glycopeptides, pool fractions that are rich in carbohydrate and free of 118-kDa glycopeptide as shown in Fig. 1. The density range of these fractions should fall between 1.38 and 1.45 g/mL.
8. To obtain the "link" glycopeptide, pool fractions that are low in carbohydrate (PAS) and contain a protein band at 118 kDa on a 7.5% SDS-PAGE following silver/Coomassie staining (Fig. 1). The density of these fractions usually ranges from 1.26–1.35 g/mL.

3.2. Deglycosylation of Mucin Components

Methods are given for achieving significantly deglycosylated or totally deglycosylated high-mol-wt GP, as well as totally deglycosylated link GP. We have monitored deglycosylation by following hexosamine and protein content (Table 1).

3.2.1. Preparation of the Highly Deglycosylated High-Mol-Wt Glycopeptides (GP-hd)

This method removes over 99% of the GlcNAc and 74% of the GalNAc from the high-mol-wt GP (Table 1).

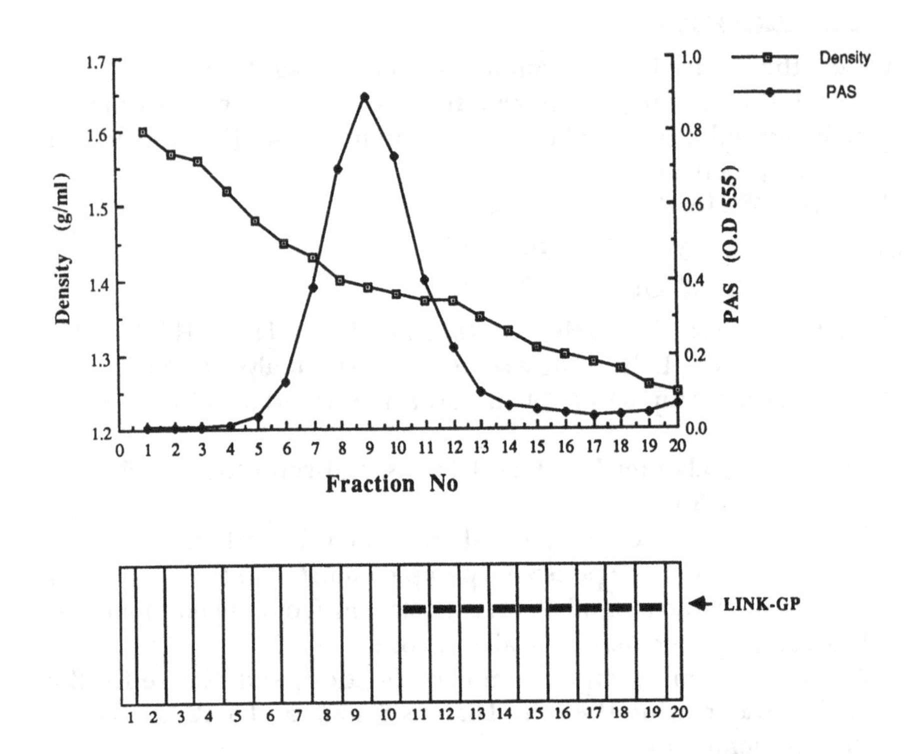

Fig. 1. Composite picture of PAS, SDS-PAGE, and density patterns of fractions following CsCl density gradient ultracentrifugation of reduced and alkylated mucin. High-mol-wt GP is free of link GP in fractions 6–10, within a density range of 1.38–1.45 g/mL. Fractions 13–19 within a density range of 1.26–1.35 g/mL, contain link GP free of high-mol-wt GP.

Table 1
Deglycosylation of High-Mol-Wt and Link Glycopeptide

nmol Sugar, µg protein	Native mucin	High-mol-wt GPs	GP-hd[a]	GP-td[b]	Link GP	d-Link[c]
GalNAc	3.95	4.23	1.07	0.005	2.12	0.00
GlcNAc protein	3.60	4.09	0.051	0.00	3.72	0.00

[a]GP-hd represents highly deglycosylated glycopeptides.
[b]GP-td represent totally deglycosylated glycopeptides.
[c]d-Link represents totally deglycosylated link glycopeptide.

1. Dry 5–10 mg lyophilized high-mol-wt GP in acid-washed pyrex brand borosilicate glass tubes by placing over P_2O_5 in a dessicator flask for 2 d. Carry out the remaining steps in a fume hood.
2. Add 3 mL TFMS: Anisole (2:1 [v/v]), mix by vortexing, fill the tube with a jet of nitrogen, and cap tightly. Keep at 20°C for 3 h while slowly stirring with a magnetic bar.
3. Cool a can (100 mL) of diethyl ether to –40°C by placing in a 95% ethanol–dry ice bath. Cool the tube from step 2 by placing in crushed dry ice, and while holding in crushed ice, uncap and add 2 vol of cooled diethyl ether.
4. Keeping the reaction tube cool in an ice-water bath, add dropwise 1 vol ice-cold (50% [v/v]) aqueous pyridine. A precipitate of pyridinium salts will form with each drop, but dissolves quickly on gentle agitation by hand. The reaction is exothermic. Therefore, proceed carefully making sure that the tube is cooled between drops.
5. Cap and centrifuge the tube at 4°C for 10 min at 2000*g*.
6. Remove the overlying ether phase with a Pasteur pipet and discard. Dialyze the remaining aqueous phase against 5 m*M* pyridine acetate buffer, pH 5.5, at 4°C for 48 h with six changes. The dialysis should be performed with low-mol-wt (5 kDa cutoff) Spectrapor cellulose dialysis tubing to avoid losses, since the deglycosylation step may cause some peptide breakdown.
7. Remove the dialyzed sample from the dialysis tubing, and wash the inside of the dialysis bag with a small volume of the dialysis buffer to remove residual protein on the bag.
8. Concentrate the deglycosylated sample by partial lyophilization to ~300 µg protein/mL, adjust to 1 m*M* with PMSF and 0.05% with sodium azide, and store at –20°C.

3.2.2. Preparation of Totally Deglycosylated Glycopeptides (GP-td)

1. To obtain totally deglycosylated glycopeptides, subject the high-mol-wt GP to two successive treatments with TFMS exactly as described in Section 3.2.1., except use the lower temperature (0°C, 2 h or 0°C, 4 h) (*see* Note 6). Carry out the first TFMS reaction for 2 h at 0°C, followed by a second TFMS reaction at 0°C for 4 h.
2. Achieve additional deglycosylation as described in Section 3.2.2.1. and 3.2.2.2. by oxidation-elimination reaction *(12)* and a final treatment with *O*-glycanase (*see* Note 7).

3. Dialyze the TFMS-treated sample against water at 4°C for 48 h with six changes. The dialysis should be performed with low-mol-wt (5 kDa cutoff) dialysis tubing. Transfer the dialysate to a 5-mL polypropylene tube, and concentrate to dryness by lyophilization.

3.2.2.1. Oxidation-Elimination ("Reverse" β-Elimination)

1. Dissolve TFMS-treated high-mol-wt GPs (~2 mg in 2 mL water) in a 10-mL polypropylene test tube.
2. Add 0.2 mL of 0.1 *M* acetic acid, and adjust pH to 4.5 with 0.1 *M* NaOH.
3. Add sodium metaperiodate to a final concentration of 200 m*M*; add a small magnetic stirring bar, cap, and incubate overnight while stirring at 4°C.
4. Destroy excess sodium metaperiodate by adding 1 mL of Quenching solution. The pH of the solution should rise to 10.5; if not, adjust with 0.1 *N* NaOH.
5. Incubate overnight at 4°C. Dialyze against distilled water using low-mol-wt (1 kDa cutoff) dialysis tubing. Transfer to a 5-mL polypropylene test tube, and concentrate to dryness by lyophilization. Store at –20°C.

3.2.2.2. Digestion with *O*-Glycanase

1. Dissolve ~2 mg protein in 2 mL of 10 m*M* Tris maleate buffer, pH 6.0, containing 1 m*M* PMSF and 0.05% sodium azide in a 5.0-mL test tube. Add 75 mU of *O*-glycanase.
2. Incubate at 37°C for 16 h in a water bath shaker at low speed.
3. Dialyze the sample against 10 m*M* phosphate buffer at 4°C using a 1 kDa cutoff dialysis tubing.
4. Transfer to a 5.0-mL polypropylene tube, and concentrate the sample to ~300 µg protein/mL. Store at –20°C in the presence of 1 m*M* PMSF and 0.05% sodium azide.

3.2.3. Deglycosylation of "Link" Glycopeptide

Since the link GP is less glycosylated, it does not require strong deglycosylating conditions. Two rounds of mild TFMS reaction, each carried out at 0°C for 2 h, completely eliminate all the sugars (Table 1). The deglycosylated link GP is referred to as the "d-Link" peptide.

3.3. Preparation of Antibodies to Native Mucin and to Its Deglycosylated Components

3.3.1. Immunization

1. Mix together 200 µg (protein) of mucin antigen with Freund's complete adjuvant in a ratio of 1:1 (v/v), and inject sc into two New Zealand rabbits.
2. Repeat the injection at 2, 6, and 10 wk with 100 µg antigen, mixed together with Freund's incomplete adjuvant.

Table 2
Relative Reactivity of Antibodies Prepared to Different Mucin Fractions[a]

	High-mol-wt	GP-hd	GP-td	Link GP	d-Link
Antibody to native mucin (1:5000)	+++++	++	±	++	±
Antibody to GP-hd (1:1000)	++	++++	+	±	–
Antibody to GP-td (1:1000)	+	+	++++	–	–
Antibody to d-Link (1:1000)	–	–	–	++++	+++

[a]100 ng Antigen was slotted onto NC paper and the relative reactivities of the antibodies determined by ELISA at the antibody titer shown.

3. Bleed the rabbit 2 wk after each injection, and monitor the antiserum for antibody production by an ELISA. The comparable reactivity of antibodies raised against mucin antigens, prepared as described in this chapter, is summarized in Table 2.

3.3.2. Monitoring Antibody Development

1. Spot 100 ng of mucin antigens onto a nitrocellulose (NC) sheet. This can be done conveniently using a Manifold or similar apparatus.
2. Cut out each spot, and transfer it to a 5-mL polypropylene test tube and cap (5-mL tissue culture wells will do equally well).
3. Block the remaining binding sites around the spot by adding 3 mL of 1% BSA in TBS. Incubate for at least 30 min at room temperature with shaking.
4. Prepare three serial dilutions of each antiserum (1:100, 1:1000, 1:10,000) in TBS containing 1% BSA. Discard the blocking solution from each tube from step 3, and incubate one slot in each dilution by adding 3 mL of the diluted antiserum for 2 h at room temperature or for overnight at 4°C.The tubes should be kept shaking at all times.
5. Remove the antibody solution, and wash by shaking in 3 mL TBST with five changes every 10 min.
6. Discard the TBST after the final wash from step 5, and to each tube, add 3 mL of goat antirabbit alkaline phosphatase at a dilution of 1:3000 (v/v) in TBS containing 1% BSA. Incubate with shaking for 1–2 h at room temperature.
7. Discard the conjugate solution from step 6, and wash by shaking in 3 mL TBST with five changes every 10 min.
8. Mix alkaline phosphatase color developing reagents NBT and BC1P with 0.1*M* Tris buffer, pH 9.0, according to the manufacturer's

instructions. Discard the TBST after the final wash from step 7, and add to each tube 1 mL of the color developing reagent.

9. Make sure that the spot is completely immersed in the solution. Incubate for 10–20 min at room temperature (in dark) to allow color development. Then rinse each spot with distilled water onto filter paper and dry.
10. Define antibody strength as the highest dilution that gives a visible color reaction.

4. Notes

1. Prepare a 100-m*M* solution of PMSF in absolute ethanol. Add this solution dropwise to the homogenization buffer to achieve a final concentration of 1 m*M* PMSF. Since an aqueous solution of PMSF is inactivated very rapidly, add fresh aliquots at each stage of purification.
2. The hollow-fiber cartridge concentrator is extremely useful when concentrating large volumes of sample. For example, about 2 L of the homogenate can be concentrated to ~500 mL in <8 h.
3. Nucleic acid can be conveniently estimated spectrophotometrically at OD 260. Protein can be estimated either by Lowry et al. *(9)* or BCA protein assay *(10)*. Total carbohydrate can be assayed using the periodic acid-Schiff (PAS) reagent *(11)*. Dialysis is necessary before measuring protein or carbohydrate by these methods since CsCl interferes with the assays.
4. In practice, it is best to discard fractions with a density >1.5 g/mL to avoid nucleic acid contamination, since it is possible for trace amounts of nucleic acid to escape detection in the unconcentrated samples.
5. Native mucins prepared by two density gradient ultracentrifugation runs have no detectable uronic acid, chondroitin sulfate, ribose, deoxyribose, triglyceride, IgA, IgA secretory component, or fibronectin. Monitor purity by SDS-PAGE using 7.5% separating gel followed by silver or Coomassie staining. Since intestinal mucin barely enters a 7.5% gel, no stained bands should be seen below the stacking gel.
6. The lower temperature was necessary because the method described in Section 3.2.1. produces low-mol-wt products, some of which may be lost during the dialysis steps required for subsequent "reverse" β-elimination and *O*-glycanase digestion.
7. In the oxidation-elimination reaction *(12)*, cleavage of the core GalNAc-Ser (Thr) linkage requires an unsubstituted 3' (OH) on the GalNAc. Some 3' hydroxyls may remain substituted (i.e., by galactose) after the TFMS treatment and, in this situation, "reverse" β-elimination will not remove the remaining Galβ1-3GalNAc residues. These residues are removed efficiently by *O*-glycanase.

References

1. McCool, D. J., Marcon, M. A, Forstner, J. F., and Forstner, G. G. (1990) The T-84 human colonic adenocarcinoma cell line produces mucin in culture and releases it in response to various secretogogues. *Biochem. J.* **267,** 491–500.
2. Dekker, J., Van Beurden-Lamers, W. O., and Strous, G. J. (1989) Biosynthesis of gastric mucus glycoprotein of the rat. *J. Biol. Chem.* **264,** 10,431–10,437.
3. Fahim, R. E. F., Specian, R. D., Forstner, G. G., and Forstner, J. F. (1987) Characterization and localization of the 'putative' link component in rat small intestinal mucin. *Biochem. J.* **243,** 631–640.
4. Mantle, M., Forstner, G. G., and Forstner, J. F. (1984) Antigenic and structural features of goblet-cell mucin of human small intestine. *Biochem. J.* **217,** 159–167.
5. Gendler, S., Taylor-Papadimitriou, J., Duhig, T., Rothbard, J., and Burchell, J. (1988) A highly immunogenic region of a human polymorphic epithelial mucin expressed by carcinomas is made up of tandem repeats. *J. Biol. Chem.* **263,** 12,820–12,823.
6. Gum, J. R., Byrd, J. C., Hicks, J. W., Toribara, N. W., Lamport, T. A., and Kim, Y. S. (1989) Molecular cloning of human intestinal mucin cDNAs. *J. Biol. Chem.* **264,** 6480–6487.
7. Forstner, G., Forstner, J., and Fahim, R. (1989) Small intestinal mucin: polymerization and the "Link Glycopeptide," in *Mucus and Related Topics* (Chantler, E. and Ratcliffe, N. A., eds.), Society for Experimental Biology, The Company of Biologists, Ltd., Cambridge, UK, pp. 251–271.
8. Laemmli, U. K. (1970) Cleavage of structural proteins during assembly of the head of the bacteriophage T4. *Nature (London)* **227,** 680–685.
9. Lowry, O. H., Rosebrough, N. J., Farr, A. L., and Randall, R. J. (1951) Protein measurement with the Folin phenol reagent. *J. Biol. Chem.* **193,** 265–275.
10. Redinbaugh, M. J. and Turley, R. B. (1986) Adaptation of the Bicinchoninic acid protein assay for use with microtiter plates and sucrose gradient fractions. *Anal. Biochem.* **153,** 267–271.
11. Mantle, M. and Allen, A. (1978) A colorimetric assay for glycoproteins based on the periodic acid Schiff base stain. *Biochem. Soc. Trans.* **6,** 607–609.
12. Gerken, T. A., Gupta, R., and Jentoft, N. (1992) A novel approach for chemically deglycosylating O-linked glycoproteins. The deglycosylation of submaxillary and respiratory mucins. *Biochemistry* **31,** 639–648.

CHAPTER 19

The Use of Immunoperoxidase to Characterize Gastrointestinal Mucin Epitopes

Jacques Bara and Rafael Oriol

1. Introduction

Two main types of glycoconjugates are expressed by gastrointestinal (GI) epithelial cells: glycoproteins and glycolipids. Among glycoproteins, mucins are quantitatively well respresented throughout the GI tract.

Mucins are composed of very high-mol-wt components, forming a protective gel, in vivo, that covers the GI epithelium. They are produced by mucus cells. There are different types of mucous cells: In the stomach, there are columnar mucus cells of surface epithelium, fundic neck mucus cells, and pyloric mucus cells. In the intestine, there are mucus cells of Brünner glands and goblet cells.

It is not easy to determine the biochemical nature of a mucin epitope recognized by an antibody, since it can be a peptide, a saccharide, or both. We describe an immunoperoxidase method (including chemical and enzymatic pretreatments of deparaffinized sections) that allows to define the saccharide nature of epitopes, especially those related to ABH and Lewis histo-blood group systems (ref. *1* and Note 1). Indeed, GI mucin can be regarded as a mosaic of epitopes that are made of three main building blocks:

1. Carbohydrate residues associated to saccharide chains bearing the well known histo-blood-group-related epitopes;
2. Peptide components, associated to the peptide core; and
3. Mixed peptide and carbohydrate components.

From: *Methods in Molecular Biology, Vol. 14: Glycoprotein Analysis in Biomedicine*
Edited by: E. F. Hounsell

In this last case, the mucin epitopes cover both saccharide and peptide moities.

1.1. Carbohydrate Epitopes

These epitopes are under the control of genes coding for several glycosyltransferases, one for each glycosidic linkage. Some of these genes are genetically monomorphic as the genes encoding for the glycosyltransferases responsible for the synthesis of core chains. Considering only the terminal disaccharide of these chains, there are at least three different types of precursor chains in ABO mucin histogroup-related antigens *(2)*:

Type I	Galβ1-3GlcNAcβl-R
Type II	Galβ1-4GlcNAcβ1-R
Type III	Galβ1-3GalNAcαl-R

On top of the percursor chains, several genetically polymorphic glycosyltransferases can add terminal carbohydrate residues, which give the ABH or Lewis (Le) antigenic properties to the final epitope. The H- and Lewis-related epitopes result from the epistatic interaction of the products of at least four genes (Se, Le, H, and X genes) encoding for at least five fucosyltransferases *(3,4)*. These enzymes can transfer fucose molecules onto the terminal Galβ unit or onto the penultimate GlcNAcβ unit of the different histo-blood group precursor chains (e.g., fucose α1–2 to Gal gives H type I, II, and III). Then, in the GI mucus cells, according to the type of the precursor chain and the expression levels of these enzymes, which are inherited as simple Mendelian traits, a great number of epitopes can be found related to Le^a, Le^b, Le^x, Le^y, Tn, T, H type I–III, the I, i antigens, and in addition, their branched, extended, or sialosyl forms (*see* the Glossary and Chapter 20). Moreover, the Le^b, Le^y, and H type I–III are precursors of the A and B blood group antigens generated by a GalNAcα-transferase and Galα-transferase, respectively, encoded by the ABO locus on chromosome 9, and confer in turn a great number of different epitopes according to the spatial configuration induced by the addition of these terminal sugars *(5)*. In addition, such configurations can be observed on other glycoproteins and glycolipids that show different branching precursor chains inducing, in turn, other different structures generating different epitopes that can be expressed in absorptive cells.

1.2. Peptide Epitopes

These epitopes are present whatever the ABO and Lewis status. On the other hand, they can show organ-specificity and be restricted to peculiar parts of the gastrointestinal tract: stomach *(6,7)* or large intestine *(8)*. They are specific with regard to mucus cells, and they are not expressed in other organs devoid of mucus cells. Unlike this, histo-blood-group-related antigens have a large pattern of distribution throughout the human body.

1.3. Mixed Peptide and Saccharide Epitopes

Such epitopes are composed of very short saccharide chains (one or two sugars) interacting with the peptide core *(9)*, such as Tn (GalNAcα1-Ser/Thr) and T (Galβ1-3GalNAcα1-Ser/Thr). Such immunoreactivity remains unchanged after partial deglycosylation.

In order to study these epitopes, different methods have been used, such as ELISA, using artificial histo-blood group antigens produced by chemical synthesis *(10)* or hemagglutination.

The immunoperoxidase method described in this chapter can be useful in analyzing the different histo-blood-group-related epitopes associated with mucins taking into account their natural tissue environment (*see* Note 2). The method allows the study of mucin epitope patterns throughout the normal human GI tract of individuals of different ABH and Lewis phenotypes (*see* Note 1).

2. Materials

1. Human normal GI tracts obtained from cadaveric donors (age range: 15 to 48 yr old) just after the kidneys were removed for transplantation.
2. A glass box (20 × 45 × 6 cm) with a cork plank covering the ground for tissue fixation.
3. A plastic box (10 × 15 × 2 cm) containing a wet plastic sponge covering the ground for immunoperoxidase incubations.
4. Phosphate-buffered saline (PBS): Phosphate buffer 0.15*M* containing 0.9% NaCl, pH 7.4.
5. Ethanol 95%, kept at room temperature, xylene, and Tween 20.
6. Monoclonal antibodies, MAbs (*see* Note 1), made up at a dilution of 1/2 or 1/10 for culture supernatants; 1/100 or 1/1000 for ascites fluids.
7. Anti-IgG mouse antibodies linked to peroxidase (Diagnostic Pasteur, Marnes-la-Coquette, France).
8. Amino-ethylcarbazole (AEC) reagent: 60 mg of AEC were diluted in 5 mL of *N,N*-dimethylformamide and then diluted in 95 mL acetate

buffer 50 m*M*, pH 5. Then 100 µL of H_2O_2 were added to this solution. The reagent was prepared just before use. Care should be taken with this reagent, since AEC is toxic and a possible carcinogen.
9. Hematein solution (0.5%) (light sensitive).
10. Kaiser's glycerol gelatine (Merck, Darmstadt, Germany).
11. 2*M* Acetate buffer, pH 5.6, diluted 40 times in injectable H_2O just before use to give a final concentration of 50 m*M*.
12. Injectable H_2O.
13. Periodic acid ($NaIO_4$) kept dry and made up fresh in 50 m*M* acetate buffer, pH 5.0.
14. Glycine.
15. Neuraminidase from *Vibrio cholerae* (Behringwerke, Marburg, Germany).

3. Methods

3.1. Immunoperoxidase Labeling

1. Wash the GI tracts with PBS, and take tissue samples measuring 10 × 5 cm at different histological areas of the gut (*see* Note 3).
2. Immerse the mucosae in the glass tissue box containing 95% ethanol, and cleave with a scalpel to remove the muscularis mucosae.
3. Pin the tissue samples on the cork, and fix in an ethanol bath for 48 h (*see* Note 4).
4. Cut longitudinal strips of mucosae measuring 1 × 20 cm with a scalpel, coil into "swiss rolls" (*see* Note 5), and bind with cotton yarn.
5. Include the samples in paraffin, and cut 3-µm sections (*see* Note 6).
6. Deparaffinize sections using three successive baths of xylene and ethanol incubating for 10 min in each bath.
7. Rinse with water for 10 min, and then incubate the sections for 1 min with PBS and for 30 min with antimucin MAbs (Section 2., step 6).
8. Wash the sections with PBS containing 0.1% of Tween 20 three times for 1 min, add peroxidase-labeled anti-IgG antibodies (diluted 1/50), and incubate for 30 min.
9. Wash the sections three times with PBS containing 0.1% of Tween 20 and then incubate for 4 min with the AEC reagent (Section 2., step 8).
10. Rinse the sections with water, incubate for 1 or 2 min in hematein solution, and mount using Kaiser's glycerol gelatin.

3.2. Periodic Acid Pretreatment

1. Before immunoperoxidase labeling, incubate deparaffinized sections with different concentrations (0.1–80 m*M*) of $NaIO_4$ in acetate buffer for 1 h at room temperature and then 1% glycine for 30–60 min at room temperature *(13)*.

2. Incubate with PBS-Tween for 10 min, and then perform immunoperoxidase labeling using antimucin MAbs. When immunoreactivity is abolished (at concentrations between 1 and 5 m*M* of periodic acid), this suggests a carbohydrate nature of the epitope (*see* Note 7).
3. Set up a control of a deparaffinized section of ileal-colonic mucosae pretreated with 5 m*M* of periodic acid, and incubated with antisialosyl Le^a MAb (MAb NS 19-9). Negative immunoreactivity shows the efficacy of periodate treatment.

3.3. Enzyme Pretreatments

1. Incubate deparaffinized tissue sections with 1 U/mL of neuraminidase for 1 h at 37°C in 50 m*M* acetate buffer, pH 6, containing 2% $CaCl_2$.
2. After enzymatic treatment, wash the sections with PBS-Tween, and then stain with antimucin MAbs by immunoperoxidase.
3. Set up a positive control of enzymatic degradation using either MAb NS 19-9 incubated on the ileal mucosa near the ileal-colonic junction *(14)* or antisialosyl-Tn MAb-MAb TKH2 *(15)* incubated on the small intestinal mucosa *(16)*.

4. Notes

1. The ABH and Lewis status of tissue donors was determined by this immunoperoxidase method using MAbs against A, B, Le^a: MAb 7LE *(12)* (quoted as 34W6 in this ref.), Le^b: MAb 2-25LE *(6)* and MAb LM 129/181 from Glasgow West of Scotland B.T. Service, Carluke, UK, anti-H-type II *(12)* (quoted as 34W4 in this ref.), and T epitope: MAb T001 from Biocarb AB, Lund, Sweden. The individuals are classified according to the immunoreactivity of the surface gastric epithelium as:
   a. Nonsecretors, Le(a + b–); if anti-T and anti-Le^a MAbs strongly stain surface gastric epithelium, which does not react with anti-H type II;
   b. As secretors, Le(a – b+) or Le(a – b–), if the anti-Le^b and/or anti-H type II MAbs strongly stain the surface gastric epithelium respectively; or
   c. Lewis negative, Le(a – b–), if the anti-Le^b and anti-Le^a MAbs do not stain this epithelium.

   In our laboratory, we obtained a library of 78 GI tracts belonging to the following phenotypes: ALe(a + b-) = 4 cases, ALe(a – b+) = 28 cases, ALe(a – b–) = 2 cases, BLe(a + b–) = 2 cases, BLe(a – b+) = 8 cases, OLe(a + b–) = 6 cases, OLe(a – b+) = 24 cases, Ole(a – b–) = 1 case, ABLe(a – b+) = 3 cases.
2. The tissue location of an epitope can suggest its association to a particular precursor chain. For instance, in the pylorus, the blood group type I epitopes are mainly expressed in the surface gastric epithelium, and the

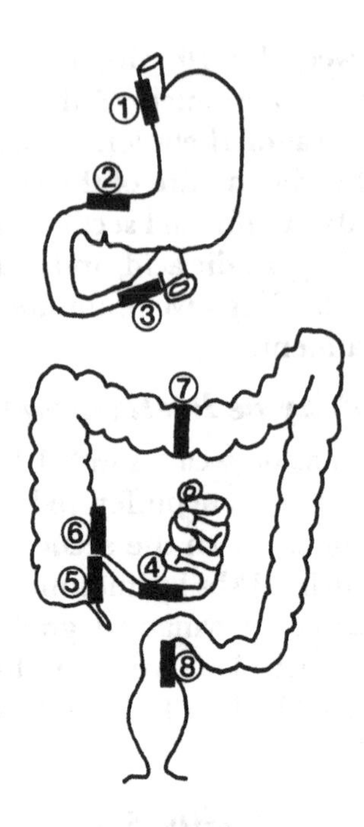

Fig. 1. Localization of different samples of gastrointestinal mucosae: 1 = esophagus and cardia junction, 2 = pylorus–duodenum junction; 3 = jejunum; 4 = ileum; 5 = cecum; 6 = right colon; 7 = transverse colon; 8 = sigmoid.

type II chains are found in the deep glands exclusively *(19)*. On the other hand, MAbs that do not react on the Le(a – b–) mucosae could be related to type I Lewis antigens *(14)*; those reacting only on the surface gastric epithelium of the nonsecretor individuals Le(a + b–) could be related to Le^a or T antigens, and those reacting in addition with the gastric glands could be related to type II antigens. Some immunoreactivity is dependent on the expression of the A or B blood group substance. For instance, the anti-Le^y MAb (12-4LE) reacts better in the O than A or B individuals *(20)*. With this immunoperoxidase method, it is possible to distinguish small differences between the specificity of MAbs against neighboring carbohydrate epitopes and determine eventually overlapping specificity areas. Indeed, each MAb shows a peculiar immunoperoxidase pattern throughout the GI tract, as a fingerprint, and MAbs showing identical fingerprints are rare *(17)*. The method is very sensitive to differences between closely related anticarbohydrate MAbs.

3. The different histological areas are: esophagus–cardia junction, pylorus–duodenal junction, jejunal and ileal mucosae, ileal–right colonic junction, and transverse and sigmoid mucosae (Fig. 1).

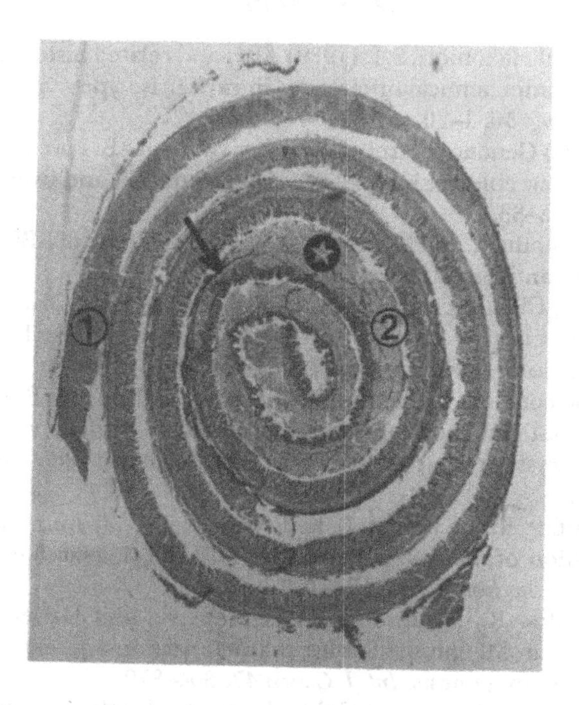

Fig. 2. Swiss roll sections of the pylorus duodenum junction. 1 = pylorus; 2 = duodenum, arrow = junction between two mucosae; ✪ = Brünner's glands.

4. It is important to use ethanol to fix tissue, because this fixative does not destroy or mask antigenic determinants. It acts by reversible precipitation, has no denaturing effects, and leads to excellent conservation of both saccharide and peptidic mucin epitopes. In addition, after such fixation, the strips of tissue samples are flexible enough to coil the mucosae easily into Swiss rolls as described by Magnus 50 yr ago.
5. The original method of Magnus *(11)*, who coiled the GI mucosae into "Swiss rolls," allows study of strips of mucosae of 10–20 cm on the same histological section (Fig. 2).
6. After the cuts are done, the slides with the paraffin-embedded histological sections can be kept in the cold room for 2–3 mo without modification of their immunoreactivity.
7. Not all the histo-blood group saccharide epitopes are destroyed using periodic acid treatment *(18)*. On the contrary, some cryptic peptidic epitopes can become accessible to MAbs after partial deglycosylation, and consequently, immunoperoxidase immunoreactivity can become positive after such pretreatment.

References

1. Clausen, H. and Hakomori, S. I. (1989) ABH and related histo-blood group antigens; immunochemical differences in carrier isotypes and their distribution. *Vox Sang.* **56,** 1–20.
2. Oriol, R. (1990) Genetic control of the fucosylation of ABH precursor chains. Evidence for new epistatic interaction in different cells and tissues. *J. Immunogenetics* **17,** 23–33.
3. Oriol, R., Le Pendu, J., and Mollicone, R. (1986) Genetics of ABO, H, Lewis, X and related antigens. *Vox Sang.* **51,** 161–171.
4. Mollicone, R., Gibaud, A, François, A., Ratcliffe, M., and Oriol, R. (1990) Acceptor specificity and tissue distribution of three human α-3-fucosyltransferases. *Eur. J. Biochem.* **191,** 169–176.
5. Bara, J., Mollicone, R., Le Pendu, J., Oriol, R., and Burtin, P. (1985) Evidence of an antigen common to human intestine, endocervix and mucinous ovarian cysts present exclusively in ALe^b patients. *Bull. Cancer (Paris)* **72,** 104–107.
6. Bara, J., Gautier, R., Le Pendu, J., and Oriol, R. (1988) Immunochemical characterization of mucins: polypeptide (MI) and polysaccharide (A and Le^b) antigens. *Biochem. J.* **254,** 185–193.
7. Bara, J., Gautier R., Mouradian P., Decaens C., and Daher N. (1991) Oncofetal mucin MI epitope family: characterization and expression during colonic carcinogenesis. *Int. J. Cancer* **47,** 304–310.
8. Shochat, D., Pank, K. D., and Goldenberg, D. M. (1982) Colon specific antigen-p (CSAp) II: further characterization in colorectal and pancreatic cancer. *Cancer (Phila.)* **50,** 927–931.
9. Gerken, T. A., Butenhof, K. J., and Shogren, R. (1989) Effects of glycosylation on the conformation and dynamics of O-linked glycoproteins: carbon-13 NMR studies of ovine submaxillary mucin. *Biochemistry* **28,** 5536–5543.
10. Gane, P., Mollicone, R., Rouger, Ph., and Oriol, R. (1987) Inhibition of heamagglutination with synsorbs and salivas of anti-A monoclonal antibodies. *Blood Transfusion and Immunohaematology* **30,** 435–442.
11. Magnus, H. A. (1937) Observations of the presence of intestinal epithelium in the gastric mucosa. *J. Pathol. Bacteriol.* **44,** 389–398.
12. Bara, J., Daher, N., Mollicone, R., and Oriol, R. (1987) Immunohistological patterns of 20 monoclonal antibodies against non-A non-B glycoconjugates in normal human pyloric and duodenal mucosae. *Blood Transfusion Immunohaematology* **30,** 685–692.
13. Woodward, M. P., Young, W. W., and Bloodgood, R. A. (1985) Detection of monoclonal antibodies specific for carbohydrate epitopes using periodic oxidation. *J. Immunol. Methods* **78,** 143–153.
14. Bara, J., Herrero Zabeleta, E., Mollicone, R., Nap, M., and Burtin, P. (1986) Distribution of GICA in normal gastrointestinal and endocervical mucosae and mucinous ovarian cysts using antibody NS 19-9. *Am. J. Clin Pathol.* **85,** 152–159.

15. Kjeldsen, T., Clausen, H., Hirohashi, S., Ogawa, T., Iijima, H., and Hakomori, S. (1988) Preparation and characterization of monoclonal antibody directed to the tumor-associated O-linked sialosyl-2-6-*N*-acetylgalactosaminyl (Sialosyl-Tn) epitope. *Cancer Res.* **48,** 2214–2220.
16. Kordari, P., Zamora, P. O., Kjeldsen, T., Gautier, R., and Bara, J. (1990) Comparative study of 4 MAbs against sialyl-Tn- related antigens in normal, fetal and cancerous colonic mucosae. *Tumor Biol.* **11 (abstract),** 111.
17. Bara, J., Mollicone, R., Mouradian, P., and Oriol, R. (1987) Immunohistologic patterns of 32 anti-A monoclonal antibodies in normal human pyloric and duodenal mucosae. *Blood Transfusion and Immunohaematology* **30,** 455–463.
18. Bara, J. Decaens, C., Loridon-Rosa, B., and Oriol, R. (1992) Immunohistological characterization of mucin epitopes by pre-treatment of gastrointestinal sections with periodic acid. *J. Immunol. Methods* **149,** 105–113.
19. Mollicone, R., Bara, J., Le Pendu, J., and Oriol, R. (1985) Immunohistologic pattern of type 1 (Lea, Leb) and type 2 (X,Y,H) blood group-related antigens in the human pyloric and duodenal mucosae. *Laboratory Investigation* **53,** 219–227.
20. Bara, J., Mollicone, R., Herrero-Zabeleta, E., Gautier, R., Daher, N., and Oriol, R. (1988) Ectopic expression of the Y (Ley) antigen defined by monoclonal antibody 12-4LE in distal colonic adenocarcinomas. *Int. J. Cancer* **41,** 683–689.

CHAPTER 20

The Application of Lectins to the Study of Mucosal Glycoproteins

Jonathan M. Rhodes and Chi Kong Ching

1. Introduction

Mucosal surfaces are the most vulnerable parts of any animal, and their defense to a large extent depends on carbohydrates: the carbohydrates of the glycocalyx and the carbohydrates of the heavily glycosylated mucus glycoproteins, which coat the mucosa. In mucus, the carbohydrates that constitute the oligosaccharide side chains (of which there are about 150/mucin molecule) confer on the mucus its well-known physical properties as a lubricant and physical barrier, and also help to protect its protein core from attack by proteases. Carbohydrates both on mucus and on the mucosal surface also act as receptors for bacteria and protozoa via lectin–carbohydrate interactions and constitute an essential part of many growth factor receptors. Carbohydrate structures are also thought to be of critical importance in determining cell–cell interactions. Complete sequencing of the oligosaccharide chains in a glycoprotein can be a very lengthy process, whereas simple quantitative analysis of the individual carbohydrates in tissue extracts gives us no information about their likely function.

1.1. Lectin Histochemistry

Lectins have proved very useful tools for obtaining valuable information about carbohydrate structures quickly and simply. Much of this work has been histochemical *(1)* and has involved the use of

From: *Methods in Molecular Biology, Vol. 14: Glycoprotein Analysis in Biomedicine*
Edited by: E. F. Hounsell

lectins that have been tagged with a suitable label, such as peroxidase or fluorescein. Such work has, for example, demonstrated that some carbohydrate antigens may behave as oncofetal antigens, for example Galβ1-3GalNAcα1, which is the Thomsen-Friedenreich (T) blood group antigen. It can be detected by binding with the lectin peanut agglutinin (PNA) in fetal *(2)* and neoplastic colonic mucosa *(3,4)*, and in such conditions as ulcerative colitis where there is an increased risk for malignancy *(5)*, but not in normal adult colon. There are many other examples of altered blood group expression by mucosae in malignant diseases *(6)*. Much remains to be learned about the functional significance of these changes in carbohydrate expression, but it has already been shown that the pattern of lectin binding of malignant cells may correlate with their metastatic potential *(7)* and with the prognosis of the disease *(8)*. We are currently very interested in the possibility that dietary lectins may have important mitogenic effects on the intestinal mucosa, either by a direct action *(9)* or by interaction with other growth factors *(10)*. Some lectins, such as the red kidney bean lectin (phytohaemagglutinin, PHA), are of course well known for their toxic effects on normal intestinal epithelium, but there is also the interesting possibility that changes in the lectin-binding profile of the mucosa, such as those that have been observed in coeliac disease *(11)*, might result in lectins that are normally nontoxic having a harmful effect. The pattern of carbohydrate expression by the mucosa is also of great potential importance in the interaction among the mucosa and pathogenic bacteria and protozoans, many of which bind to the mucosa by lectin–carbohydrate interactions *(12)*.

Care should be taken when interpreting the result of lectin-binding studies. It must be realized that different lectins vary considerably in their degree of specificity *(13)*. Many lectins will only bind to a saccharide if it is expressed as a terminal residue, some will only bind to the saccharide if it is in the correct configuration (e.g., ricin [RCA1] to β, but not to α galactose), or if it is linked in a particular way to the more proximal saccharide in the oligosaccharide chain (e.g., *Ulex europaeus* [UEA1] to fucoseα on a Galβ1-4GlcNAcβ1 (Type II) side chain, but not to fucoseα on a Galβ1-3GlcNAcβ1 (Type I) chain, and some will only bind to a sugar that has a particular pattern of substitution, e.g., *Cancer antennarius* lectin to *O*-acetylated sialic acids *(14)*.

1.2. Oligosaccharide Structural Analysis of Glycoproteins in Polyacrylamide Gels by Sequential Degradation and Lectin Analysis

Further information about carbohydrate structure can be obtained if sequential lectin binding analysis is used in combination with sequential degradation, a technique proposed by Irimura and Nicholson *(15)* for use on electrophoresis gels, which can be also applied to electroblotted glycoproteins *(16)* (*see* Figs. 1–6). A panel of peroxidase-tagged lectins is selected for their differing carbohydrate specificities and is used to assess carbohydrate expression by the electroblotted glycoprotein that is under investigation. A "replica" of the electroblotted glycoprotein is treated on the blot by mild acid hydrolysis, which releases sialic acid and fucose, and is then reassessed using the same panel of lectins. Further electroblotted replicas are then subjected to mild acid hydrolysis followed by one, two, or three Smith degradations. Smith degradation consists of sequential oxidation by periodate, reduction by borohydride, and mild acid hydrolysis with sulfuric acid. Each degradation step cleaves off successive monosaccharides, so that the subterminal carbohydrate revealed by each degradation can be identified by its pattern of lectin binding. Smith degradation, which depends on the availability of vicinal diols for periodate oxidation, destroys all monosaccharides at the nonreducing terminal of side chains and, in addition, will destroy any nonterminal monosubstituted saccharides unless they are substituted at the C3 position for hexopyranoses and C3 or C4 for 2-acetamido hexopyranoses *(17)*. Since most glycoproteins will have more than one oligosaccharide side chain, this technique gives a picture of the degree of variability of the side chain structures rather than full sequence information.

1.3. Enzyme-Linked Lectin-Binding Assays (ELLA) for Quantification of Soluble Glycoproteins

Enzyme-linked lectins can also be used in binding assays (ELLAs) for the quantification of soluble glycoproteins in a way analogous to enzyme-linked antibodies *(18)*. The specificity of binding will of course be much less than that for an antibody, but this may be useful in certain situations, for example, the quantification of mucins *(19)* (*see* Figs. 7 and 8) or sialylated as compared with desialylated glycoproteins.

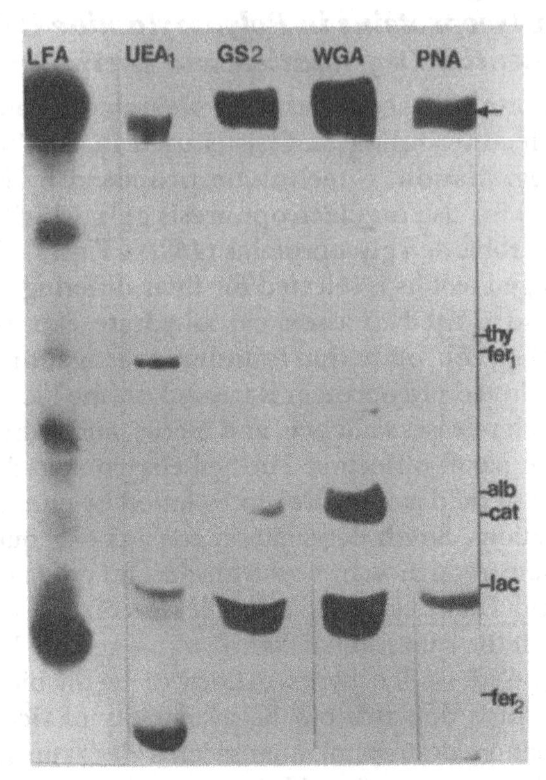

Fig. 1. Lectin blotting of five aliquots of a pancreatic cancer patient's serum following SDS-polyacrylamide (2–16%) gel electrophoresis and high-intensity transfer onto nitrocellulose paper. The high-mol-wt cancer-related glycoprotein under investigation is arrowed. Lectins used were *Limax flavus* (LFA), *Ulex europaeus* (UEA$_1$), *Griffonia simplicifolia* (GS2), wheat germ agglutinin (WGA), and peanut agglutinin (PNA). Molecular-weight standards: thy = thyroglobulin 330, fer $_1$ = ferritin 1 220, alb = albumin 67, cat = catalase 60, lac = lactate dehydrogenase 32, Fer$_2$ = ferritin 2 18, kDa. From ref. *16*, reprinted with permission.

2. Materials

2.1. Lectin Histochemistry

1. Peroxidase-tagged lectins (various commercial suppliers) is made up as described in Section 2.2., step 11. Peanut agglutinin (PNA) and concanavalin A (Con A) both exist as polymers at pH 7.4, and they are best dissolved in PBS buffer at pH 6.8, when they both become dimers. Other lectins can be dissolved in PBS. Most lectins depend

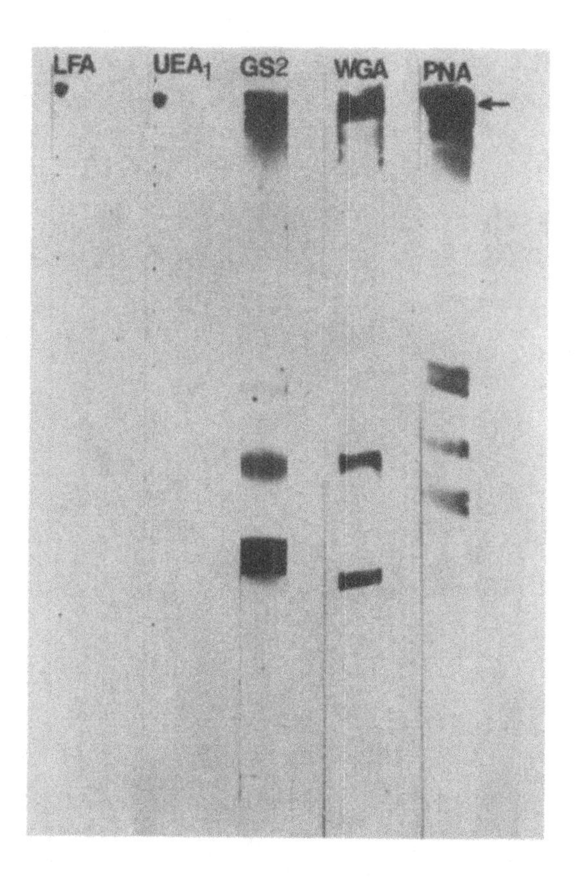

Fig. 2. Lectin blotting after mild acid hydrolysis showing the absence of LFA (sialic acid) and UEA_1 (fucose) binding. From ref. *16*, reprinted with permission.

on cations for their binding activities *(20)*, and 10 m*M* calcium chloride, magnesium chloride, and manganese chloride should therefore be added to the lectin solutions. This is particularly important with Con A. Some lectins, for example, PNA and Con A, are mitogenic, and others are associated with potent toxins (e.g., *Ricinus communis I* and *Abrus pecatorius)*, so they should be handled with care.

2. Tris-buffered saline (TBS): 10 m*M* Tris-HCl, 150 m*M* NaCl, pH 8.0.
3. Hydrogen peroxide.
4. 3,3-diaminobenzidine solution: 0.6 mg/mL in TBS containing 0.1% hydrogen peroxide (freshly prepared) (*see* Note 1).
5. 0.01*M* Phosphate-buffered saline (PBS) (0.01*M*) pH 7.2.

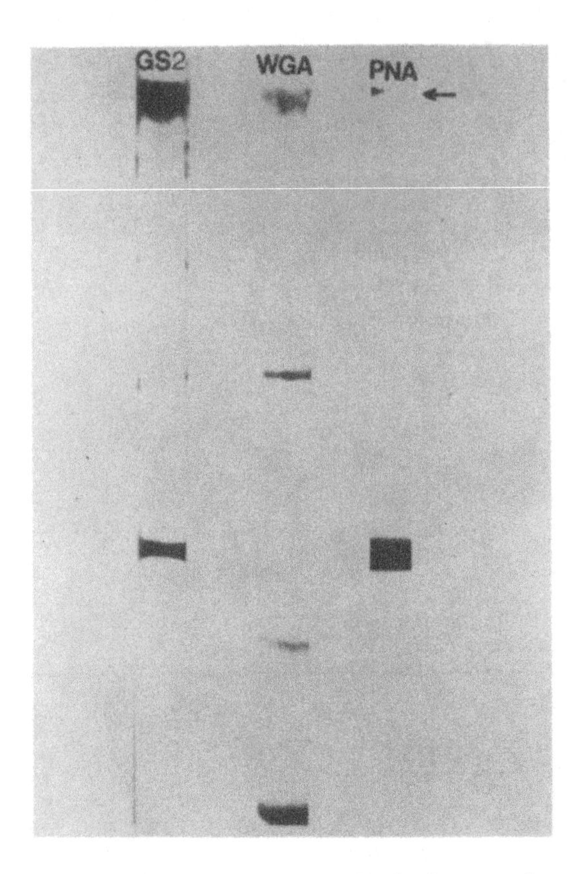

Fig. 3. Lectin blotting after mild acid hydrolysis and one Smith degradation showing disappearance of PNA positivity and persistence of GS2 positivity. From ref. *16*, reprinted with permission.

2.2. Sequential Degradation and Lectin Blotting

1. Polyacrylamide gels (2–16% gradient gels used for mucin studies; *see* Note 2).
2. Whatman No. 1 filter papers.
3. Nitrocellulose membrane (slightly larger than the gels).
4. Tris/glycine buffer: 1.25 m*M* Tris and 96 m*M* glycine (pH 8.3).
5. PBS/Tween 20 (0.1%) in 0.01*M* phosphate-buffered saline (PBS), pH 7.2, prepared fresh. (*See* Note 3.)
6. Ponceau S (0.2%) in distilled water.
7. 50 m*M* and 25 m*M* Sulfuric acid.

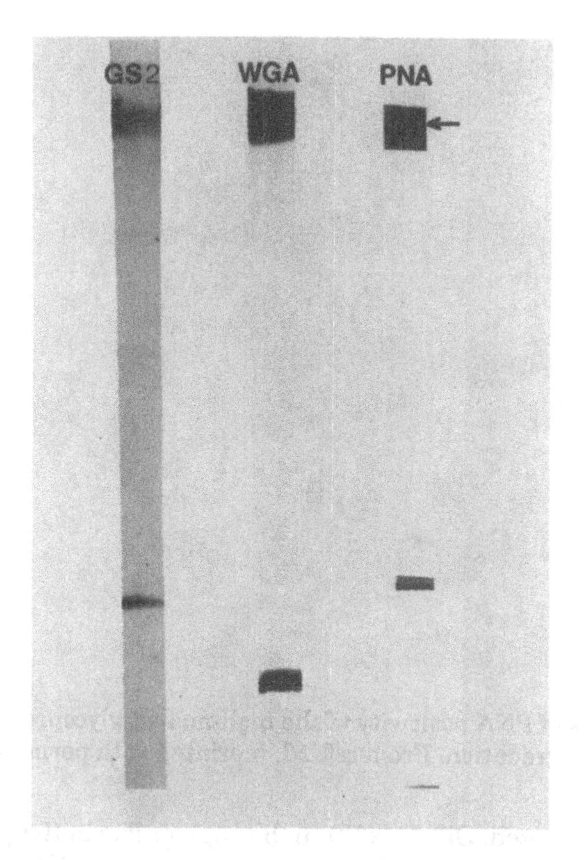

Fig. 4. Reappearance of PNA binding to the high-mol-wt glycoprotein, and persistence of GS2 and WGA binding after a second Smith degradation. From ref. *16*, reprinted with permission.

8. 75 m*M* Sodium periodate in 50 m*M* sodium acetate, pH 4.0—made up fresh avoiding strong sunlight.
9. 1,2 ethanediol, 0.1% in distilled water.
10. 0.1*M* Sodium borohydride in 0.1*M* sodium borate buffer, pH 8.0.
11. For the lectin solutions, prepare PBS buffer (pH 7.2) containing 10 m*M* Ca^{2+}, Mg^{2+}, and Mn^{2+}. Separately prepare PBS buffer, pH 6.8, containing the same amount of Ca^{2+}, Mg^{2+}, and Mn^{2+} ions by titrating with 1*N* HCl. Prepare a stock solution (at 10X the final lectin concentration) for each lectin in the appropriate buffer (PNA and ConA in PBS, pH 6.8; UEA I, WGA, *Limax flavus* [LFA], and *Griffonia simplicifolia* [GS2] in PBS, pH 7.2 for example). Store the lectin solutions in aliquots at

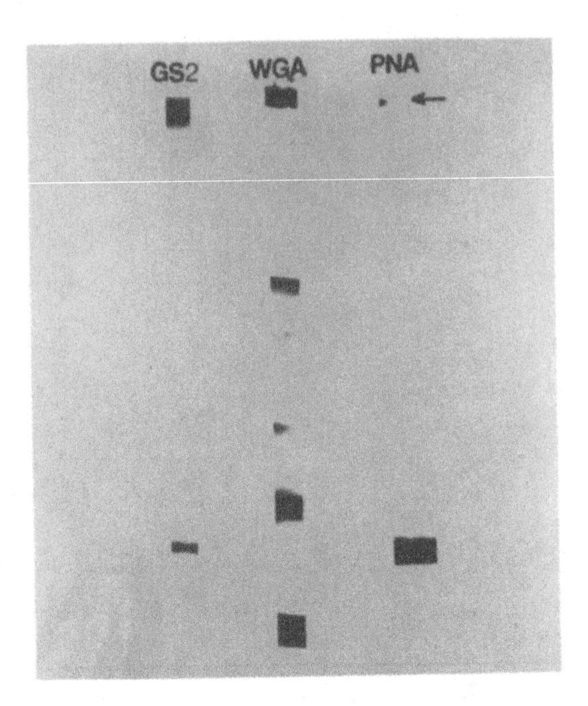

Fig. 5. Loss of PNA positivity of the high-mol-wt glycoprotein after the third Smith degradation. From ref. *16*, reprinted with permission.

–20°C until used. Dilute (x10) in the appropriate buffer prior to use. Suggested lectin–peroxidase concentrations are: PNA 6.25 µg/mL, LFA 6.25 µg/mL, WGA 6.25 µg/mL, UEAI 25 µg/mL, GS2 25 µg/mL.

12. 4-Chloro-1-naphthol in hydrogen peroxide prepared fresh just before use by mixing 1 part of 3 mg/mL 4 chloro-1-naphthol in methanol and 4 parts of H_2O_2 in PBS (50 µL 30% H_2O_2 in 40 mL PBS, pH 7. 2). **Caution:** 4-Chloro-1-naphthol is a carcinogen and should be handled with care.
13. Methanol.

2.3. Enzyme-Linked Lectin Binding Assay

1. Cobalt-irradiated ELISA plates (e.g., M 124B Dynatech).
2. Methanol.
3. Ortho phenylenediamine 5 mg/1% hydrogen peroxide (75 µL) in 12.5 mL 0.2*M* phosphate/0.1*M* citrate buffer, pH 5.0.
4. Sulfuric acid (4*M*)
5. Glycoprotein solution under assay mixed 1:1 with carbonate buffer containing 50 m*M* Na_2CO_3, 50 m*M* $NaHCO_3$, pH 9.6.

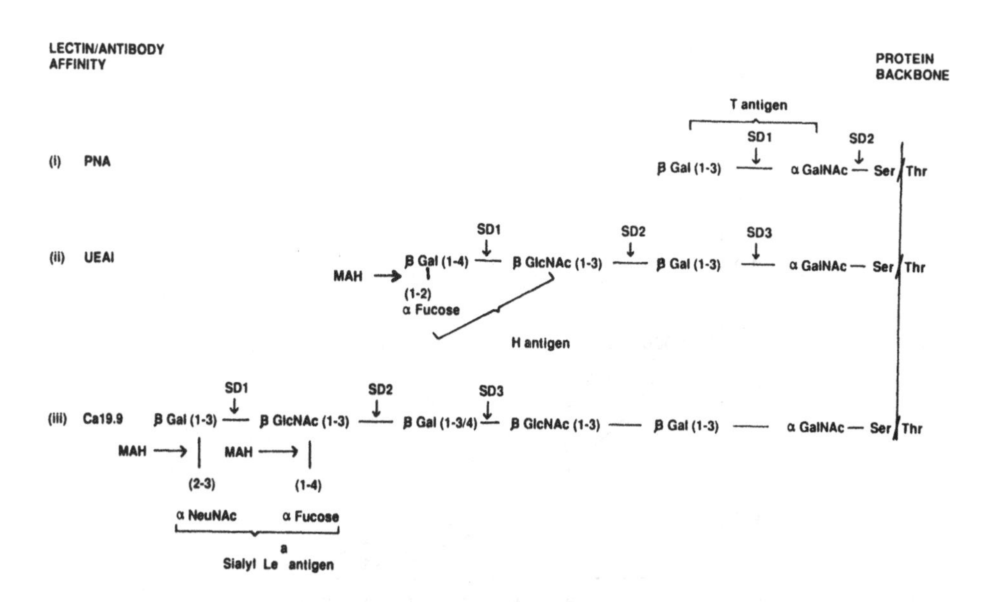

Fig. 6. Postulated minimum variation in oligosaccharide structures of the high-mol-wt glycoprotein studied in Figs. 1–5. (The glycoprotein was also known to express sialylated Lewis[a]). From ref. *16*, reprinted with permission.

3. Methods

3.1. Lectin Histochemistry of Formalin-Fixed Tissue (see Note 4)

1. Mount 5-µm tissue sections on acetone-cleaned glass slides.
2. Deparaffinate the sections in xylene, and pass them from absolute alcohol into PBS.
3. Inhibit endogenous peroxidase by treatment with 1% hydrogen peroxide for 1 h.
4. Incubate for 16 h at 4° C with lectin–peroxidase conjugate (usually at 0.002 mg/mL, but may vary between lectins, *see* Note 5).
5. Wash off unbound lectin by 3 × 5 min washes in PBS.
6. Incubate for 10 min with 3,3-diaminobenzidine 0.6 mg/mL in TBS containing 0.01% hydrogen peroxide to reveal bound lectin/peroxidase (*see* Note 6).
7. Repeat the incubation in the presence of the appropriate inhibitory sugar at 0.2*M* (*see* Note 7).

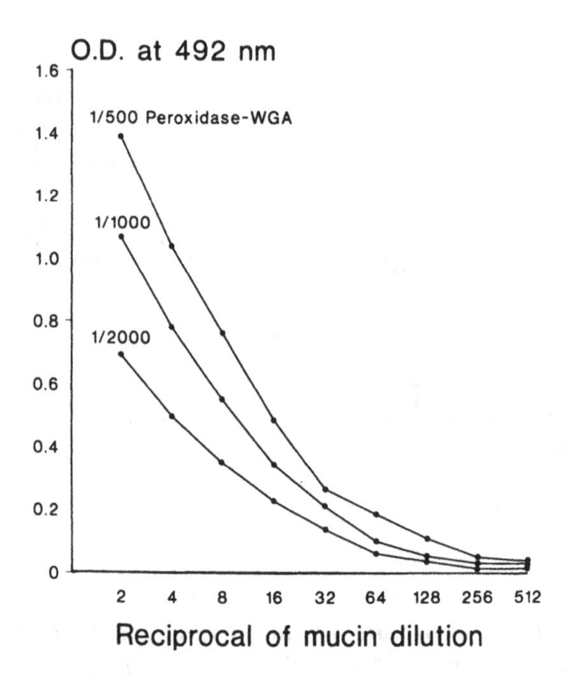

Fig. 7. Serial dilutions of human colonic mucin assayed by WGA-peroxidase binding assay (using three different concentrations of lectin).

3.2. Structural Analysis by Sequential Degradation and Lectin Blotting (see *Note 8*)

1. Prepare an SDS-PAGE gel with several lanes (or one broad lane) containing the glycoprotein under study, and equilibrate this in Tris/glycine buffer for 30 min. Enough lanes will have to be run to allow a panel of at least five lectins to be tested after each of four degradation steps, i.e., 20 lanes, not allowing for duplicates.
2. Soak Whatman No. 1 filter papers and a nitrocellulose membrane (slightly larger than the gel) in Tris/glycine buffer (pH 8.3) for 30 min, and then use them to form a "sandwich" with the electrophoresed and equilibrated gel.
3. Electroblot at 100 V for 2 h in a transblot cell with the temperature constantly kept below 10°C by a supercooling coil containing ice-chilled water.
4. After high-intensity transfer, carefully separate the nitrocellulose membrane (blot) from the polyacrylamide gel.
5. Quench the blot by incubation in PBS/Tween 20 for 1 h with two changes of buffer.

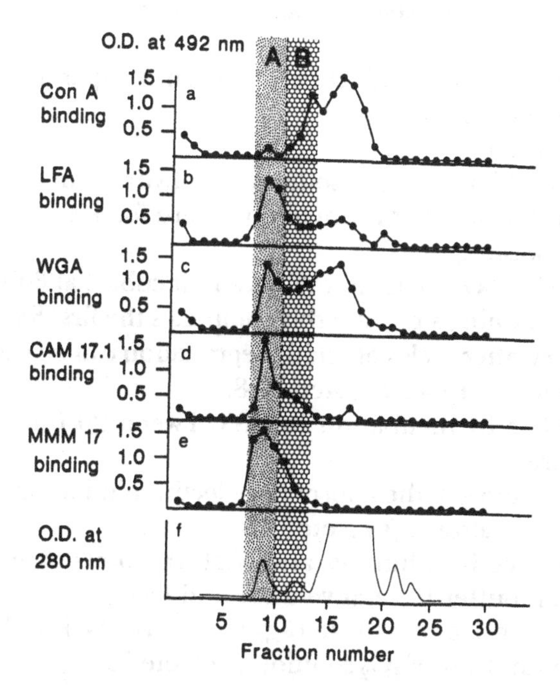

Fig. 8. Elution profile of human colonic mucosal glycoproteins identified by lectin binding assays following gel filtration of soluble glycoproteins extracted from a single endoscopic biopsy. CAM 17.1 and MMM 17 are antimucin mouse monoclonal antibodies.

6. Reveal the position of the glycoprotein bands on the nitrocellulose blot using Ponceau S as a temporary stain.
7. Cut replica lanes from the blot, creating a series of "mini blots" (or 0.25 cm wide strips if one broad lane has been run), and mark the site of the glycoprotein under study on each blot with a pencil or ball point pen. Remove the Ponceau S stain by repeated washings in PBS/Tween 20 until clear.
8. Incubate the miniblots in 50 m*M* H_2SO_4 at 80°C for 3 h (*see* Notes 9 and 10).
9. Wash (5 × 3 min) with PBS/Tween 20 buffer.
10. Store at –20°C a sufficient number of miniblots for analysis by the complete range of lectins. The remaining miniblots are used for Smith degradation (*see* Note 11).
11. Incubate the miniblots with sodium periodate/sodium acetate in the dark at 4°C for 48 h (*see* Note 9).

12. Discard the periodate solution, and wash the blots with PBS/Tween 20 buffer (3 × 3 min).
13. Add 1,2 ethanediol, and incubate at room temperature for 1 h to stop the oxidation process.
14. Repeat the washing step.
15. Incubate the blots in 0.1*M* sodium borohydride in sodium borate buffer (0.1*M*, pH 8.0) at room temperature for 4 h.
16. Repeat the washing step.
17. Add 25 m*M* H_2SO_4 to the miniblots and incubate at 80°C for 1 h.
18. Repeat the washing step, and this completes the first Smith degradation cycle. Further cycles of Smith degradation can be carried out on the same blots by repeating steps 1–8.
19. Quench blots by incubation in PBS/Tween 20 for 1 h at room temperature.
20. Incubate the blots in the appropriate lectin solutions at 4°C for 16 h on a rotating platform (*see* Note 12).
21. Discard the lectin solutions, and wash the blots three times with PBS/Tween buffer to remove unbound lectin.
22. Identify bound peroxidase-tagged lectin by incubation in 4 chloronaphthol and H_2O_2 solution (*see* Note 13).

3.3. Enzyme-Linked Lectin Binding Assay (see Note 14)

1. Incubate 100 µL of the glycoprotein solutions under assay in each well of a cobalt-irradiated ELISA plate for 16 h at 4°C (*see* Note 15). At least three wells on each plate are incubated with buffer alone as blanks.
2. Remove unbound glycoprotein by washing three times with PBS/Tween 20.
3. Quench by incubation in PBS/Tween 20 for 1 h at room temperature.
4. Add 50 µL of a 2 µg/mL solution of peroxidase-tagged wheat germ agglutinin to each well, and incubate for 2 h at 37°C (*see* Note 16).
5. Tap out wells, and wash three times with PBS/Tween 20 to remove unbound lectin.
6. Add 100 µL orthophenyldiamine/H_2O_2 and incubate for 10 min.
7. Stop the reaction by addition of 100 µL 4*M* H_2SO_4.
8. Read the plate at 450 nm in a suitable ELISA plate reader.

4. Notes

1. 3,3-Diaminobenzidine is carcinogenic, and 3-amino,9-ethylcarbazole may be used as a less carcinogenic alternative.
2. We have used 2–16% polyacrylamide gels, because we have been studying mucins that are only able to enter gels with a low polyacrylamide

concentration. The 2% end of the gel is rather soft and becomes adherent to the nitrocellulose blot after high-intensity transfer. The adherent gel can be carefully rubbed off with a presoaked tissue while immersed in the Tris/glycine buffer. Every attempt should be made to remove the gel completely; otherwise, it will cause unwanted artifacts.

3. Fresh buffer is required for every high-intensity transfer procedure. The use of old buffer results in rapid temperature rise and distortion of the gel. The addition of SDS (0.01%) to the transfer buffer has been claimed to enhance transfer, but it causes excessive foaming during the high-intensity transfer and should be avoided. Methanol is commonly included in transfer buffers for Western blotting, but we have avoided it because it reduces the pore size of gels and reduces the speed and efficiency of the electrophoretic transfer.
4. Many lectins (e.g., PNA, WGA, UEAI) bind similarly to formalin-fixed and frozen tissue, but there are exceptions (e.g., *Cancer antennarius [14]*, which will only detect *O*-acetylated sialic acids in frozen sections), and if in doubt a pilot study should be performed to compare the staining characteristics on frozen and formalin fixed tissue. Lectin histochemistry can be performed on frozen sections using the same technique as described for fixed tissue if the sections are mounted onto adhesive slides (coated with poly-l-lysine), air-dried for 12 h, and fixed in 50:50 acetone:methanol for 10 min prior to lectin histochemistry.
5. The technique described here, which is essentially that described by Kuhlmann and colleagues *(21)*, uses a very low concentration of lectin with a long incubation. This has the advantages that there is very good specificity and low cost, but many workers use higher concentrations of lectin-conjugate, e.g., 0.05 mg/mL with a 20–30 min incubation.
6. There are a wide range of methods available for identifying bound lectins in histochemical preparations. These include fluorescein, avidin-biotin systems, and antilectin antibodies. Peroxidase-tagged lectins are widely available and usually represent the simplest option. The sensitivity may vary considerably with different systems, and this can cause confusion. Biotinylated PNA, for example, will demonstrate binding to normal adult colon, whereas peroxidase-tagged PNA will not *(3)*. Antilectin antibodies should be used with caution, since many lectins will bind to oligosaccharide chains on immunoglobulins. This is a particular problem with galactose-binding lectins, such as *Ricinus communis I* (RCAI) and PNA. Alkaline phosphatase-conjugated lectins should be avoided in the study of small intestinal mucosa, because this contains a high concentration of endogenous alkaline phosphatase, which is difficult to eradicate by the conventional levamisole (1*M*) preincubation step.

7. Some lectins are easier to inhibit than others (wheat germ agglutinin tends only to be partly inhibited by 0.2*M* GlcNAc). Although some lectins bind to specific disaccharides (e.g., peanut agglutinin and Galβ1-3GalNAc), it is sufficient to use the terminal carbohydrate of the binding site (i.e., D-galactose for peanut agglutinin) as the inhibitor when checking the specificity of lectin binding.
8. The technique as originally described by Irimura and Nicolson *(15)* was carried out on the polyacrylamide gel using radiolabeled lectins. We find the nitrocellulose blots much easier to handle, and they allow repeated cycles of chemical degradation to be carried out on the same blot, a procedure that is impossible on a polyacrylamide gel. Peroxidase-tagged lectins are also easier to use and are more readily available than radiolabeled lectins.
9. The duration of mild acid hydrolysis and each Smith degradation step may be shortened for some glycoproteins *(15)*.
10. Exo-glycosidases, e.g., α-fucosidase and neuraminidase, can be used instead of mild acid hydrolysis to remove fucose and sialic acid.
11. Thaw out stored miniblots at room temperature prior to study.
12. The lowest possible concentration of each lectin solution should be used to avoid nonspecific binding.
13. Developed blots can be stored between sheets of filter paper at –20°C for future photography. The blots should be thawed out slowly, reimmersed in deionized water, and kept at 4°C prior to photography.
14. This assay will not allow absolute quantification of glycoprotein concentration, since the degree of glycosylation of the glycoprotein may vary. We have found it particularly useful for estimating relative concentrations of very small amounts of glycoprotein in elution fractions obtained after chromatography of soluble glycoproteins from mucosal biopsy samples. When used in this way to study mucus glycoproteins in fractions obtained following gel filtration or ion-exchange chromatography, it gives a coefficient of variation between plates of 8.7% and is able to detect >0.2 μg of mucin.
15. The choice of ELISA plate is important, since some are not very adherent for heavily glycosylated glycoproteins. We have found that cobalt-irradiated plates tend to be better in this respect.
16. Wheat germ agglutinin, which binds *N*-acetylglucosamine and sialic acid (*N*-acetylneuraminic acid), will bind most if not all mucosal glycoproteins and is particularly useful as a broad screen for glycoprotein content. Similar assays can be performed using other peroxidase-tagged lectins, such as Con A and *Limax flavus*, to obtain further information about the relative concentrations of mannose-containing or sialylated glycoproteins, respectively.

References

1. Damjanov, I. (1987) Biology of disease. Lectin cytochemistry and histochemistry. *Lab. Invest.* **57,** 5–20.
2. Coapman, R. A. and Cooper, H. S. (1986) Peanut lectin binding sites in human fetal colon. *Arch. Pathol. Lab. Med.* **110,** 124–127.
3. Cooper, H. S. (1982) Peanut lectin binding sites in large bowel carcinoma. *Lab. Invest.* **47,** 383–390.
4. Rhodes, J. M., Black, R. R., and Savage, A. (1986) Glycoprotein abnormalities in colonic carcinomata, adenomata and hyperplastic polyps shown by lectin peroxidase histochemistry. *J. Clin. Pathol.* **39,** 1331–1334.
5. Rhodes, J. M., Black, R. R., and Savage, A. (1988) Altered lectin binding by colonic epithelial glycoconjugates in ulcerative colitis and Crohn's disease. *Dig. Dis. Sci.* **33,** 1359–1363.
6. Hakomori, S. (1984) Philip Levine Award Lecture. Blood group glycolipid antigens and their modifications as human cancer antigens. *Am. J. Clin. Pathol.* **82,** 635–648.
7. Kellokumpu, I. H. (1986) Differences in lectin reactivities of cellular glycoconjugates between primary human colorectal carcinomas and their metastases. *Cancer Res.* **46,** 4620–4625.
8. Veerman, A. J. P., Hogeman, P. H. G., Huismans, D. R., Van Zantwijk, C. H., and Bezemer, P. D. (1985) Peanut agglutinin, a marker for T-cell acute lymphoblastic leukemia with a good prognosis. *Cancer Res.* **45,** 1890–1893.
9. Ryder, S. D., Rhodes, E. G. H., Smith, J. A., and Rhodes, J. M. (1989) Lectins modulate growth in HT29 colon cancer cells. *Gut* **30,(abstr)A,** 1449.
10. Feizi, T. and Childs, R. A. (1985) Carbohydrate structures of glycoproteins and glycolipids as differentiation antigens, tumour-associated antigens, and components of receptor systems. *Trends Biochem. Sci.* **10,** 24–29.
11. Vecchi, M., Torgano, G., de-Franchis, R. Tronconi, S., Agape, D., and Ronchi, G. (1989) Evidence of altered structural and secretory glycoconjugates in the jejunal mucosa of patients with gluten sensitive enteropathy and subtotal villous atrophy. *Gut* **30,** 804–810.
12. Chadee, K., Petri, W. A., Innes, D. J., and Ravdin, J. I. (1987) Rat and human colonic mucins bind to and inhibit adherence lectin of *Entamoeba histolytica. J. Clin. Invest.* **80,** 1245–1254.
13. Goldstein, I. J. and Poretz, R. D. (1986) Isolation, physicochemical characterization, and carbohydrate-binding specificity of lectins in *The Lectins: Properties, Functions, and Applications in Biology and Medicine* (Liener, I. E., Sharon, N., and Goldstein, I. J., eds.), Academic, London and New York, pp. 35–250.
14. Ravindranath, M. H., Higa, H. H., Cooper, E. L., and Paulson, J. C. (1985) Purification and characterization of an *O*-Acetylsialic acid-specific lectin from a marine crab *Cancer antennarius. J. Biol. Chem.* **260,** 8850–8856.
15. Irimura, T. and Nicolson, G. L. (1983) Carbohydrate chain analysis by lectin binding to mixtures of glycoproteins, separated by polyacrylamide slab-gel electrophoresis, with *in situ* chemical modifications. *Carbohydrate Res.* **115,** 209–220.

16. Ching, C. K. and Rhodes, J. M. (1990) Purification and characterization of a peanut-agglutinin-binding pancreatic cancer-related serum mucus glycoprotein. *Int. J. Cancer* **45,** 1022–1027.
17. Beeley, J. G. (1985) Structural analysis, in *Glycoprotein and Proteoglycan Techniques.* (Burdon, R. H. and van Knippenberg, R. H., eds.), Elsevier, Amsterdam, p. 285.
18. McCoy, J. P., Varani, J., and Goldstein, I. J. (1983) Enzyme-linked lectin assay (ELLA): Use of alkaline phosphatase-conjugated *Griffonia simplicifolia* B4 isolectin for the detection of D-galactopyranosyl end groups. *Anal. Biochem.* **130,** 437–444.
19. Raouf, A., Parker, N., Iddon, D., Ryder, S., Langdon-Brown, B., Milton, J. D., Walker, R., and Rhodes, J. M. (1991) Ion-exchange chromatography of purified colonic mucus glycoproteins in inflammatory bowel disease; absence of a selective subclass defect. *Gut,* in press.
20. Lonnerdal, B., Borrebaeck, C. A. K., Etzler, M. E., and Erssoon, B. (1983) Dependence on cations for the binding affinity of lectins as determined by affinity electrophoresis. *Biochem. Biophys. Res. Comm.* **115,** 1069–1074.
21. Kuhlmann, W. D., Peschke, P., and Wurster, K. (1983) Lectin-peroxidase conjugates in histopathology of gastrointestinal mucosa. *Virchows Arch. Pathol. Anat.* **398,** 319–328.

CHAPTER 21

Detection of Tumor-Associated Expression of Carbohydrate-Binding Proteins (Lectins)

The Use of Neoglycoproteins and Neoglycoenzymes in Glycohistochemical and Glycocytological Studies

Hans-Joachim Gabius and Sigrun Gabius

1. Introduction

Plant lectins and carbohydrate-specific monoclonal antibodies play a significant role in the characterization of carbohydrate structures of cellular glycoconjugates. It is important to emphasize that the remarkable specificity of these tools should not only be considered to be exploitable analytically. Since their binding behavior may mimic the respective capacity of sites in the tissue, they can serve as models to infer the presence of defined ligands for potential endogenous sugar receptors. Successful application of such tools, e.g., by uncovering developmental regulation of certain carbohydrate structures, encourages experimental attempts to prove the presence of endogenous receptors for these carbohydrate ligands. Detection of their expression in the tissue is a consequent step in verifying the assumed significance of protein–carbohydrate interactions. Indeed, the notion that endogenous carbohydrate-binding proteins, such as lectins, participate in a variety of physiological processes, such as intracellular routing and serum circulation of glycoproteins or cell interactions, is receiving increasing attention *(1–4)*. Within the class of carbohydrate-binding proteins, lectins are strictly delimited from antibodies or

From: *Methods in Molecular Biology, Vol. 14: Glycoprotein Analysis in Biomedicine*
Edited by: E. F. Hounsell

enzymes *(5)*. Concomitant monitoring of expression of carbohydrate determinants and endogenous lectins with respective specificity in tissue sections and on cells will provide evidence as to whether the assumption of a physiologically meaningful glycobiological interplay is reasonable in the experimental system under investigation. In this respect, tumors offer a readily available source of tissue. Work on this material can also guarantee that any progress in basic science can translate into clinical benefit. Explicitly, the measurement of expression of sugar receptors in tissue sections and on cells may aid in refining tumor classification, in uncovering correlations of this parameter to tumor progression, stage, and spread, and in leading to biochemically defined targets for therapy *(6)*.

The lack of enzymatic activity restricts the localization of lectins to assays that employ labeled ligands for receptor detection. Because of the specificity of lectins to sugar structures, it is obvious that such ligands are ideally suited for this purpose. Chemical conjugation of carbohydrate ligands to a labeled, itself sugar-free carrier establishes a neoglycoprotein, rendering the desired tool available for the search for carbohydrate-specific receptors *(7–10)*. The chosen synthetic procedure allows the production of an array of probes that only differ in the nature of the carbohydrate ligand. When starting with commercially available *p*-aminophenyl glycosides, conjugation to a carrier protein by the diazonium or phenylisothiocyanate reactions has been shown to yield ideal probes *(7–10)*. The carrier protein should combine absence of inherent ligand properties with easy access and presence of further functional groups for label incorporation. By taking advantage of commercially available standardized kits for highly sensitive histochemical detection of biotinylated probes, this modification is selected for the neoglycoproteins. Since paraffin-embedded tissue specimens are commonly processed, the glycohistochemical protocol is described for this type of material. The glycocytological procedure on fixed cells is similarly based on application of biotinylated neoglycoproteins. When binding studies with native cells are performed, chemically glycosylated enzymes, termed neoglycoenzymes, are employed in a two-step procedure as nonradioactive, yet highly sensitive tools instead of iodinated neoglycoproteins to assess the presence of accessible binding sites quantitatively *(11,12)*. The sugar incorporation into all types of carrier protein can be determined by a resorcinol-sulfuric acid micromethod *(13)*. Ascertainment of a constant level of modification ensures reproducibility of lectin detection.

The described protocols thus serve as a means to visualize binding of carbohydrate ligands, immobilized on a labeled carrier protein, to fixed sections and cells. Moreover, neoglycoenzymes allow binding of defined sugar structures to native cells to be evaluated quantitatively. Following this type of measurement as a function of the neoglycoenzyme concentration, transformation of binding data can be carried out by routine procedures to yield the dissociation constant and an estimation of the number of bound probe molecules per cell at saturation *(14)*.

2. Materials

1. *p*-Aminophenyl glycosides.
2. Bovine serum albumin (carbohydrate-free; *see* Note 1).
3. β-galactosidase *(E. coli, Aspergillus oryzae)*.
4. Gel-filtration columns (Sephadex G-50, 3.7 × 50 cm, and Sepharose 4B, 1.8 × 60 cm).

2.1. Preparation of Neoglycoprotein Using HNO_2

1. Ice-cold 0.05*M* $NaNO_2$ (prepare fresh solution).
2. HCl (1*M*, 0.1*M*, and 0.05*M*).
3. 0.9% NaCl adjusted to pH 9 with 0.05*M* NaOH.

2.2. Preparation of Neoglycoprotein Using Thiophosgene

1. 0.1*M* Sodium carbonate buffer, pH 8.6 and pH 9.5.
2. 1% Thiophosgene in CH_2Cl_2 (v/v). **Caution:** Handle with extreme care in a fume hood.
3. Thin-layer chromatography (TLC) silica plates.
4. TLC solvent system ($CHCl_3$:CH_3OH:H_2O, 13:8:2 [v/v]).

2.3. Analysis of Hexose Incorporation into Neoglycoproteins

1. 6 mg/mL Aqueous solution of resorcinol.
2. Ninety-six-well microtiter plate (U-shaped wells).
3. 75% Sulfuric acid.
4. Pristane.

2.4. Biotinylation of Neoglycoproteins and Their Detection

1. Biotinyl-*N*-hydroxysuccinimide ester (store at –20°C) 10 mg dissolved in 1.8 mL DMF.
2. PBS containing 1% BSA.
3. Biotinylated BSA.

4. ABC (Avidin Biotin Conjugate) kit (various commercial suppliers).
5. Streptavidin- or avidin-peroxidase (1 µg/mL).
6. 3,3'-diaminobenzidine hydrochloride (DAB; prepared fresh 30 mg/mL in PBS) containing 0.01% hydrogen peroxide or a less carcinogenic alternative 3-amino,9-ethylcarbazole (0.2 mg/mL in 0.1*M* sodium acetate buffer, pH 5.2).
7. Enhancing buffer (freshly prepared and filtered): Ammonium paramolybdate (0.5 mL of a 1% solution), Nickel chloride (0.5 mL of a 1% solution), and Cobalt chloride (0.5 mL of a 1% solution) in 100 mL 0.15*M* Sorensen phosphate buffer (pH 6.6) containing 30 mg DAB.
8. Hematoxylin or fast green.
9. Entellan (Merck, Darmstadt, FRG).

2.5. Preparation of Neoglycoenzymes

1. Phosphate buffered saline (20m*M*) (PBS, pH 7.4).
2. 1-Ethyl-3-(3-dimethylaminopropyl)carbodiimide (prepare fresh solution).
3. PBS containing 60% glycerol.
4. Gel-filtration column Sepharose 4B (1.8 × 60 cm) equilibrated in PBS.

2.6. Preparation of Tissues and Cells

1. Buffered formalin or Bouin's solution.
2. Ethanol, xylol, 2% methanolic H_2O_2.
3. 0.1–2% BSA in PBS.
4. Hank's balanced salt solution.
5. Periodic acid (0.23%) in PBS.
6. 20 m*M* HEPES, pH 7.5 containing 0.1% BSA (Hepes I).
7. 100 m*M* HEPES, pH 7.0 containing 0.5% Triton X-100, 150 m*M* NaCl, 2 m*M* $MgCl_2$, 0.1% NaN_3, 0.1% BSA, and 1.5 m*M* chlorophenolred-β-D-galactopyranoside (Hepes II).
8. 0.2*M* Glycine, pH 10.5.

3. Methods

3.1. Preparation of Neoglycoprotein Using HNO_2 (see Note 2)

1. Dissolve 0.1 g *p*-aminophenyl glycoside (monosaccharide) in 5 mL of ice-cold 0.1*M* HCl.
2. Acidify the $NaNO_2$ solution by addition of five drops of 1*M* HCl, and add dropwise with stirring to the glycoside solution until a slight excess of HNO_2 is detected by color formation with starch-iodide indicator paper.
3. Immediately add this solution dropwise to an ice-cold solution of 0.5 g bovine serum albumin in 0.9% NaCl solution (pH 9), and let

the reaction continue in an ice bath for 2 h, whereas constantly maintaining the pH at 9 with additions of 0.5*M* NaOH.

4. Neutralize with 0.05*M* HCl, extensively dialyze against distilled H_2O prior to concentration by ultrafiltration, and further separate reagents from the neoglycoprotein by gel filtration on Sephadex G-50 and lyophilize.

3.2. Preparation of Neoglycoprotein Using Thiophosgene (see Note 2)

1. Dissolve 0.1 g *p*-aminophenyl glycoside (monosaccharide) in 8 mL 0.1*M* sodium carbonate buffer (pH 8.6).
2. Add 80 µL thiophosgene in 8 mL CH_2Cl_2 under stirring, and monitor the progress of the reaction by TLC. Remove residual thiophosgene by a constant stream of nitrogen, adjust pH to 6.0 by addition of 0.2*M* NaOH, and subsequently concentrate the solution to near dryness. Wash product with cold water.
3. Add the resulting *p*-isothiocyanatophenyl glycoside in small portions to a solution of 175 mg bovine serum albumin in 10 mL 0.1*M* sodium carbonate buffer (pH 9.5), and let the reaction continue for 16 h, while constantly maintaining the pH at 9.5 with additions of 0.2*M* NaOH.
4. Add 0.1*M* HCl to neutralize, remove unreacted carbohydrate derivative by gel filtration on Sephadex G-50, and extensively dialyze against distilled water prior to lyophilization.

3.3. Analyis of Hexose Incorporation into Neoglycoprotein

1. Place 1–100 nmol of neutral sugar, attached to the neoglycoprotein, in a volume of 20 µL in a microtiter well. Add 20 µL of resorcinol solution, 100 µL of 75% sulfuric acid, and 50 µL of pristane.
2. Cautiously shake the plate to mix them and heat at 90°C for 30 min.
3. Read the absorbance at 430 or 480 nm in a plate reader after keeping the plate in the dark for 30 min. Comparison to a standard curve will reveal sugar incorporation into the carrier protein.

3.4. Biotinylation of Neoglycoprotein

1. Dissolve 100 mg neoglycoprotein in 70 mL 0.1*M* sodium carbonate buffer (pH 8.0) containing 0.9% NaCl.
2. Add a solution of 10 mg biotinyl-*N*-hydroxysuccinimide ester in 1.8 mL dimethylformamide dropwise with stirring.
3. Let the reaction continue for 16 h at room temperature.
4. Remove reagents by gel filtration (*see* Section 3.2., step 4) and by extensive dialysis against distilled water, and lyophilize the biotinylated neoglycoprotein (*see* Note 3).

3.5. Preparation of Neoglycoenzymes

1. Dissolve 8 mg of β-galactosidase in 1 mL of 20 m*M* phosphate-buffered saline (PBS, pH 7.4) at 4°C.
2. Add dropwise with stirring a solution of 16 μmol *p*-aminophenyl glycoside and 32 μmol 1-ethyl-3-(3-dimethylaminopropyl)carbodiimide at 4°C, and let the reaction continue for 16 h.
3. Dialyze the mixture against PBS for 2 d, and remove residual reagents by gel filtration on Sepharose 4B, equilibrated with PBS.
4. Combine enzyme-containing fractions, dialyze against PBS containing 60% glycerol for 16 h, and store at –20°C.

3.6. Glycohistochemical Staining of Tissue Sections

1. Fix the tissue specimen in buffered formalin or in Bouin's fluid, dehydrate in graded ethanol solutions, and embed in paraplast at a temperature not higher than 56°C (*see* Note 4).
2. Cut sections of appropriate thickness, dewax in xylol, rehydrate in graded ethanol solutions, and incubate for 10–30 min with 2% methanolic hydrogen peroxide to block endogenous peroxidase activity prior to rinsing with PBS.
3. Incubate sections for 1 h with 0.1–2% bovine serum albumin (BSA) solution in PBS to saturate protein-binding sites, and blot this solution off cautiously with filter paper.
4. Incubate sections with 5–150 μg/mL biotinylated neoglycoprotein, dissolved in PBS + 0.1% BSA, for 1–12 h at 4°C, room temperature, or 37°C in an incubator. In parallel, control sections are incubated with the same concentration of nonglycosylated, but biotinylated BSA to exclude binding of the probes by protein–protein interaction. Further control sections are incubated in the presence of an excess of homologous, but unlabeled neoglycoprotein to ascertain sugar specificity of binding (*see* Note 5).
5. Rinse thoroughly with PBS, and incubate with ABC kit reagent for 1 h at room temperature or with streptavidin- or avidin-peroxidase (1 μg/mL) overnight at 4°C (*see* Note 6).
6. Rinse thoroughly, and incubate in DAB or 3-amino,9-ethylcarbazole solution (Section 2.4., step 6).
7. Enhance a weak signal in the DAB reaction by serially incubating the section for 10 min each with the enhancing buffer (Section 2.4., step 7) in the absence and then in the presence of 300 μL hydrogen peroxide.

8. Rinse thoroughly with water, counterstain with hematoxylin (without intensification), or fast green (with intensification) and cover with Entellan prior to light microscope evaluation (*see* Notes 7–10).

3.7. Glycocytochemical Staining of Cells

1. Wash cells carefully with Hank's balanced salt solution containing 1% BSA to remove serum components.
2. Prepare specimen of 5×10^4 cells/slide with a cytospin centrifuge for cells in suspension, or use coverslips with washed adherent cells, dry the samples, fix in 80% ice-cold acetone for 10 min, and wash with PBS containing 0.1% BSA.
3. Incubate preparations for 15–60 min at room temperature with PBS containing 2% BSA to saturate protein-binding sites and blot this solution off cautiously with filter paper.
4. Perform the same steps 4–8, as described in the previous section, with one addition after the incubation with the labeled probe: Abolish endogenous peroxidase activity by treatment with 0.23% periodic acid in PBS for 45 s and wash thoroughly.

3.8. Cell-Surface Binding of Neoglycoenzymes

1. Wash cells thoroughly with 10 m*M* PBS (pH 7.2) or Hank's balanced salt solution with HEPES I (Section 2.6., item 6) to remove serum components.
2. Incubate cells in suspension with a certain amount of neoglycoenzyme in 400 µL of buffer solution for various periods, e.g., 5–120 min, at 4°C with constant shaking. Incubate adhesive cells that were grown in 24-well culture plates for at least 24 h after trypsinization with a certain amount of neoglycoenzyme in 200 µL of buffer solution for various periods, e.g., 5–120 min, at 4°C with constant shaking.
3. Wash thoroughly and quickly three times with buffer solution (800 µL for cells in suspension, 300 µL for adhesive cells) to remove unbound neoglycoenzyme.
4. Add 200 µL freshly prepared HEPES II (Section 2.6., item 7) to the wells for adhesive cells or to the test tubes for cells in suspension.
5. Stop enzymatic reaction by addition of 200 µL of 0.2*M* glycine (pH 10.5), and determine development of the chromogenic product in a microtiter plate reader at 590 nm.
6. Perform binding studies as a function of neoglycoenzyme concentration with an incubation time, when plateau binding is invariably achieved, concomitantly evaluating the extent of noninhibitable binding that is referred to as nonspecific (Fig. 1).

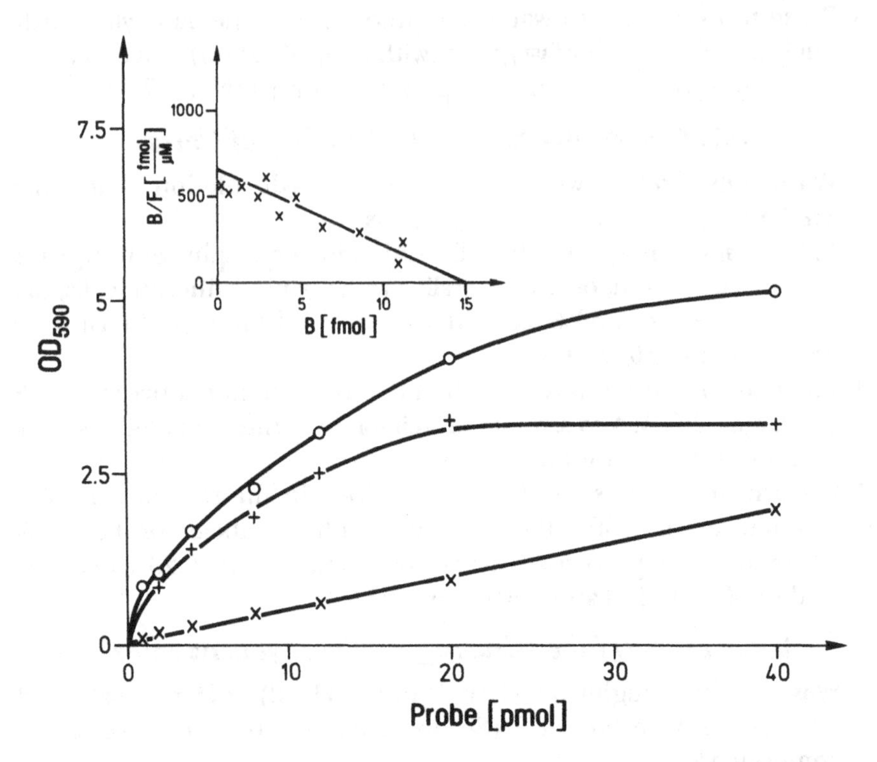

Fig. 1. Determination of specific binding (+) of lactosylated *E. coli* β-galactosidase to 3.8 × 10^4 human promyelocytic leukemia cells (HL-60) after 15 min at 4°C as a function of ligand concentration. Total binding (o) was reduced by the extent of nonspecific binding (x). Scatchard plot analysis (inset) revealed a single, noninteracting class of binding sites (K_D = 23 n*M*; B_{max} = 2.4 × 10^5 enzymes/cell).

7. Determine the enzymatic activity of the used neoglycoenzyme batch by activity graphs with the same substrate solution in order to enable algebraic transformation of extent of probe binding in femtomoles (*see* Note 11).

4. Notes

1. Since BSA is used as carrier of the glycohistochemically crucial ligands, as blocking substance for protein-binding sites and as supplement to the probe-containing solution, it is of pivotal importance to use carbohydrate-free preparations. Any carbohydrate contamination can evidently cause masking of the endogenous binding sites. BSA (4 g/100

mL 0.1*M* sodium acetate buffer at pH 4.5) should thus be treated with 10 m*M* periodic acid for 6 h at room temperature, followed by elimination of excess periodate by adding glycerol to a final concentration of 10 m*M* and extensive dialysis against PBS *(15)*. Moreover, biotinylated BSA of each batch should initially be used to check the extent of tissue binding. Lot-to-lot variability in the level of nonspecific binding has been reported *(16)*.

2. The properties of the probe, e.g., degree of glycosylation and biotinylation, and type of linker between sugar moiety and carrier protein, have an impact on quantitative aspects of receptor detection, as illustrated in Table 1. It is thus advisable to test different types of a certain neoglycoprotein, derived from various synthetic procedures, to elucidate the optimal nature of the probe for the tissue type under investigation.
3. It is necessary to avoid repetitive cycles of freezing and thawing as well as prolonged storage at 4°C for neoglycoprotein-containing solutions.
4. The conditions for tissue fixation should be adapted and optimized for the individual tissue type. At any rate, the conditions should be as gentle as possible to maintain the ligand-binding properties of the binding sites without impairing the tissue structure.
5. The control reaction in the absence of the labeled probe will ascertain whether any component of the detection system causes signal generation by specific binding. The glycoproteins avidin and horseradish peroxidase can in principle bind to mannose-specific tissue sites. This unwanted side reaction needs to be excluded. Similarly, the carbohydrate chains of the β-galactosidase of *Aspergillus oryzae* may themselves serve as ligands, when this enzyme is employed for cell-binding studies. Application of streptavidin conjugates of a nonglycosylated enzyme, performance of the binding reaction in the presence of mannose and periodate treatment of the enzyme can address these issues.
6. Application of nonglycosylated, but biotinylated BSA will contribute to prove the absence of nonspecific protein–protein interaction for probe binding. Additionally, it excludes participation of biotin-binding tissue sites, such as biotinidase, in probe binding. The negative control reaction, too, ascertains lack of influence of endogenous biotin-containing proteins to binding of the kit reagents (avidin or streptavidin). In the case of notable background staining owing to one of these possibilities, this interference must be eliminated by saturation of such sites prior to incubation with the labeled probe and the kit reagents.
7. The participation of enzymes to probe binding can be assessed by prolonged incubation or by addition of nucleotides, e.g., 20 m*M* GTP, UDP, or CDP, to the solution that contains the labeled neoglyco-

Table 1
Synopsis of Binding of Neoglycoproteins, Derived from Four Different Types of Synthesis, to Tumor Cells of 20 Cases of Invasive Ductal Mammary Carcinoma[a]

	Lac			Gal	α-GalNAc		β-galNAc		GalNAc
	BSA_{10}	BSA_{17}	BSA_{30}	BSA_{22}	BSA_{10}	BSA_{30}	BSA_{10}	BSA_{30}	BSA_{22}
Cytoplasm	3.3/2.2	1.0/0.6	2.4/1.5	2.8/1.6	3.7/2.7	1.9/1.0	2.5/1.6	2.2/1.1	3.1/1.9
Nuclei	2.5/1.6	1.3/1.0	2.8/1.9	2.1/1.3	3.2/1.9	2.5/1.4	2.4/1.3	2.4/1.5	2.8/1.3

[a]The categories for the percentage of positive cells (first value) and the categories for staining intensity (second value) are expressed in numerical values: 0 = 0% or no staining; 1 = 0–5% or weak, but significant staining; 2 = 5–20% or medium staining; 3 = 20–50% or strong staining; 4 = 50–100% or very strong staining. The number of carbohydrate moieties (lactose or galactose) is given for each BSA derivative. The method of synthesis is indicated by its yield of carbohydrate incorporation: BSA_{10} = diazotization, BSA_{17} = reductive amination, BSA_{30} = thionylation, BSA_{22} = monosaccharide incorporation via an aliphatic spacer; labeled probes were applied at a concentraion of 100 μg/mL, as described in detail elsewhere *(17)*.

protein. In both cases, reduction of the staining reaction should be noticed. Glycosidases will hydrolyze the crucial carbohydrates off the labeled carrier, preferably at acidic pH values, whereas nucleotides serve as inhibitors for binding of the probe to glycosyltransferases.

8. To evaluate the presence of Ca^{2+}-dependent lectins, the binding reaction should be performed in Tris- or HEPES-based buffer solutions in the presence of 20 m*M* $CaCl_2$.
9. The labeled probes allow the visualization of binding sites that are neither harmed by the fixation procedure nor masked by their endogenous ligands. Concomitant visualization of a predominant lectin in serial sections of a certain tissue by its specific antibody as well as by the respective neoglycoprotein indicates the validity of the glycohistochemical approach *(18–20)*.
10. Inhibition studies with heterologous mixtures of labeled probe and nonlabeled neoglycoprotein can assess the carbohydrate-binding specificity of the lectin *in situ*.
11. Neoglycoenzymes can bind to more than one cell surface receptor. Consequently, the apparent number of bound enzymes per cell will not necessarily reflect the total amount of actually present binding sites on the cell surface. When the spatial extension of the enzyme is lowered, e.g., with β-galactosidases from *E. coli* (tetrameric) and *Aspergillus oryzae* (monomeric), without reduction of carbohydrate incorporation per monomeric subunit, the number of measurable total binding sites increases. Concomitantly, the dissociation constant increases in agreement with the assumption that binding of one molecule of the smaller probe is mediated by less surface sites than binding of one molecule of the tetrameric *E. coli* enzyme *(12)*. In aggregate, the study of neoglycoenzyme binding inherently assesses the cell's capacity to interact specifically with carbohydrate ligands of the same type in spatial proximity. Evidence for receptor clustering and, thereby, for generation of tight binding to relatively large probes may be of relevance as a crude model for interaction of cell-surface lectins with ligands on other cells.

References

1. Barondes, S. H. (1981) Lectins: Their multiple endogenous cellular functions. *Ann. Rev. Biochem.* **50,** 207–231.
2. Ashwell, G. and Harford, J. (1982) Carbohydrate-specific receptors of the liver. *Ann. Rev. Biochem.* **51,** 531–554.
3. Gabius, H.-J. (1988) Mammalian lectins: Their structure and their glycobiological and glycoclinical roles. *ISI Atlas Sci.: Biochem.* **1,** 210–214.

4. Sharon, N. and Lis, H. (1989) Lectins as cell recognition molecules. *Science* **246,** 227–234.
5. Barondes, S. H. (1988) Bifunctional properties of lectins: lectins redefined. *Trends Biochem. Sci.* **13,** 480–482.
6. Gabius, H.-J. (1988) Tumorlectinology: At the intersection of carbohydrate chemistry, biochemistry, cell biology and oncology. *Angew. Chem. Int. Ed. Engl.* **27,** 1267–1276.
7. McBroom, C. R., Samanen, C. H., and Goldstein, I. J. (1972) Carbohydrate antigens: Coupling of carbohydrates to proteins by diazonium or phenylisothiocyanate reactions. *Methods Enzymol.* **28,** 212–219.
8. Stowell, C. P. and Lee, Y. C. (1980) Neoglycoproteins: The preparation and application of synthetic glycoproteins. *Adv. Carbohydr. Chem. Biochem.* **37,** 225–281.
9. Schrevel, J., Gros, D., and Monsigny, M. (1981) Cytochemistry of cell glycoconjugates. *Progr. Histochem. Cytochem.* **14,** 1–269.
10. Gabius, H.-J. and Bardosi, A. (1991) Neoglycoproteins as tools in glycohistochemistry. *Progr. Histochem. Cytochem.* **22(3),** 1–66.
11. Gabius, S., Hellmann, K. P., Hellmann, T., Brinck, U., and Gabius, H. -J. (1989) Neoglycoenzymes: A versatile tool for lectin detection in solid-phase assays and glycohistochemistry. *Anal. Biochem.* **182,** 447–451.
12. Gabius, S., Schirrmacher, V., Franz, H., Joshi, S. S., and Gabius, H.-J. (1990) Analysis of cell surface sugar receptor expression by neoglycoenzyme binding and adhesion to plastic-immobilized neoglycoproteins for related weakly and strongly metastatic cell lines of murine tumor model systems. *Int. J. Cancer* **44,** 500–507.
13. Monsigny, M., Petit, C., and Roche, A. C. (1988) Colorimetric determination of neutral sugars by a resorcinol-sulfuric acid micromethod. *Anal. Biochem.* **175,** 525–530.
14. Scatchard, G. (1949) The attractions of proteins for small molecules and ions. *Ann. NY Acad. Sci.* **51,** 660–672.
15. Glass, W. F., Briggs, R. C., and Hnilica, L. S. (1981) Use of lectins for detection of electrophoretically separated glycoproteins transferred onto nitrocellulose sheets. *Anal. Biochem.* **115,** 219–224.
16. Howlett, B. J. and Clarke, A. E. (1981) Detection of lectin binding to glycoconjugates immobilized on polyvinylchloride microtitre plates. *Biochem. Int.* **2,** 553–560.
17. Gabius, H.-J., Gabius, S., Brinck, U., and Schauer, A. (1990) Endogenous lectins with specificity to β-galactosides and α- or β-*N*-acetyl-galactosaminides in human breast cancer: Their glycohistochemical detection in tissue sections by synthetically different types of neoglycoproteins, their quantitation on cultured cells by neoglycoenzymes and their usefulness as targets in lectin-mediated phototherapy in vitro. *Path. Res. Pract.* **186,** 597–607.
18. Bardosi, A., Dimitri, T., Wosgien, B., and Gabius, H.-J. (1989) Expression of endogenous receptors for neoglycoproteins, especially lectins, that allow fiber typing on formaldehyde-fixed, paraffin-embedded muscle biopsy speci-

mens. A glycohistochemical, immunohistochemical and glycobiochemical study. *J. Histochem. Cytochem.* **37**, 989–998.
19. Kuchler, S., Zanetta, J. P., Vincendon, G., and Gabius, H.-J. (1990) Detection of binding sites for biotinylated neoglycoproteins and heparin (endogenous lectins) during cerebellar ontogenesis in the rat. *Eur. J. Cell Biol.* **52**, 87–97.
20. Bardosi, A., Bardosi, L., Hendrys, M., Wosgien, B., and Gabius, H.-J. (1990) Spatial differences of endogenous lectin expression within the cellular organization of the human heart: A glycohistochemical, immunohistochemical and glycobiochemical study. *Am. J. Anat.* **188**, 409–418.

CHAPTER 22

Scanning Tunneling Microscopy of Biopolymers

Terence J. McMaster and Victor J. Morris

1. Introduction

Scanning tunneling microscopy (STM), introduced by Binnig et al. *(1)* in 1982, has proven to be of immense use in determining the surface structures of semiconductors *(2,3)*, metals *(4)*, and superconductors *(5)*. The application of this technique to the visualization of the structures of biological molecules has until recently lagged behind. However, the principal advantages of STM—superior resolution of 1 Å or less in the sample plane and 0. 1 Å vertically, and the ability to image at ambient pressures or in liquid environments with minimal specimen damage—are of great potential for biomolecular structure investigation.

Consideration of its application to studies of biological molecules is timely for several reasons. First, there are a growing number of relatively inexpensive commercial instruments available. The technique is thus becoming more widespread in biological laboratories, but it is not yet routine, and details on its practical use in biology are not generally available. Second, and more importantly, considerable success has been achieved in imaging a broad range of biopolymers. Recent work has reported the near-atomic-scale resolution of DNA structure *(6)*, images of proteins in air *(7–9)*, lipid bilayer structure *(10)*, and polysaccharide deposition *(11)* and conformation *(12)*.

Although STM for biological studies is still being developed, it is very important that results from it be checked with those obtained from other sources. This has been done in studies of vicilin using

From: *Methods in Molecular Biology, Vol. 14: Glycoprotein Analysis in Biomedicine*
Edited by: E. F. Hounsell

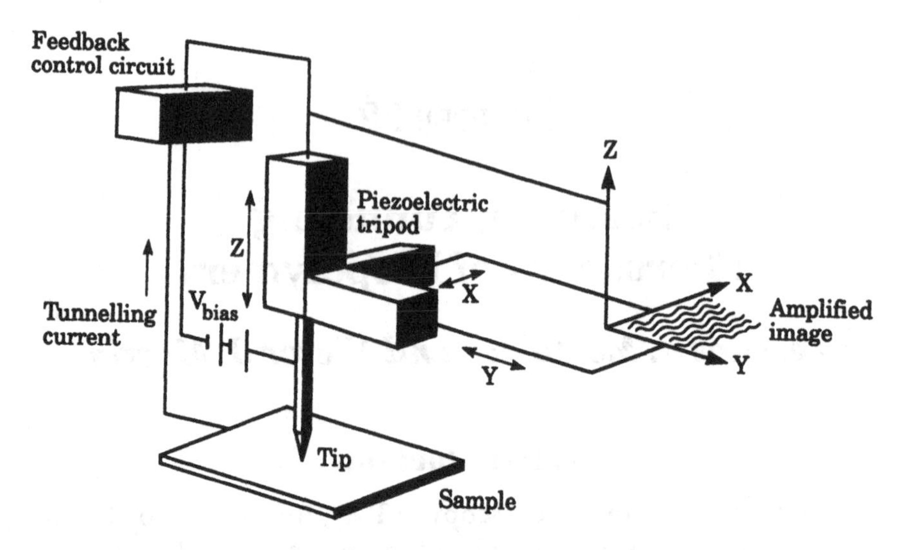

Fig. 1. Schematic diagram of an STM.

STM and X-ray scattering *(7)*, correlation of circular dichroism, computer structure predictions and viscometry with STM data in studies of wheat subunit protein *(8)*, and TEM and STM experiments on the HPI-layer protein *(13)*.

At present, emphasis has been placed on imaging biopolymers with known shapes and sizes in order to test the applicability of STM. The ability to image polysaccharides and proteins suggests that the technique may be used in the future to image complex glycoconjugates or molecular interactions where information on molecular size and shape is unknown and difficult to obtain directly from other biophysical methods.

An understanding of the basic principles governing the STM is essential for its use. The instrument is shown schematically in Fig. 1. When the conducting probe, or tip, which is held at a potential of approx ± 1 V with respect to the substrate, is positioned very close to the substrate, typically several tens of angstroms or less, then a tunneling current of the order of 10^{-9}A will flow between the two conductors. This is a quantum mechanical phenomenon, whereby electrons traverse, or "tunnel through," a potential barrier. A more detailed treatment of the theory of STM may be found elsewhere in the literature *(14,15)*. The particular theoretical aspects of imaging biomolecules have been helpfully outlined by Welland and Taylor *(16)*.

The remarkable vertical sensitivity of STM arises from the strong exponential dependence of the tunnel current on the separation of the tip and substrate.

The motion of the tip in x, y, and z is controlled either with a single piezo tube scanner or, as depicted, with a set of three orthogonal piezoelectric crystals. In the plane of the sample and the substrate, voltages applied to x and y piezos cause the tip to scan a square area line by line in a raster pattern entirely analogous to the motion of an electron beam in a scanning electron microscope (SEM). The scan area is typically made up of a 256 × 256 square array of tunneling points. The magnification of the STM is determined by the ratio of the area scanned in this way to the area of display on a monitor screen. The z motion of the tip is most commonly controlled via a feedback circuit that operates to maintain the tunnel current at a constant value. This means that at each discrete point of a single x-line scan, the tip moves vertically to keep the current constant, and topographic information is obtained by monitoring the motion of the tip.

The image acquisition time is commonly of the order of a minute, and most microscopes utilize a computer for storage and display of image data. Image processing is normally carried out postacquisition, and the methods used in other microscopies for image treatment are generally applicable in STM.

2. Materials

2.1. Substrates

An atomically flat conducting substrate upon which the biomolecule is deposited is a fundamental requirement for STM. For work in air, the substrate must also be inert, because formation of an oxide layer would inhibit tunneling. Two types of inert, conducting substrate have been commonly used in biological STM.

2.1.1. Highly Oriented Pyrolytic Graphite (HOPG)

HOPG (Union Carbide, Cleveland, OH; ref. *17*) is a layered material, and the ease with which fresh surfaces can be prepared has made it a popular choice for biomolecule studies.

2.1.2. Atomically Flat Gold (III) Crystal Faces

Prepared from 0.5 mm diameter, 99.9% pure gold wire.

2.2. Tips

The tunneling tip should ideally be atomically sharp, i.e., terminating in a single atom. A tunneling tip is fabricated from thin (typically 0.5 mm diameter) metal wire, commonly platinum-iridium alloy ($Pt_{0.7}Ir_{0.3}$ or $Pt_{0.8}Ir_{0.2}$), gold, or tungsten.

2.3. STM Environment

The STM head assembly, namely the piezo element, tip, sample/substrate support, and associated structures, is especially vulnerable to acoustic interference, mechanical vibration, shock, and thermal gradients *(14)*. An acoustically insulated cabinet with thermal insulation, mounted on an optical-style antivibration table or suspended by springs from a rigid support will give very good results.

3. Methods

3.1. Substrate Preparation

3.1.1. HOPG

1. Place graphite crystal on a glass slide, and cover HOPG and surrounding area with a piece of sticky tape.
2. Remove the tape with a slow, steady motion. The first few crystal layers will adhere to the tape, exposing a fresh crystal surface for biomolecule deposition. The disadvantages of HOPG are that it is relatively hydrophobic and that it displays atomic-step edge structures, which may be misleading (*see* Note 1).

3.1.2. Gold (111) Surfaces

Au (111) crystal faces may be grown epitaxially on very flat mica at elevated substrate temperatures (350°C) and under vacuum (10^{-6} mm Mg) *(18)*, but this is time-consuming, and may be beyond the resources or capabilities of many biology laboratories. A simpler procedure developed initially for reflection electron microscopy is more useful *(19,20)*.

1. Melt the gold wire in an oxygen-acetylene flame, set midway between oxygen-rich and acetylene-rich.
2. As the molten gold forms a ball at the end of the wire, feed more wire into the flame. A length of 4 cm of 0.5 mm diameter wire should form a sphere approx 2 mm in diameter.
3. Remove from the flame, and allow to cool in air to room temperature. The gold ball should exhibit atomically flat (111) facets.

3.2. Tip Preparation

Two techniques have been used: mechanical cutting or electrochemical etching *(21,22)*.

3.2.1. Mechanically Cut Tips

1. Wash a 5-mm piece of wire in acetone. Additional washing in alcohol in an ultrasonic bath may be helpful.
2. Mount wire in tip holder, and cut at a shallow angle using pliers (for Pt-Ir tips). It is necessary to use a corundum disk when cutting tungsten wire to avoid splitting; however, a razor blade is sufficient when cutting gold wire.
3. Mount tip in STM and use.

3.2.2. Electrochemically Etched Tips

1. Wash as for mechanically cut tips.
2. Mount wire in tip holder, and immerse typically 0.5 mm into etchant solution. This solution may vary in composition and concentration according to tip material (*see* Note 2).
3. Apply an ac voltage to the tip, and continue etching until current between tip and counterelectrode drops to between 0 and 10% of its initial value. The cutoff, ac peak-to-peak voltage, frequency, shape, and composition of the counterelectrode may all be variable (*see* Note 2).
4. Remove tip from etchant, and wash with distilled water. Remove excess water by spotting onto filter paper, taking care to avoid touching the point of the tip.
5. Mount in STM and use.

3.3. Sample Preparation and Deposition

A wide variety of techniques have been developed, and they are briefly summarized with examples of the particular molecules imaged.

3.3.1. Air-Drying

This is particularly useful for proteins such as vicilin *(7)* from aqueous buffered solution (Fig. 2A) or polypeptides from organic solvents *(23)*. The concentration of the solution will vary according to mol wt and shape, and the coverage of the substrate that is desired. A good starting point is a concentration of 1 mg/mL.

1. Deposit a 1–2 µL drop of the biomolecule solution onto the substrate.
2. Cover and leave until it has evaporated to dryness. When using gold sphere substrates, dip the sphere into the solution, remove, and leave to dry.

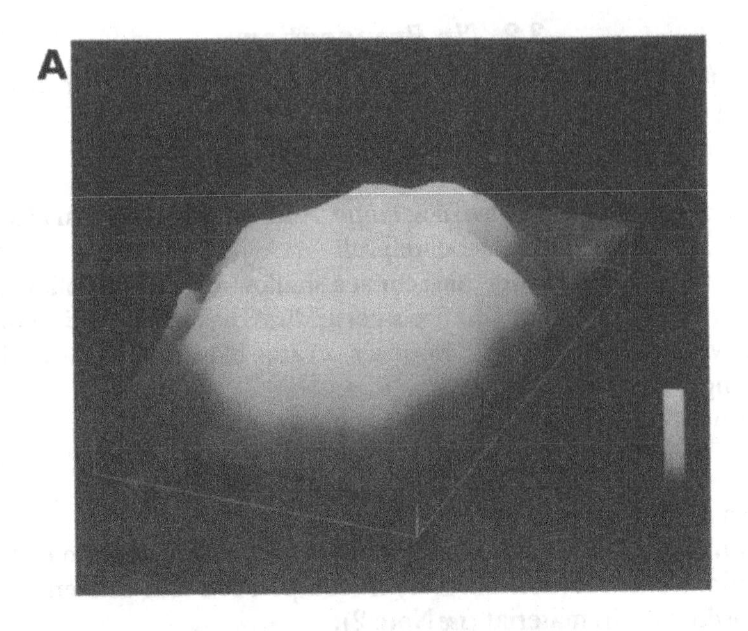

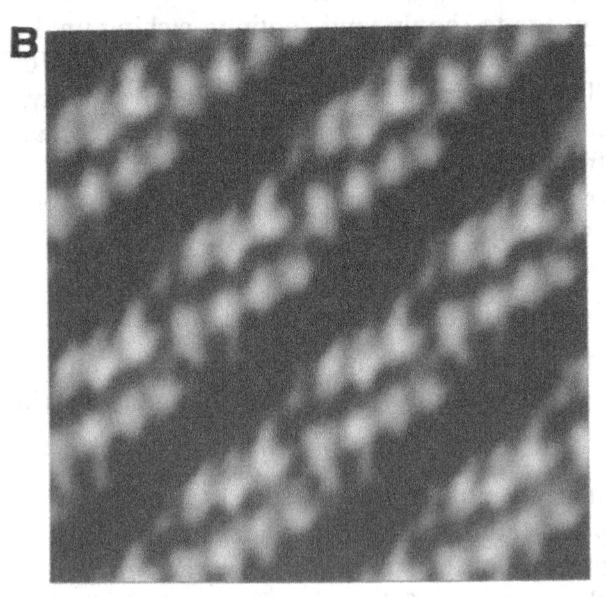

Fig. 2A. STM image of a single molecule of the pea protein, vicilin. The molecule is approx 100 Å across *(7)*. B. Surface ordering of the liquid crystal 4-*n*-octyl-4'-cyanobiphenyl (8CB). The bright spots correspond to individual phenyl rings.

C

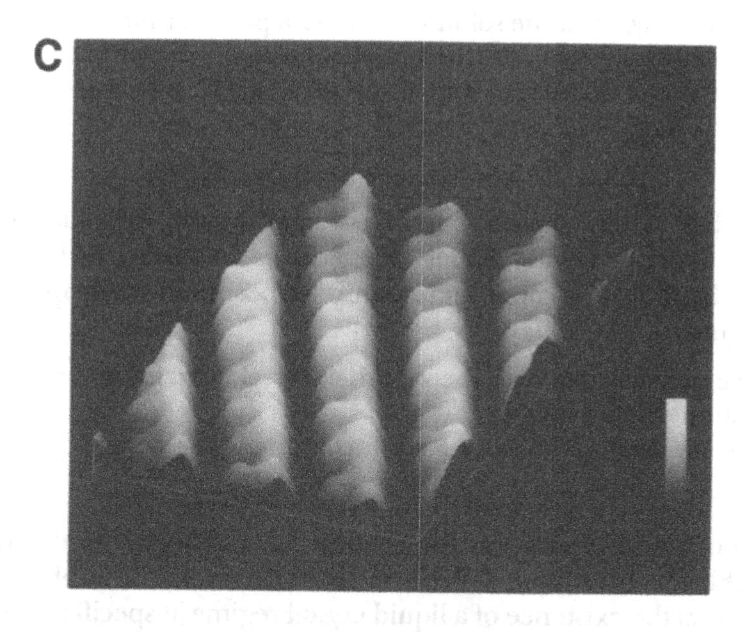

D

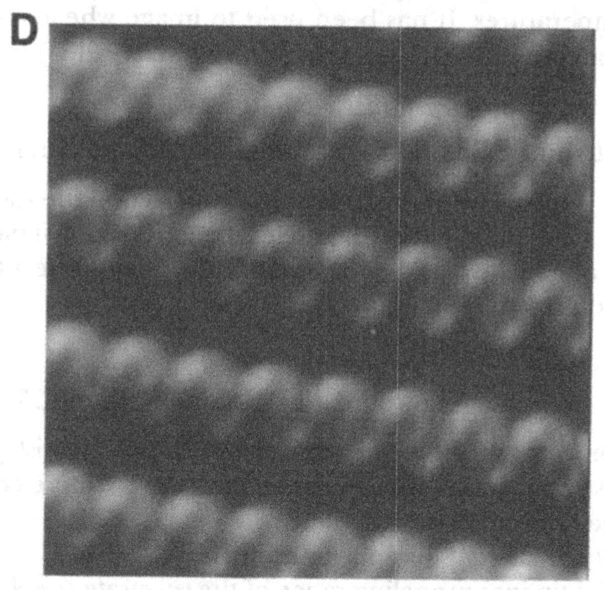

C. Several rod-like molecules of the high-mol-wt subunit protein from wheat. The periodicity along the molecule is 14.9 Å, molecule width is 19 Å, and sideways spacing is 31.0 Å *(8)*. D. The helical polypeptide, poly (γ-benzyl-L-glutamate). Periodic repeat along molecule is 16.5 Å *(22)*.

3. Alternatively, wash the solution off after a predetermined time, and allow substrate and adsorbed molecules to dry either naturally or by the use of vacuum drying, freeze-drying, or forced air-flow drying (*see* Note 3).

3.3.2. Sublimation / Vapor Deposition

This has been successful with STM of small organic molecules, such as the liquid crystal 8CB *(24)*. Figure 2B shows the surface ordering of 8CB on HOPG after room temperature deposition of a drop of the pure liquid.

1. Place the substrate over the mouth of vessel containing the sample solution or pure liquid.
2. Heat the vessel to produce a vapor containing the sample.

3.3.3. Liquid Crystal Deposition

Deposit and prepare as for Section 3.3.1. This method is only applicable for anisotropic molecules, whose thermodynamic phase diagram shows the existence of a liquid crystal regime at specific concentrations and temperatures. It has been used to image wheat subunit protein *(8)* and an α-helical polypeptide *(23)* (Fig. 2C and D).

3.3.4. Langmuir-Blodgett

This is used to image bilayers of cadmium arachidate as in ref. *10.*

1. Prepare an L-B trough with the molecule of interest on the surface.
2. If a flat substrate is initially above liquid level, then position it parallel or perpendicular to the plane of the liquid. Dip the substrate into the trough, and then remove (*see* Note 4).

3.3.5. Electrodeposition Methods

These are used in imaging RNA and DNA under water *(25)*.

1. Form the substrate as the base of a cylindrical electrochemistry cell.
2. With the substrate positive, raise the potential of the counterelectrode until hydrogen evolution is observed.
3. Pass a small (μA) current for approx 1 min.
4. Move the STM tip into tunneling range of the substrate (*see* Note 5).
5. As an alternative to steps 1–4, dip the tip into a solution containing the biomolecule of interest, remove, and position very close to the substrate, i.e., within tunneling range. Raise the tip potential to about 4 V for 10 μs to eject adsorbed molecules onto the surface. This is a recent technique, and success has been claimed for work with DNA *(26)*.

3.3.6. Metal Coating

For example, this is used to obtain images of recA-DNA complexes *(27)*, protein layers and collagen *(28)*, bacteriophage polyhead surface structure *(29)*, and freeze-fracture replicas of lipid structures *(30)*.

1. Prepare the samples using any of Sections 3.3.1.–3.3.5.
2. Coat the sample with metal particles as per standard electron microscope procedures. This method and air-drying (Section 3.3.1.) are probably the most common methods currently used for biomolecules. Their relative merits are discussed in Note 3.

3.4. Setting Up the STM

1. Set the tunnel current at a relatively low value, for example, 0.1 nA, and the bias voltage at ~1 V for initial scans in the topographic mode. This will ensure that the tip sample distance is relatively large at the outset, and the risk of a tip "crash" (impact of tip onto the sample) is reduced (*see* Note 6).
2. Set the scan area at a large value, for example, 5000 × 5000 Å. Move this area across the sample until molecules are observed, and then increase magnification (*see* Note 7).
3. Set the scan time for a 5000 × 5000 Å area at about 3 min. Adjust as necessary for optimum resolution (*see* Note 6).

3.5. Scanning Methods

3.5.1. Constant Current Mode

For the constant current or topographic mode, maintain the current at a constant value (Fig. 3A) using a feedback circuit. This mode is best for an irregular surface, or where the topography is unknown and there is a risk of the tip crashing into the substrate or sample. (*See* Note 8 for a variation of this method for use with large or uneven structures.) This is the most common mode of imaging biomolecules.

3.5.2. Constant Height Mode

In this mode of operation, keep the STM tip at a constant voltage as it is scanned across the surface, and measure the changes in tunneling current (Fig. 3B). This mode is significantly faster than the constant current mode, although it is generally less favored for biological studies, since it is most suitable for very smooth surfaces. Scanning artifacts and their treatment are discussed in Note 9.

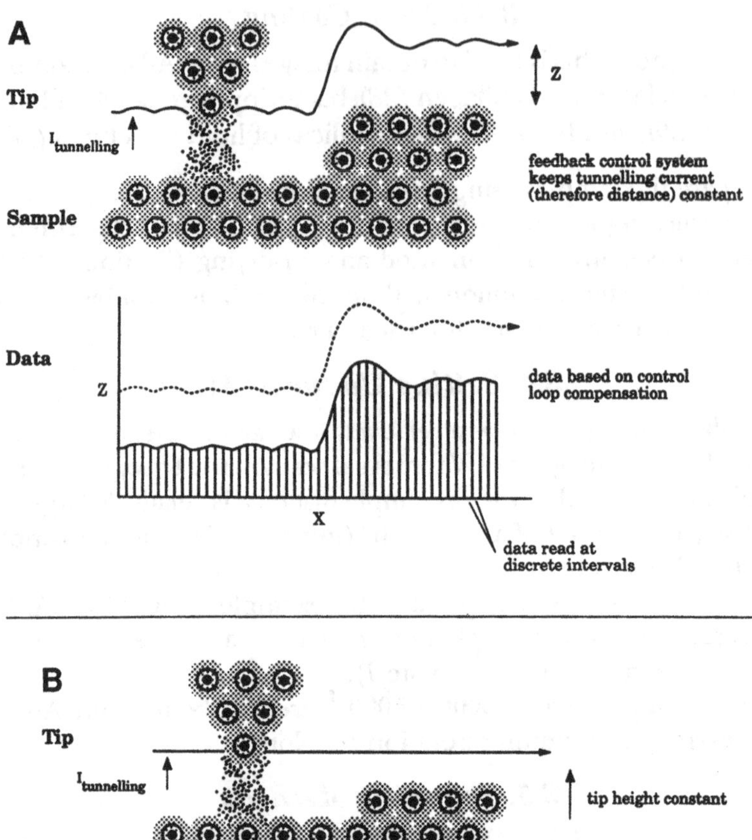

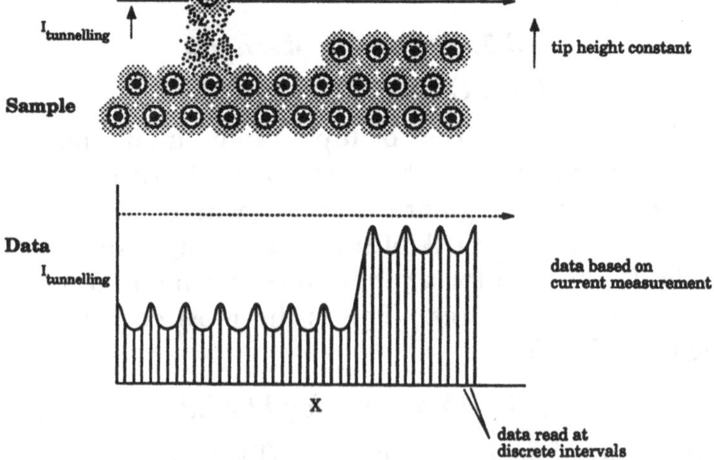

Fig. 3A. Schematic of constant current imaging mode. B. Schematic of constant height imaging mode. (Reproduced from Burleigh Instruments' Scanning Probe Microscopy Book, with permission.)

3.6. Spectroscopy

Using the STM, it is possible to vary the bias voltage and the tip to sample distance at any point, and by utilizing its sensitivity to electron energy states of the sample, it is possible to obtain spectroscopic data with atomic-level resolution. There are two related types of measurement:

1. I-V or current–voltage spectroscopy.
2. I-Z or current–height spectroscopy.

3.6.1. I-V Spectroscopy

1. Disable the feedback loop, and fix the tip-sample distance.
2. Measure the tunnel current, I, as a function of bias voltage, V, typically –1 to + 1V, at a defined pixel of a topographic image.
3. Plot (dI/dV) (I/V) against V to provide data on the local density of electronic states. Such data have been correlated with molecular properties, e.g., vibrational modes, for copper phthalocyanine *(31)*.

Specific interpretation of spectra obtained for proteins is much more difficult. Despite this, however, I-V spectroscopy may be used to determine if biomolecules are present or not by comparison of spectra obtained before and after deposition of the molecules.

3.6.2. I-Z Spectroscopy

1. Measure the tunnel current as a function of tip-sample separation.
2. Calculate the sample work function (the energy required to remove an electron from its normal state, approximately proportional to d lnI/dZ). In conjunction with more conventional topographic imaging, this information can delineate molecular features, e.g., the phosphate backbone from base pairs in the DNA helix *(6,32)*.

4. Notes

1. HOPG is the most convenient of substrates for biological use. However, its hydrophobicity may cause problems in obtaining good adsorption of some biomolecules, particularly proteins. Gold has already been mentioned as an alternative. Platinum-carbon-coated mica has also been used *(27)*, but the corrugation of the surface should be small enough so as not to obscure the deposited molecules.

   Care is required in the interpretation of biomolecule images obtained on HOPG substrates, especially since molecules may adsorb more readily at step edges of crystal surfaces. Observed surface structures of HOPG at crystal plane edges may in some cases resemble features of biomolecules and may mimic the periodicities expected in helical molecules *(33)*.

2. Tip production: The exponential dependence of current on tip-substrate separation renders the practical achievement of an atomically sharp tip more feasible. The literature reveals that STM research groups tend to develop their own tip-making "recipes." However, typical etchants include 32% HCl for Au tips, 3*M* NaCN and 1*M* NaOH for Pt-Ir tips, and 1*M* KOH for W tips *(21,22)*. The counterelectrode may be in the form of a carbon rod, or as a circular loop or foil surrounding the etching tip. Voltages are usually in the range 0–20 V ac peak-to-peak. The end point of the etching process is marked by an abrupt drop in current, since the tip begins to neck prior to the lower portion falling off. An electronic circuit to switch off the current as this happens is often useful.

   Mechanically cut tips often have a large radius of curvature and/or multiple protrusions. The former property may inhibit optimum resolution of edge structures, for example, whereas the latter may lead to the production of multiple or "ghost" images (*see* Note 9). Electrochemical etching produces a tip of much smaller radius of curvature. The choice of tip material and tip production method is determined largely by experiment.
3. Deposition: Air-drying of molecules onto substrates works well in most cases, although there are possible problems owing to structure collapse arising from surface tension effects. Freeze-drying followed by metal coating is good for tunneling and immobilization of the biomolecule, but some loss of resolution is inevitable. Obviously, uncoated samples or those examined under liquid will be most like the native hydrated state; however, the theoretical understanding of tunneling in biological material is poor, and height information is difficult to quantify.
4. Langmuir-Blodgett deposition: For some molecule— substrate combinations, depending on relative hydrophobicities—it may be helpful to coat the substrate by drawing it out of the trough.
5. Electrochemical STM: Tips for use in electrochemical environment may be made using any of the methods described above (Section 3.2.). However, the tip may require electrical insulation with an inert substance to minimize unwanted Faradaic currents. The shank of the tip is covered with an insulator, leaving just the end exposed *(21)*.
6. Setting up the STM: Precise starting conditions of tunnel current, bias voltage, scan time, and feedback circuit variables are difficult to

define, since optimum STM performance depends on both individual instrument characteristics and on the particular biomolecule system that is being studied. A low (≤0.1 nA) tunnel current is a good starting point, although with biomolecules in their native, uncoated state, too low a current may give very poor resolution. In general, increasing the bias voltage will move the tip away from the substrate. For STM of semiconductors, images at different bias voltages will display different electron energy states of the surface atoms *(34)*; however, biomolecules do not show the same striking effect. High bias (≥5V) should be avoided, since the microscope would then operate in the field-emission regime because of the extremely high field gradient between the tip and the substrate. Feedback circuit variables, such as gain and bandwidth, may also be adjusted to improve the electronic performance of the microscope.

Many microscopes have an automatic tip-to-sample approach. If this is not available, extreme care is required at this stage. The tip has to be advanced gradually until tunneling is observed.

Should the tip crash into the sample, it may continue to function normally. If this is not so, then the tip may sometimes be reconditioned *in situ* by application of a high bias voltage (>10 V). However, this is not always reliable and may also damage the specimen.

7. Locating the molecules: It is possible to tag molecules to help in locating them on the substrate: e.g., antibodies tagged with gold spheres, fluorescently labeled proteins, or benzyl-substituted side chains of polypeptides *(23)*. Other techniques can then be used to monitor adsorption onto the substrate before STM studies commence. It is possible to operate an STM inside an SEM *(35)* or in conjunction with a light microscope *(36)* to locate molecules and perform parallel experiments.
8. Hopping mode: The aim of this technique, first suggested by Jericho et al. *(37)*, is to reduce mechanical interaction between the tip and the sample, which might lead to sideways motion of the molecules, sample deformation, or tip damage. The tip is periodically withdrawn from the substrate and then brought back into tunneling range by a fixed distance in the z direction during each x scan. For each hopping cycle, the tip is withdrawn for ~75% of the time and is tunneling for ~25%. It has been used in imaging large bacterial cell wall sheaths and pepsin molecules *(38)*.
9. Scanning artifacts and their treatment: Double images may be recorded (Fig. 4A) arising from two (or more) tunneling sites on the tip. Although not totally invalidating the data, tip replacement is inevitable.

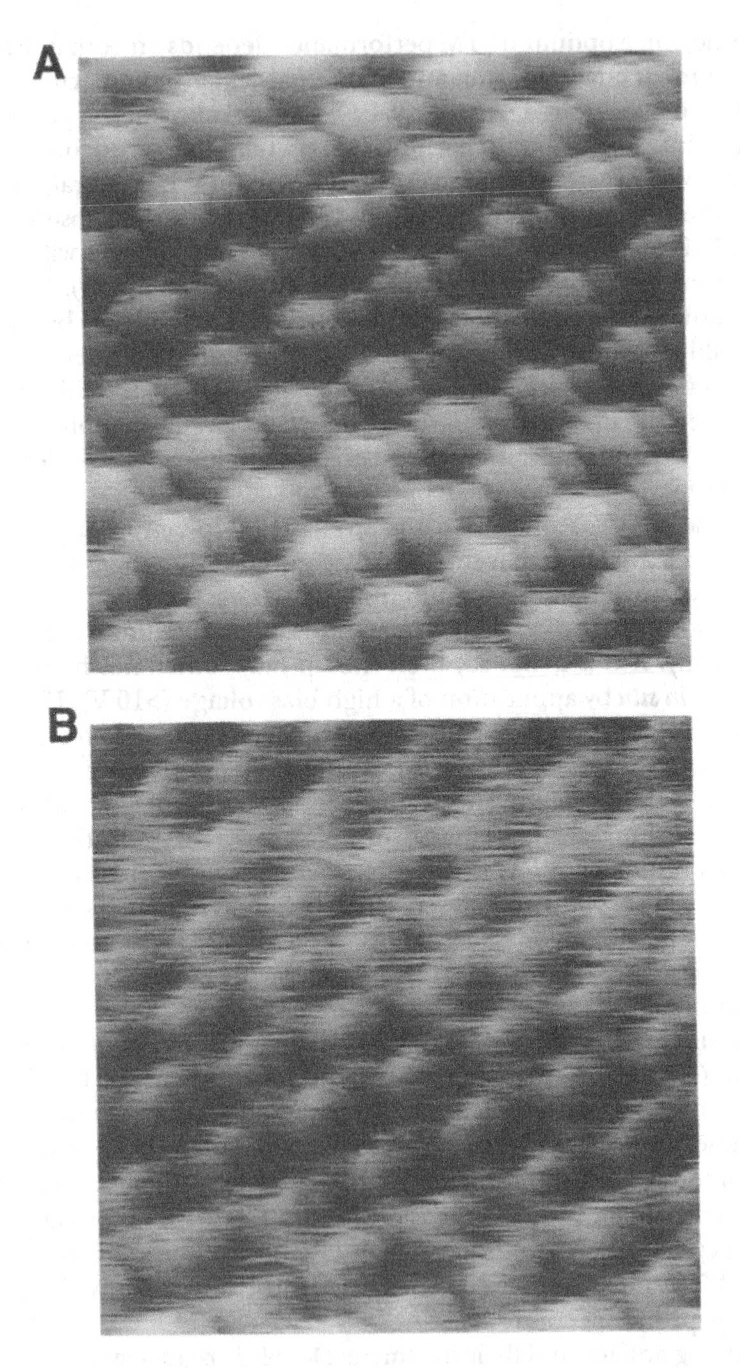

Fig. 4. Scanning artifacts. A. Double images in a 15 × 15 Å scan of graphite. Note the smaller "ghost" images to the right of each carbon atom. B. "Streaky" 15 × 15 Å scan of graphite.

C
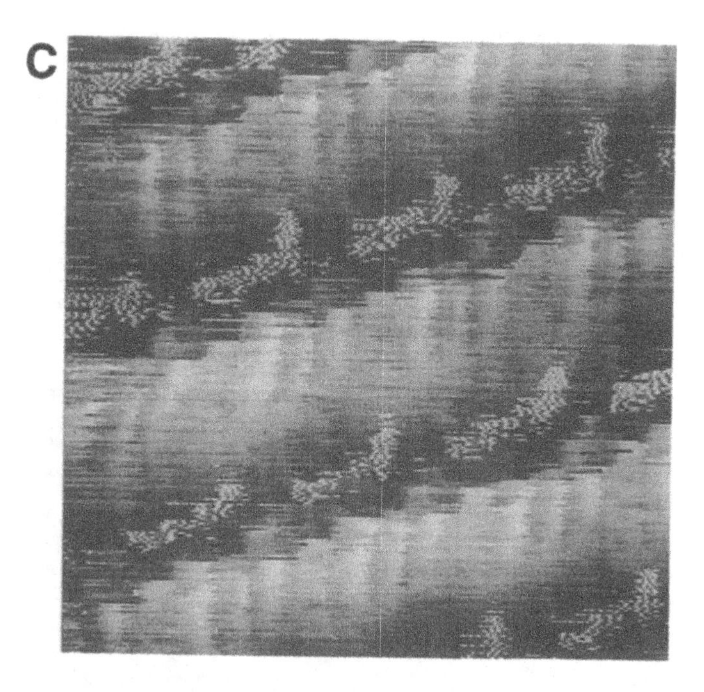

D
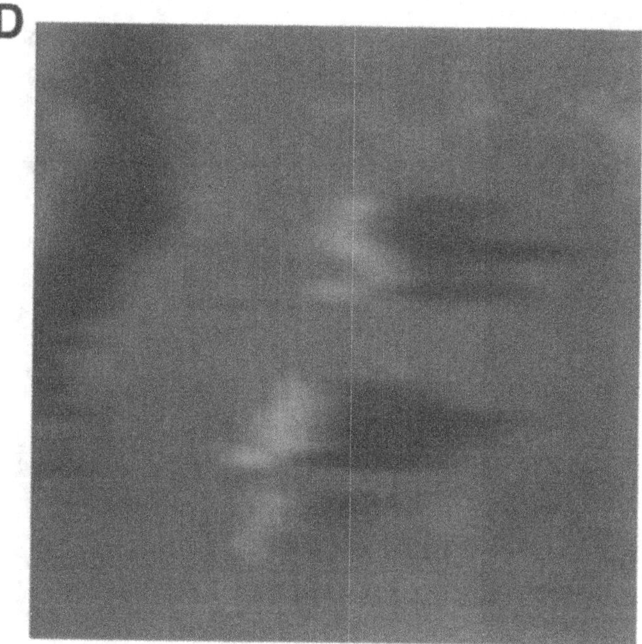

C. Low-resolution 72 × 72 Å scan of 8CB (*see* Fig. 2B). Note the periodic noise between the biphenyl bands. D. Tunneling shadow effect demonstrated on a 345 × 345 Å scan of protein fragments. The scan direction was left to right.

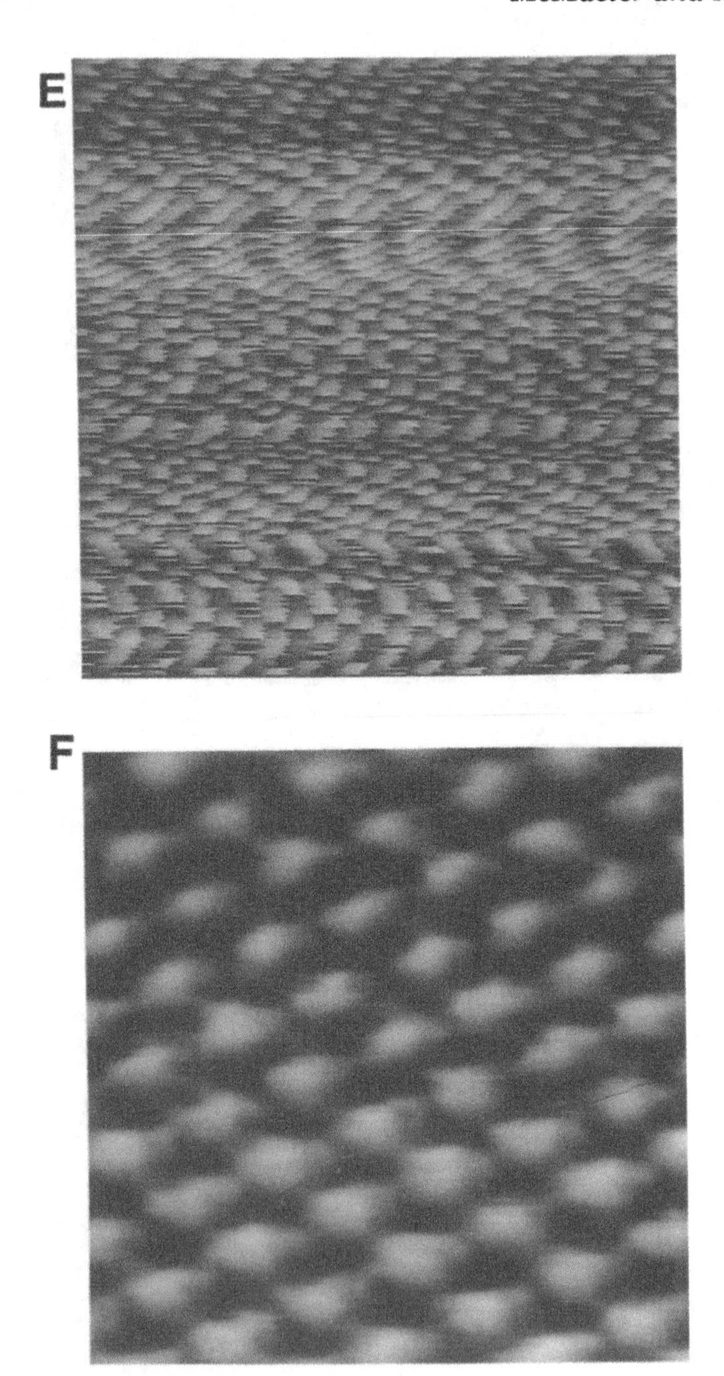

E. 50 × 50 Å scan of graphite showing variable drift. F. Virtually distortion-free 15 × 15 Å image of graphite lattice.

Streakiness in the image (Fig. 4B) indicates dragging of the tip across the sample or extraneous matter on the tip bridging it and the sample. A rapidly varying tunnel current is commonly also associated with this phenomenon. Moving the tip out of tunneling range and then back in again, or changing scan area may help. Reconditioning of the tip out of tunneling range may be helpful (*see* Note 6). An alternative cause of this problem may be too fast a scan speed.

Sets of periodic lines or dots (Fig. 4C) indicate an electronic noise problem. Inadequate shielding of the head, producing interference, or unstable conditions for the feedback servo may be the cause. Again, it may be a scan speed effect.

Scans of some biological structures (Fig. 4D) show a characteristic tunneling "shadow" or depression on the downscan side of the structure. Changing the scan direction should result in movement of this shadow around the molecule. This effect arises from work function differences between the substrate and the sample. Altering the scan speed and tunnel current may reduce it. The data are not invalidated, although care should be exercised in measurement and analysis. Such shadows may be useful in indicating the geometric shape of molecules.

Sudden changes in temperature or mechanical vibrations can cause drift in image data (Fig. 4E). Appropriate insulation and environmental control are required. Images 4A–E may be helpfully compared to a good graphite image (Fig. 4F).

To check if an observed structure is not an artifact produced by the scanning mechanism, the following tests may be used:

1. Change scan speed	No effect
2. Alter magnification	Size change
3. Change scan direction	Basic shape unaltered
4. Alter tunnel current and bias voltage	Basic shape unaltered, although resolution may be affected

Where possible, blank experiments should be carried out under similar conditions, i.e., substrate alone, and solvent without molecules present deposited onto the substrate.

References

1. Binnig, G., Rohrer, H., Gerber, Ch., and Weibel, E. (1982) Surface studies by scanning tunneling microscopy. *Phys. Rev. Lett.* **49,** 57–61.
2. Binnig, G., Rohrer, H., Gerber, Ch., and Weibel, E. (1983) 7 x 7 reconstruction on Si (111) resolved in real space. *Phys. Rev. Lett.* **50,** 120–123.
3. Tromp, R. M., Hamers, R. J., and Demuth, J. E. (1985) Si (001) dimer structure observed with scanning tunneling microscopy. *Phys. Rev. Lett.* 55, 1303–1306.

4. Kuk, Y., Silverman, P. J., and Nguyen, H. Q. (1988) Study of metal surfaces by scanning tunneling microscopy with field ion microscopy. *J. Vac. Sci. Technol.* **A6,** 524–528.
5. Hess, H. F., Robinson, R. B., Dynes, R. C., Valles, J. M., Jr., and Waszczak, J. V. (1990) Spectroscopic and spatial characterization of superconducting vortex core states with a scanning tunneling microscope. *J. Vac. Sci. Technol.* **A8,** 450–454.
6. Driscoll, R. J., Youngquist, M. G., and Baldeschwieler, J. D. (1990) Atomic-scale imaging of DNA using scanning tunneling microscopy. *Nature* **346,** 294–296.
7. Welland, M. E., Miles, M. J., Lambert, N., Morris, V. J., Coombs, J. H., and Pethica, J. B. (1989) Structure of the globular protein vicilin revealed by scanning tunneling microscopy. *Int. J. Biol. Macromol.* **11,** 29–32.
8. Miles, M. J., Carr, H. J., McMaster, T. J., I'Anson, K. J., Belton, P. S., Morris, V. J., Field, J. M., Shewry, P. R., and Tatham, A. S. (1991) Scanning tunneling microscopy of a wheat storage protein reveals details of an unusual supersecondary structure. *Proc. Natl. Acad. Sci. USA* **88,** 68–71.
9. Guckenberger, R., Wiegräbe, W., Hillebrand, A., Hartmann, T., Wang, Z., and Baumeister, W. (1989) STM of a hydrated bacterial surface protein. *Ultramicroscopy* **31,** 327–332.
10. Smith, D. P. E., Bryant, A., Quate, C. F., Rabe, J. P., Gerber, Ch., and Swalen, J. D. (1987) Images of a lipid bilayer at molecular resolution by scanning tunneling microscopy. *Proc. Natl. Acad. Sci. USA* **84,** 969–972.
11. Miles, M. J., McMaster, T. J., Carr, H. J., Tatham, A. S., Shewry, P. R., Field, J. M., Belton, P. S., Jeenes, D., Hanley, B., Whittam, M., Cairns, P., Morris, V. J., and Lambert, N. (1990) Scanning tunneling microscopy of biomolecules. *J. Vac. Sci. Technol.* **A8,** 698–702.
12. Yang, X., Miller, M. A., Yang, R., Evans, D. F., and Edstrom, R. D. (1990) Scanning tunneling microscopic images show a laminated structure for glycogen molecules. *Fed. Am. Soc. Exp. Biol. J.* **4,** 3140–3143.
13. Amrein, M., Wang, Z., and Guckenberger, R. (1991) Comparative study of a regular protein layer by STM and TEM. *J. Vac. Sci. Technol.* **89,** 1276–1281.
14. Kuk, Y. and Silverman, P. J. (1989) Scanning tunneling microscope instrumentation. *Rev. Sci. Instrum.* **60,** 165–180.
15. Hansma, P. K. and Tersoff, J. (1987) Scanning tunneling microscopy. *J. Appl. Phys.* **61,** R1–23.
16. Welland, M. E. and Taylor, M. E. (1990) Scanning tunneling microscopy, in *Modern Microscopies: Techniques and Applications* (Duke, P. J. and Michette, A. G., eds.), Plenum, London, pp. 231–254.
17. Union Carbide Corporation, Advanced Ceramics Division, PO Box 94637, Cleveland, OH 44101.
18. Hallmark, V. M., Chiang, S., Rabolt, J. F., Swalen, J. D., and Wilson, R. J. (1987) Observation of atomic corrugation on Au(lll) by scanning tunneling microscopy. *Rhys. Rev. Lett.* **59,** 2879–2882.
19. Hsu, T. and Cowley, J. M. (1983) Reflection electron microscopy of fcc metals. *Ultramicroscopy* **11,** 239–250.

20. Schneir, J., Harary, H. H., Dagata, J. A., Hansma, P. K., and Sonnenfeld, R. (1989) Scanning tunneling microscopy and fabrication of nanometre scale structures at the liquid-gold interface. *Scanning Microscopy* **3,** 719–724.
21. Nagahara, L. A., Thundat, T., and Lindsay, S. M. (1989) Preparation and characterization of STM tips for electrochemical studies. *Rev. Sci. Instrum.* **60,** 3128–3130.
22. Stemmer, A., Hefti, A., Aebi, U., and Engel, A. (1989) Scanning tunneling and transmission electron microscopy on identical areas of biological specimens. *Ultramicroscopy* **30,** 263–280.
23. McMaster, T. J., Carr, H. J., Miles, M. J., Cairns, P., and Morris, V. J. (1991) STM of Poly (γ-benzyl-L-glutamate). *Macromolecules* **24,** 1428–1430.
24. Smith, D. P. E., Hörber, J., Gerber, Ch., and Binnig, G. (1989) Smectic liquid crystal monolayers on graphite imaged by scanning tunneling microscopy. *Science* **245,** 43–45.
25. Lindsay, S. M., Nagahara, L. A., Thundat, T., Knipping, U., Rill, R. L., Drake, B., Prater, C. B., Wiesenhorn, A. L., Gould, S. A. C., and Hansma, P. K. (1989) STM and AFM images of nucleosome DNA under water. *J. Biomolecular Structure and Dynamics* **7,** 279–287.
26. Allen, M. J., Tench, R. J., Mazrimas, J. A., Balooch, M., Siekhaus, W. J., and Balhorn, R. (1991) A pulse-deposition method for STM analysis of DNA on graphite. *J. Vac. Sci. Technol.* **B9,** 1272–1275.
27. Amrein, M., Stasiak, A., Gross, H., Stoll, E., and Travaglini, G. (1988) Scanning tunneling microscopy of recA–DNA complexes coated with a conducting film. *Science* **240,** 514–516.
28. Guckenberger, R., Wiegräbe, W., and Baumeister, W. (1989) Scanning tunneling microscopy of biomacromolecules. *J. Microsc.* **152,** 795–802.
29. Amrein, M., Durr, R., Winkler, H., Travaglini, G., Wepf, R., and Gross, H. (1989) STM of freeze-dried and platinum-iridium-carbon coated bacteriophage T4 polyhead. *J. Ultrastruct. Mol. Struct. Res.* **102,** 170–177.
30. Zasadzinski, J. A. N., Schneir, J., Gurley, J., Eilings, V., and Hansma, P. K. (1988) Scanning tunneling microscopy of freeze-fracture replicas of biomembranes. *Science* **239,** 1013–1015.
31. Mizutani, W., Sakakibara, Y., Ono, M., Tanishima, S., Ohno, K., and Toshima, M. (1989) Measurement of copper phthalocynanine ultrathin films by STM and spectroscopy. *Jap. J. Appl. Phys.* **28,** L1460–L1463.
32. Cricenti, A., Selci, A., Felici, A.C., Generosi, R., Gori, E., Djaczenko, W., and Chiarotti, G. (1986) Molecular structure of DNA by scanning tunneling microscopy. *Science* **245,** 1226–1227.
33. Clemmer, R. C. and Beebe, T. P. Graphite: a mimic for DNA and other molecules in scanning tunneling microscope studies. *Science* **251,** 640–642.
34. Hamers, R. J., Tromp, R. M., and Demuth, J. E. (1987) Electronic and geometric structure of Si(lll) -(7x7) and Si(001) surfaces. *Surf. Sci.* **181,** 346–355.
35. Fuchs, H. and Laschinski, R. (1990) Surface investigations with a combined scanning electron-scanning tunneling microscope. *Scanning* **12,** 126–132.

36. Emch, R., Clivaz, X., Taylor-Denes, C., Vaudaux, P., and Descouts, P. (1990) Scanning tunneling microscope for studying the biomaterial–biological tissue interface. *J. Vac. Sci. Technol.* **A8,** 655–658.
37. Jericho, M. H., Blackford, B. L., and Dahn, D. C. (1989) Scanning tunneling microscope imaging technique for weakly bonded surface deposits. *J. Appl. Phys.* **65,** 5237–5239.
38. Jericho, M. H., Blackford, B. L., Dahn, D. C., Frame, C., and Maclean, D. (1990) Scanning tunneling microscopy imaging of uncoated biological material. *J. Vac. Sci. Technol.* **A8,** 661–666.

Glossary

A. Monosaccharide Nomenclature

Monosaccharides are in the pyranose form unless otherwise stipulated. There are two alternative forms for portraying monosaccharides, e.g., β-D-*N*-acetylglucosamine (GlcNAc).

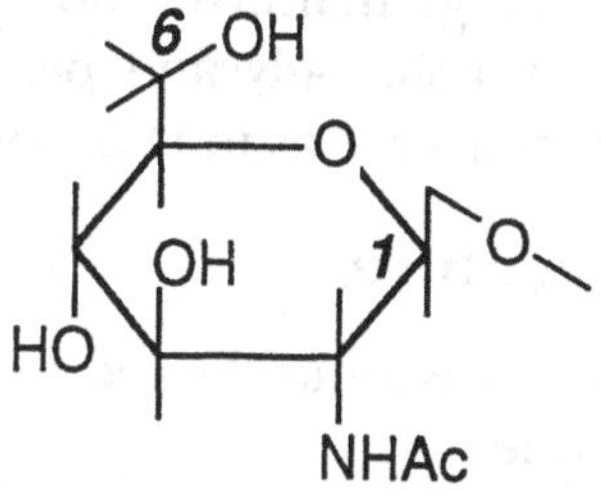

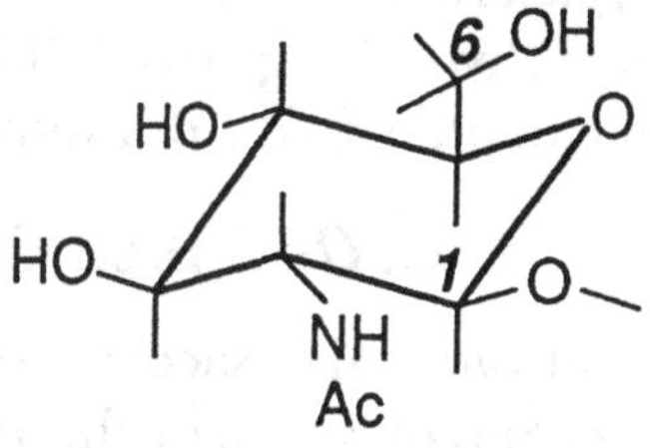

Different monosaccharides vary by the number and orientation of their functional groups, i.e., OH, NHAc, and the like. Linkage can be to any of the hydroxyl groups with either β or α anomeric configuration (above and below the plane of the ring, respectively, at C1 for D-pyranosides for example).

The Sialic Acids

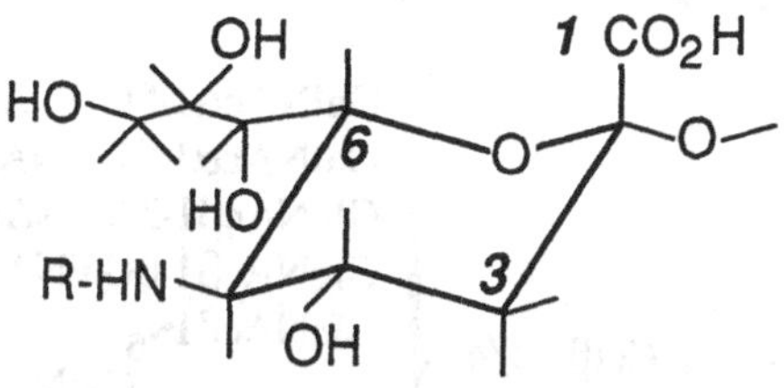

R = CH_3-CO- (*N*-acetylneuraminic acid) or CH_2OH-CO- (*N*-glycolylneuraminic acid); the hydroxyl groups can be substituted with various acyl substituents and those at C8 and C9 by additional sialic acid residues.

B. *N*-Linked Chain Structures

All *N*-linked oligosaccharides have a common pentasaccharide core ($Man_3GlcNAc_2$) originating from a common biosynthetic intermediate. They differ in the number of branches and the presence of peripheral sugars such as fucose and sialic acid. They can be categorized according to their branched constituents, which may consist of mannose only (high mannose *N*-glycans); alternating GlcNAc and Gal residues terminated by any of the sequences in D below, and with the possibility of intrachain substitutions of bisecting GlcNAc and core Fuc (complex *N*-glycans); or attributes of both high mannose and complex chains (hybrid *N*-glycans). The variety of basic *N*-glycan types are shown in Chapters 2–7. Additional glycosylations are as in D below.

C. *O*-Linked Chain Structures

The following oligosaccharide sequences are found linked to the hydroxyl group of serine or threonine residues:

Sequence	Found in
GlcNAcβ1-	Cytoplasmic glycoproteins
Fucα1- and ±Xyl_n- Glcα1-	EGF domains of coagulation and fibrinolytic proteins
Glcα1→2Galβ1-	bacterial glycoproteins
sulfated disaccharide repeats–GlcAβ1-3Galβ1-3Galβ1-4Xylβ1- (Galβ1-3Gal substituted with ±SO_4; Xyl substituted with PO_3)	proteoglycans; linked to Ser only

±(Galβ1-3/4GlcNAcβ1-3/6)$_n$Galβ1-3/4 {
- GalNAcα1-3GalNAcα1-
- GalNAcα1-6GalNAcα1-
- GlcNAcβ1-3GalNAcα1-
- GlcNAcβ1-6GalNAcα1-
- GlcNAcβ1↘6 GalNAcα1-
- GlcNAcβ1↗3

} mucins

±(Galβ1-3/4GlcNAcβ1-3/6)$_n$ {
- GlcNAcβ1↘6 GalNAcα1-
- Galβ1↗3
- Galβ1-3GalNAcα1-

} mucins and secreted and cell membrane glycoproteins

D. Examples of Chain Terminating Sequences of *O*- and *N*-Linked Oligosaccharides

NeuAcα2-6GalNAcα1-
NeuAcα2-3Galβ1-3GalNAcα1-
NeuAcα2-3/6Galβ1-4GlcNAcβ1-
NeuAcα2-3Galβ1-3GlcNAcβ1-
| 2,6
NeuAcα

GlcNAcα1-4Galβ1- (mucins only)
Fucα1-2Galβ1- (Blood group H)

GalNAcα1↘3
Galβ1- (Blood group A)
Fucα1↗2

Galα1↘3
Galβ1- (Blood group B)
Fucα1↗2

Fucα1↘4
GlcNAcβ1- (Le^{b})
Fucα1-2Galβ1↗3

Fucα1-2Galβ1↘4
GlcNAcβ1- (Le^{y})
Fucα1↗3

Fucα1↘4
GlcNAcβ1- (Sialyl Le^{a}/Le^{a})
±NeuAcα2-3Galβ1↗3

Fucα1↘3
GlcNAcβ1- (Sialyl Le^{x}/Le^{x})
±NeuAcα2-3Galβ1↗4

GalNAcβ1↘4
Galβ1- (SD^{a})
NeuAcα2↗3

Sulfate esters can occur on Gal, GalNAc, and GlcNAc, and phosphate esters on Man. NeuAc can be *O*-acetylated or replaced in many sequences by NeuGl (*N*-glycolylneuraminic acid).

Index

N

O

P